Progress in Mathematics
Volume 182

Series Editors
Hyman Bass
Joseph Oesterlé
Alan Weinstein

Sara Billey
V. Lakshmibai

Singular Loci
of Schubert Varieties

Birkhäuser
Boston • Basel • Berlin

Sara Billey
Department of Mathematics
Massachusetts Institute of Technology
Cambridge, MA 02139
U.S.A.

V. Lakshmibai
Department of Mathematics
Northeastern University
Boston, MA 02115
U.S.A.

Library of Congress Cataloging-in-Publication Data

Billey, Sara.
    Singular loci of Schubert varieties / Sara Billey, V. Lakshmibai.
      p. cm. – (Progress in mathematics ; v. 182)
    Includes bibliographical references and indexes.
    ISBN 0-8176-4092-4 (alk. paper) – ISBN 3-7643-4092-4 (alk. paper)
    1. Schubert varieties. I. Lakshmibai, V. (Venkatramani) II. Title. III. Progress in
mathematics (Boston, Mass.) ; vol. 182

QA564.B48 2000
516.3'53–dc21
                              00-062145
                              CIP

AMS Subject Classifications: 14F32, 14M12, 14M15, 14M17, 14Q99, 22E46, 51H20, 51H25, 51H30,
                                    53C30, 57P10, 57T15

Printed on acid-free paper.
© 2000 Birkhäuser Boston   *Birkhäuser*

ISBN 0-8176-4092-4   SPIN 10685357
ISBN 3-7643-4092-4

Typeset by the authors.
Printed and bound by Hamilton Printing, Rensselaer, NY.
Printed in the United States of America.

9 8 7 6 5 4 3 2 1

With fond memories of Gian-Carlo Rota.

Contents

Preface

This monograph began to take shape in late 1997 when the authors undertook to write a survey article including current results in the theory of singular loci of Schubert varieties. The article quickly became too long for a journal article because of the vast literature on the subject. We felt there was a need to have this diverse collection of results unified in a single source. Hence, we decided to extend the article into a book.

In order to give a broad treatment of the topic we have chosen to include only a limited number of proofs. The proofs included are directly related to the computations of the singular locus. We have included many other results so the reader may comprehend how this subject sits inside the more general topics of algebraic geometry, representation theory and combinatorics.

We have attempted to make this document as accessible as possible to a wide audience. A natural place to begin is with the definitions of the flag manifold, Schubert varieties, Bruhat–Chevalley order and parabolic subgroups. Our main example throughout this text will be the Schubert varieties in the flag manifold SL_n/B (or Type A case). Therefore, we have devoted Chapter 3 to concrete computations on the flag manifold SL_n/B, the Grassmannian manifold and their analogs for the other classical groups. Several of the smoothness criteria take particularly nice forms in special cases such as types A and C. We have also included extensive tables of the singular locus of a Schubert variety for the Weyl groups of types A_5, B_4, C_4, D_4 and G_2.

Even though we have included a large amount of background material, this book is not intended to serve as an introduction to root systems, representations of Lie algebras, algebraic groups, Weyl groups or even Schubert varieties. We expect the reader to be somewhat familiar with these subjects, plus have an interest in doing computations with Schubert varieties. We only give a brief review of the definitions and main theorems which will be essential for this text so that the reader should not need to refer to other texts for basic definitions.

This book can be used for a year-long course on Schubert varieties with a main focus on their singularities. The material covered in this book should serve as a good source of information on the singularities of

Schubert varieties for graduate students and researchers working in the area
of combinatorics, algebraic geometry, algebraic groups and representation
theory.

Throughout the text we have numbered the equations, subsection (or
topics), theorems, remarks, etc, in order according to their chapter and
section. Therefore, Theorem 2.1.5 is the fifth theorem in the first section
of Chapter 2.

We would like to thank the following people for their helpful comments, criticisms, and encouragement along the way: Jim Carrell, Bonnie
Friedman, Francis Fung, Nick Gonciulea, Michael Kleber, Ann Kostant,
Bertram Kostant, Victor Kreiman, Shrawan Kumar, George Lusztig, Greg
Warrington, and the 18.318 class taught in the spring of 1999 at MIT. We
would also like to thank the reviewers who gave many insightful comments
and corrections.

Finally, we would like to thank our families Paul and Allegra, Aruna,
Girish, Rakhal and Sri for their love and encouragement.

Sara Billey and Venkatramani Lakshmibai
Boston, July 1, 2000

CHAPTER 1

Introduction

The study of Schubert varieties arising out of 19th century classical projective geometry took on a more modern treatment with the work of Ehresmann, then Chevalley, and later Bernstein, Gelfand, Gelfand and Demazure. Schubert varieties are now some of the best understood examples of complex projective varieties in the literature. Therefore, they play an important role in current mathematical research. Outside of mathematics, their applications appear in physics [85] and computer graphics [154, 153].

Investigations into the questions of smoothness and singularities have engaged mathematicians since the 1970s. In particular, many researchers have focused on the important problem of determining the singular loci of Schubert varieties. The singular locus of a Schubert variety lies at the crossroads of several fields of mathematics including geometry, combinatorics and/or representation theory. In later chapters of this book, we have unified the diverse theories in the literature and modernized the notation so that this monograph can be used as a handbook to current researchers as well as an introduction to the singularity theory of Schubert varieties.

We will briefly review some of the historical developments in this area. In 1934, Ehresmann [44] showed that the cohomology ring of the Grassmannian is generated by the classes of its Schubert subvarieties, and thus established a key relationship between the geometry of flag varieties and the theory of characteristic classes. In 1956, Chevalley [36] further enhanced this relationship by showing that Schubert classes form a $\mathbb{Z}$-basis for the Chow ring of the generalized flag variety. By the early 1980's, computations in the cohomology ring of the flag variety became much better understood through the work of Bernstein, Gelfand, and Gelfand [11], Demazure [37], and Lascoux and Schützenberger [118]. This aspect of the study of the Schubert varieties is still very active; see [65, 123, 125, 50] for an overview.

In [36], Chevalley showed that Schubert varieties are nonsingular in codimension one i.e., the singular locus has codimension at least 2, in the flag manifold G/B for any semisimple Lie group G. In that same paper Chevalley states that every Schubert variety is probably smooth! In 1973, Demazure gave a seminar on his work on "Désingularisation des variétés de Schubert généralisées" (cf. [37]) while visiting the Tata Institute. During the course of the seminar, it was discovered that there are two Schubert varieties in SL_4/B, namely $X(3412)$, $X(4231)$ which do not even satisfy Poincaré duality, hence they cannot be smooth!

In the past fifteen years, many major developments have enabled more elaborate computations on the singular locus and questions of smoothness of Schubert varieties. The first result on the singular locus of a Schubert variety is due to Lakshmibai and Seshadri ([**109**], 1984), wherein the singular locus of a Schubert variety for G classical is determined using the "Standard Monomial Basis" and the Jacobian criterion for smoothness. This work has been extended by Lakshmibai in constructing explicit bases for the tangent spaces to Schubert varieties. Explicit bases such as these are very useful both for determining singularities and for computations of multiplicities at singular points. In addition, there have been numerous results on the determination of the singular locus of a Schubert variety due to Carrell–Peterson ([**32**], 1994), Lakshmibai ([**98**], 1995), Polo ([**135**], 1994).

Another key direction has been the application of intersection cohomology theory to the study of the singularities of Schubert varieties through the work of Kazhdan and Lusztig ([**78, 79**], 1980), wherein they introduce the *celebrated* Kazhdan–Lusztig polynomials. This resulted in a useful approximation to smoothness known as rational smoothness. Using results on Kazhdan–Lusztig polynomials, Carrell and Peterson developed several computationally feasible criteria for testing rational smoothness and studying curves in a Schubert variety invariant under the action of a maximal torus ([**33**], 1994). The Carrell–Peterson criteria in turn was extended by Kumar to give a unified criteria for both smoothness and rational smoothness in terms of coefficients in the nil-Hecke ring ([**91**], 1996). This is the most general test for smoothness and/or rational smoothness known at this time, extending to all Kac–Moody algebras.

More specifically, let G be a semisimple, simply connected algebraic group over an algebraically closed field K of arbitrary characteristic. Let T be a maximal torus in G, and $W (= N(T)/T$, $N(T)$ being the normalizer of T) the Weyl group. Let B be a Borel subgroup of G, where $B \supset T$. The most familiar example is $G = SL_n(K)$, T is the diagonal matrices in G, B is the upper triangular matrices in G and W is the symmetric group S_n. This example will be emphasized throughout the text, in particular in Chapter 3.

For $w \in W$, let us denote the point in the flag manifold G/B corresponding to the coset wB by e_w. Then the set of T-fixed points in G/B for the action given by left multiplication is precisely $\{e_w \mid w \in W\}$. For $w \in W$, let $X(w)$ be the associated Schubert variety; the Zariski closure of Be_w in G/B with the canonical reduced scheme structure. We have the well-known Bruhat decomposition of G/B as the disjoint union

$$G/B = \bigcup Be_w,$$

and

$$X(w) = \bigcup_{\theta \leq w} Be_\theta, \quad \theta \in W$$

where $\leq$ is the Bruhat–Chevalley order.

Let $\text{Sing} X(w)$ denote the singular locus of $X(w)$. If $X(w)$ is not smooth, then $\text{Sing} X(w)$ is a nonempty B-stable closed subvariety of $X(w)$. Given a point $x \in X(w)$, to decide if it is a smooth point or not, it suffices (in view of Bruhat decomposition) to determine if the T-fixed point e_τ of the B-orbit through x is a smooth point or not. This reduction to testing only the T-fixed points is precisely the reason that computations on the singular locus of Schubert varieties are so beautiful.

The general outline of the book is as follows. Chapter 2 covers the necessary background material on the flag manifold, Schubert varieties, Bruhat–Chevalley order and parabolic subgroups. This chapter is not meant to replace a course in Lie groups and Lie algebras, but it should suffice to review the definitions.

Our main example throughout this text will be the Schubert varieties in the flag manifold SL_n/B (or Type A case). Therefore, we have devoted Chapter 3 to concrete computations of Bruhat–Chevalley order in SL_n (and more generally in the classical groups), the Grassmannian, the flag manifold and their Schubert varieties.

In Chapter 4, we introduce the tangent space to a Schubert variety at a given point P. In view of the Bruhat decomposition, the question of whether P is a smooth point or not, will be determined by the behavior at the (unique) T-fixed point on the B-orbit through P. One may study the behavior at a T-fixed point by considering some canonical affine neighborhood of this point inside the given Schubert variety. This affine neighborhood is introduced in this chapter. We further recall the generalities on the multiplicity at a singular point on a Schubert variety.

Several of the smoothness criteria take particularly nice forms in the special case of G being classical. In Chapter 5, we first recall the results of Polo ([135]) on the singular locus of a Schubert variety. We then use Polo's results and give a root system description of the tangent space to a Schubert variety at a T-fixed point for the classical groups. Also included in this chapter are Polo's result describing the B-module structure of the tangent space at the identity [135] and two criteria for smoothness due to Carrell–Kuttler [34].

Chapter 6 is devoted to rational smoothness and Kazhdan–Lusztig theory. In this chapter we gather the known characterizations of rational smoothness of Schubert varieties in terms of Kazhdan–Lusztig polynomials, Bruhat graphs, T-stable curves, Poincaré polynomials. These results are due to Billey–Warrington [15], Brenti [26, 27], Carrell–Peterson [33], Deodhar [39], Kazhdan–Lusztig [79], Lascoux–Schützenberger [117], Lascoux [114].

Chapter 7 is completely devoted to the discussion of Kumar's results ([91]) and the nil-Hecke algebra. In this chapter, we have tried to provide some details of proofs of Kumar's results. Included here as well is Dyer's proof of Deodhar's inequality, Fan's results [46], and results on codimension one smooth Schubert varieties.

In Chapter 8, we give combinatorial criteria for smoothness and rational smoothness for Schubert varieties, G being classical, in terms of certain pattern avoidance for the corresponding permutations. These algorithms are the most efficient tests for smoothness and rational smoothness of Schubert varieties. These results are due to Billey ([13]), Lakshmibai–Sandhya ([107]), Lakshmibai–Song ([111]). We have provided details of proofs wherever possible. We have also included a conjecture due to Lakshmibai–Sandhya ([107]) on the irreducible components of the singular locus of a Schubert variety for $G = \mathrm{SL}(n)$.

Next to the projective space $\mathbb{P}^n$, the simplest examples of projective varieties are provided by the Grassmannians and more generally by the minuscule and cominuscule G/P's. In Chapter 9, we present the results for the minuscule and cominuscule G/P. We first present the results of Zelevinsky ([157]), Sankaran–Vanchinathan ([143, 144]) on small resolutions for Schubert varieties in the minuscule and cominuscule cases. We then present the results of Brion and Polo (cf. [30]) on the tangent space to Schubert varieties in the minuscule and cominuscule cases. The results of Lakshmibai–Weyman (cf. [112]) on the irreducible components of the singular loci of Schubert varieties as well as recursive formulas for the multiplicity and the Hilbert polynomial at a singular point are then presented. We have also included two closed formulas due to Kreiman–Lakshmibai ([89]), Rosenthal–Zelevinsky [142] for the multiplicity at a singular point for Schubert varieties in the Grassmannian. We have also included a closed formula for the Hilbert polynomial at a singular point due to Kreiman–Lakshmibai ([89]).

In Chapter 10, we present the results for G of rank 2 using both Kumar's criteria and computations with the tangent space basis.

In Chapter 11, we present the results relating smoothness and factoring of the Poincaré polynomial. These results are due to Billey [13], Carrell–Peterson [33], Gasharov [52], Lascoux [116]. We also present the results of Bona [19], Haiman [60] on the generating functions on the number of smooth varieties for $G = \mathrm{SL}(n)$.

Even though determinantal varieties are classically well-known, their relationship to Schubert varieties has been developed only recently (cf. [108], [132]). In Chapter 12, we present this aspect of determinantal varieties. For the sake of completeness of the treatment of this aspect, we have included proofs for all of the statements. We further review the results of [56], [103] wherein two classes of affine varieties — a certain class of ladder determinantal varieties and a certain class of quiver varieties — are shown to be normal and Cohen–Macaulay by identifying them with opposite cells in certain Schubert varieties. For these ladder determinantal varieties and also certain of these quiver varieties, we also have a description of the singular locus (cf. [56], [101]) which gives information on the irreducible components of the associated Schubert variety and verifies the conjecture of [107] for these Schubert varieties.

As mentioned earlier, the problem of the determination of the irreducible components of the singular locus of a Schubert variety is open even for $SL(n)$. We have included computer generated tables describing the irreducible components of $\text{Sing} X(w)$, $w \in W$ for types $\mathbf{A}_5$, $\mathbf{B}_4$, $\mathbf{C}_4$, $\mathbf{D}_4$. Many mathematicians have found Goresky's tables [58] very handy and useful, and this is what motivated us to generate these tables and include them in this book.

CHAPTER 2

Generalities on G/B and G/Q

This chapter reviews the background material on *generalized flag varieties* including root systems, Weyl groups, parabolic subgroups, Schubert varieties and some representation theory for semisimple algebraic groups. For more details on root systems and Weyl groups, we recommend Humphrey's *Introduction to Lie Algebras and Representation Theory* [70] and *Reflection Groups and Coxeter Groups* [69]. Other references for this chapter are Bourbaki's *Groupes et Algèbres de Lie, Chapters 4, 5 and 6* [23], Borel's *Linear Algebraic Groups* [20], Humphrey's *Linear Algebraic Groups* [68], and Jantzen's *Representations of Algebraic Groups* [76].

2.1. Abstract root systems

Fix a vector space V over $\mathbb{R}$ with a nondegenerate positive definite inner product $(\cdot, \cdot)$. Define the *reflection* relative to a nonzero vector $\alpha \in V$ to be the linear transformation on V given by

$$(2.1.1) \qquad s_\alpha(v) = v - \frac{2(v, \alpha)}{(\alpha, \alpha)} \alpha.$$

Note, s_α maps α to $-\alpha$ and fixes the hyperplane perpendicular to α. Following the literature, we will use the notation

$$(2.1.2) \qquad \langle v, \alpha \rangle = \frac{2(v, \alpha)}{(\alpha, \alpha)}.$$

$$(2.1.3) \qquad \alpha^* = \frac{2\alpha}{(\alpha, \alpha)}$$

so $(v, \alpha^*) = \langle v, \alpha \rangle$.

2.1.4. Remark. Note that α^* is also denoted $\alpha^\vee$ in the literature. In the sequel we shall use both $\langle v, \alpha \rangle$ and (v, α^*) interchangeably.

An abstract *root system* R in V is defined by the following axioms:

1. R is finite, spans V and does not contain 0.
2. If $\alpha \in R$, the only multiples of α in R are $\pm \alpha$.
3. For each $\alpha \in R$, R is stable under s_α.
4. (Integrality) If $\alpha, \beta \in R$, then $s_\alpha(\beta) - \beta = \langle \beta, \alpha \rangle \alpha$ is an integral multiple of α.

Let $l = \dim(V)$ be the *rank* of R. A subset $S = \{\alpha_1, \ldots, \alpha_l\}$ of R is called a *basis of the root system* if S is a vector space basis of V and if for any $\alpha = \sum c_i \alpha_i \in R$, all the coefficients c_i are integers with the same sign. Using this property, R can be partitioned into *positive roots* R^+ and *negative roots* R^- with respect to S. The elements of S are called *simple roots*. The integers $\langle \alpha_i, \alpha_j \rangle$, $1 \leq i, j \leq l$, are called the *Cartan integers*, and these completely determine the root system up to isomorphism.

The information given by the Cartan integers is summarized in a *Dynkin diagram*. The Dynkin diagram is a graph with vertices indexed by the simple roots and the number of edges between α_i and α_j for $i \neq j$ is $\langle \alpha_i, \alpha_j \rangle \langle \alpha_j, \alpha_i \rangle$ with an arrow pointing to the smaller of the two roots if they have different lengths. All irreducible root systems have been completely classified and their Dynkin diagrams appear on Page 207.

Example. Let V be the hyperplane in $\mathbb{R}^3$ consisting of $\{(x_1, x_2, x_3) \in \mathbb{R}^3 \mid \sum x_i = 0\}$ equipped with the usual inner product. Let $\{e_1, e_2, e_3\}$ denote the standard basis for $\mathbb{R}^3$. Let $R = \{\pm(e_1 - e_2), \pm(e_2 - e_3), \pm(e_1 - e_3)\}$. Then R is the root system of type A_2 (of rank 2) in V. The set $S := \{\alpha_1, \alpha_2\} = \{e_1 - e_2, e_2 - e_3\}$ is a basis of the root system R. The Cartan integer $\langle \alpha_i, \alpha_j \rangle$ is 2 or -1 according as to whether $i = j$ or $i \neq j$.

The *Weyl group* W of the root system R is the subgroup of $GL(V)$ generated by the s_α for $\alpha \in R$. It follows from the axioms of root systems that W is also finite. W is minimally generated by the *simple reflections* $s_{\alpha_1}, \ldots, s_{\alpha_l}$, denoted by $s_1, \ldots, s_l$ corresponding with the simple roots. For $w \in W$, the *length* of w, denoted $l(w)$ is defined to be the minimum of the length of an expression for w as a product of simple reflections; if $l(w) = p$, then an expression $w = s_{a_1} s_{a_2} \ldots s_{a_p}$ is called a *reduced expression* for w and the sequence $a_1 a_2 \ldots a_p$ is called a *reduced word* for w. Note that for a given w, there can be many reduced expressions. For instance, in the above example, we have $W = $ the symmetric group S_3 with simple reflections $s_1 := $ the transposition $(1, 2)$ and $s_2 := $ the transposition $(2, 3)$. The two reduced expressions $s_1 s_2 s_1$, $s_2 s_1 s_2$ give the same permutation (321). Here and throughout the text we will denote a permutation w by its *one-line notation* $(w_1 w_2 \ldots w_n)$ if 1 maps to w_1, 2 maps to w_2, etc.

The following two facts about Weyl groups are used often. There is a unique element in W of *longest length*, denoted by w_0. The set of all reflections in W, denoted by $\mathcal{R}$, is obtained by conjugating all simple reflections.

Any vector $\lambda \in V$ such that $\langle \lambda, \alpha \rangle \in \mathbb{Z}$ is called an *abstract weight*. The set of all weights form a lattice Λ, called the *weight lattice*, which contains the lattice generated by R as an abelian subgroup of finite index. (The lattice generated by R is usually called the *root lattice*.) If $S = \{\alpha_1, \ldots, \alpha_l\}$ is a basis of R, then Λ has a corresponding basis consisting of *fundamental weights* $\omega_1, \ldots, \omega_l$ defined by $\langle \omega_i, \alpha_j \rangle = \delta_{ij}$ (Kronecker delta). A weight $\lambda = \sum c_i \omega_i$ is *dominant* if all c_i are nonnegative integers, and is *regular* if $c_i \neq 0$, for all i. In the sequel, Λ^+ will denote the set of all dominant weights in Λ.

The *Weyl chambers* in V are the connected components in the complement of the union of hyperplanes orthogonal to the roots. These chambers are in one-to-one correspondence with the bases of the root system. The chamber containing the dominant weights with respect to S is also called the *dominant chamber*.

2.2. Root systems of algebraic groups

Let G be an algebraic group defined over an algebraically closed field K of arbitrary characteristic. Recall (cf. [**20, 68**]) the following definitions and facts from algebraic group theory. A *torus* is an algebraic group isomorphic to a product of copies of K^*. Maximal tori exist in a given algebraic group (by reason of dimension); further, all maximal tori are conjugates and hence have the same dimension. Let l be the dimension of a maximal torus; l is defined as the *rank* of G. A *Borel subgroup* in an algebraic group G is a maximal connected solvable subgroup; further, all Borel subgroups in G are conjugates (and hence have the same dimension). The *radical* of G, denoted $R(G)$, is the connected component through the identity element of the intersection of all the Borel subgroups in G. An algebraic group G is *semisimple*, if $R(G)$ is trivial. The group $SL_n(K)$ is an example of a semisimple algebraic group.

For the rest of this chapter, we shall suppose G to be a connected, semisimple algebraic group. Let us fix a maximal torus T in G.

Let $X(T) := \text{Hom}_{\text{alg. gp.}}(T, K^*)$ be the *character group* of T (written additively). Let V be any finite-dimensional T-module. Then the *complete reducibility* of T implies (in fact, equivalent to the fact) that V is the direct sum of *weight spaces* i.e., $V = \oplus_{\chi \in X(T)} V_\chi$, where $V_\chi = \{v \in V : tv = \chi(t)v, \ \forall t \in T\}$. We say $\chi \in X(T)$ is a *weight in* V if $V_\chi \neq 0$. Let $\Phi(T, V)$ be the set of nonzero weights in V.

Denote the Lie algebra of G by $\mathfrak{g}$ or Lie G. For any $g \in G$, consider $\text{Inn}(g) : G \to G$ given by $x \mapsto gxg^{-1}$. This induces the *adjoint action* on the Lie algebra $\text{Ad}(g) : \text{Lie } G \to \text{Lie } G$. For $V = \text{Lie } G$ as a T-module via the adjoint action, the corresponding $\Phi(T, \text{Lie } G)$ is called the *root system of G relative to T*; it is an abstract root system. One considers $\Phi(T, \text{Lie } G)$ to be a root system contained in the vector space $X(T) \otimes_{\mathbb{Z}} \mathbb{R}$ of rank l ($= \dim T$) (cf. [**23**]). We shall denote this root system by R. The Weyl group W associated to $\Phi(T, \text{Lie } G)$ is isomorphic to $N(T)/T$ where $N(T)$ is the normalizer of T in G. Let B be a Borel subgroup of G such that $B \supset T$. The subset $\Phi(T, \text{Lie } B)$ of $\Phi(T, \text{Lie } G)$ is called the set of *positive* *roots of R relative to B*, denoted R^+. The indecomposable elements in R^+ i.e., the set of all α such that α cannot be expressed as a positive sum of other elements in R^+, is a set of simple roots for the root system R which we shall denote by S. Thus, choosing T and B is equivalent to choosing a set of simple roots S in R. The fundamental weights $\{\omega_i : 1 \leq i \leq l\}$ are defined as in Section 2.1. We have the inclusions

$$\text{root lattice} \subseteq X(T) \subseteq \text{weight lattice}.$$

The elements in $X(T)$ that lie in the dominant chamber with respect to S are called *dominant characters*. Also, if G is simply connected, then $X(T)$ equals the weight lattice Λ [**76**].

2.2.1. Example. Let $G = SL_n(K)$ be the special linear group of rank $n - 1$, T the subgroup of diagonal matrices, and B the subgroup of upper triangular matrices. Then T, resp. B, is a maximal torus, resp. a Borel subgroup. Further, $N(T)$ is the subgroup consisting of matrices with exactly one nonzero entry in each row and column, and $W = N(T)/T$ is the symmetric group S_n. We have,

$$\text{Lie } G = \{n \times n \text{ matrices with trace } 0\}.$$

We further have

$$\text{Lie } G = \mathfrak{h} \oplus \mathfrak{n}_+ \oplus \mathfrak{n}_-,$$

where

$$\mathfrak{h} = \{ \text{ diagonal matrices in Lie } G\},$$
$$\mathfrak{n}_+ = \oplus_{1 \leq i < j \leq n}(E_{ij}), \ \mathfrak{n}_- = \oplus_{1 \leq j < i \leq n}(E_{ij}).$$

Here, E_{ij} is the elementary $n \times n$ matrix with 1 at the (i, j)-th place and zeros elsewhere. Hence the roots may be identified with $\epsilon_i - \epsilon_j, 1 \leq i, j \leq n$, $i \neq j$, where $\epsilon_i - \epsilon_j$ is the character of T that sends $\text{diag}(t_1, \dots, t_n)$ in T to $t_i t_j^{-1}$. Further, R^+ may be identified with $\{\epsilon_i - \epsilon_j, 1 \leq i < j \leq n\}$, and the fundamental weight $\omega_i, 1 \leq i \leq n - 1$ may be identified with the character of T sending $\text{diag}(t_1, \dots, t_n)$ in T to $t_1 \cdots t_i$. In this example, G is simply connected, and hence $X(T)$ coincides with the weight lattice.

2.3. Root subgroups

The Lie algebra of G, denoted by $\mathfrak{g}$, contains a one-dimensional subalgebra $\mathfrak{g}_\alpha$ for each root $\alpha \in R$. Each $\mathfrak{g}_\alpha$ is a T-weight space of weight α, called a *root space*. Furthermore, if $\mathfrak{h}$ denotes the Lie algebra of T, then $\mathfrak{g}$ decomposes as a direct sum of T-weight spaces

(2.3.1) $$\mathfrak{g} = \mathfrak{h} \oplus \bigoplus_{\alpha \in R} \mathfrak{g}_\alpha.$$

For each $\alpha \in R$, fix a basis X_α of the one-dimensional root space $\mathfrak{g}_\alpha$. The set of all $\{X_\alpha \mid \alpha \in R\}$ along with $\{H_i | 1 \leq i \leq l\}$ form the *Chevalley basis* of $\mathfrak{g}$. By Serre's theorem, $\mathfrak{g}$ is generated as a Lie algebra by the Chevalley basis along with certain relations [**70**, 18.3].

For each $\beta \in R$, there exists a unique connected T-stable subgroup U_β of G (i.e., U_β is normalized by T) with $\mathfrak{g}_\beta$ as its Lie algebra (see [**23**, **70**] for details). The subgroup U_β is called a *root subgroup*. Each U_β is a one-dimensional unipotent subgroup of G with Lie algebra $\mathfrak{g}_\beta$.

Let $\mathbb{G}_a$ denote the additive group K. For $\beta \in R$, there exists an isomorphism $\theta_\beta : \mathbb{G}_a \cong U_\beta$ such that for all $t \in T$, $x \in \mathbb{G}_a$, $t\theta_\beta(x)t^{-1} = \theta_\beta(\beta(t)x)$. Let U denote the unipotent part of B, i.e., the subgroup of B consisting of all the unipotent elements in B. Then we have an identification of varieties

$$U \cong \prod_{\beta \in R^+} U_\beta,$$

the product being taken in any fixed order. Further, U is normalized by T, and B is the semidirect product of U and T (see [**20**, **68**] for details).

2.3.2. Example. In the example of Section 2.2, $(E_{i,j})$, the one dimensional span of $E_{i,j}$ for $1 \le i, j \le n$, $i \ne j$, are all the root spaces. The root subgroup $U_{i,j}, 1 \le i, j \le n$, $i \ne j$ is the subgroup of G consisting of matrices with 1's along the diagonal and the only possible nonzero entry occurs at position (i, j). U is the subgroup of G consisting of unipotent upper triangular matrices.

2.4. Parabolic subgroups

A *parabolic subgroup* of G is any closed subgroup $Q \subset G$ such that G/Q is a projective variety (or equivalently, a complete variety, since G/Q is quasiprojective (see [**20**, **68**] for details)). A subgroup Q of G is parabolic if and only if it contains a Borel subgroup.

Let Q be a parabolic subgroup containing B. Let $R(Q)$ be the radical of Q, namely, the connected component through the identity element of the intersection of all the Borel subgroups of Q. Let $R_u(Q)$ be the unipotent radical of Q (the subgroup of unipotent elements of $R(Q)$), and let R_Q^+ be the subset of R^+ defined by $R^+ \setminus R_Q^+ = \{\beta \in R^+ \mid U_\beta \subset R_u(Q)\}$. Let $R_Q^- = -R_Q^+ (= \{-\beta \mid \beta \in R_Q^+\})$, $R_Q = R_Q^+ \cup R_Q^-$, and $S_Q = \{\alpha \in S \mid \alpha \in R_Q\}$. Then R_Q is a subroot system of R called the *root system associated to Q*, with S_Q as a set of simple roots and R_Q^+ as the set of positive roots, resp. R_Q^- as the negative roots, of R_Q relative to S_Q.

Conversely, given a subset J of S, there exists a parabolic subgroup Q of G such that Q contains B. Namely, Q is generated by B and $\{U_{-\alpha} \mid \alpha \in R^+$ and $\alpha = \sum_{\beta \in J} a_\beta \beta\}$.

Thus the set of parabolic subgroups containing B is in bijection with the power set of S. Note that for $Q = B$ (resp. G), S_Q is the empty set, resp. the whole set S.

The subgroup of Q generated by T and $\{U_\alpha \mid \alpha \in R_Q\}$ is called the *Levi subgroup* associated to S_Q, and is denoted by L_Q. We have that Q is the semidirect product of $R_u(Q)$ and L_Q (see [**20, 68**] for details).

The set of *maximal parabolic subgroups* containing B is in one-to-one correspondence with S. Namely given $\alpha \in S$, the parabolic subgroup Q where $S_Q = S \setminus \{\alpha\}$ is a maximal parabolic subgroup, and conversely. In the sequel, we shall denote the maximal parabolic subgroup Q, where $S_Q = S \setminus \{\alpha_i\}$ by P_i.

2.5. The Weyl group of a parabolic subgroup

Given a parabolic subgroup Q, let W_Q be the subgroup of W generated by $\{s_\alpha \mid \alpha \in S_Q\}$. W_Q is called the *Weyl group* of Q. Note that $W_Q \cong N_Q(T)/T$, where $N_Q(T)$ is the normalizer of T in Q.

2.5.1. Definition. $W_Q^{\min}$, **minimal representatives of** W/W_Q. In each coset $wW_Q \in W/W_Q$, there exists a unique element of minimal length (cf. [**69**, 1.10]). Let $W_Q^{\min}$ be the *set of minimal representatives of* W/W_Q. We have

$$(2.5.2) \qquad W_Q^{\min} = \{w \in W \mid l(ww') = l(w) + l(w'), \text{ for all } w' \in W_Q\}.$$

In other words, each element $w \in W$ can be written uniquely as $w = u \cdot v$, where $u \in W_Q^{\min}$, $v \in W_Q$ and $l(w) = l(u) + l(v)$. The set $W_Q^{\min}$ may be also be characterized as

$$(2.5.3) \qquad W_Q^{\min} = \{w \in W \mid w(\alpha) > 0, \text{ for all } \alpha \in S_Q\}.$$

(Here by a root β being > 0 we mean $\beta \in R^+$.) In the literature, $W_Q^{\min}$ is also denoted by W^Q; in the sequel, we shall use both of the notations.

2.5.4. Definition. $W_Q^{\max}$, **maximal representatives of** W/W_Q. In each coset wW_Q there exists a unique element of maximal length. Let $W_Q^{\max}$ be the set of maximal length representatives of W/W_Q:

$$W_Q^{\max} = \{w \in W \mid w(\alpha) < 0 \text{ for all } \alpha \in S_Q\}.$$

Further, if we denote by w_Q the unique element of maximal length in W_Q, then we have

$$W_Q^{\max} = \{ww_Q \mid w \in W_Q^{\min}\}.$$

2.6. Schubert varieties

Let T, B, Q, W, etc. be as above. Given any $w \in W$ there exists a well-defined coset wQ in G/Q which we will denote by $e_{w,Q}$. Then the set of T-fixed points in G/Q for the action given by left multiplication is precisely $\{e_{w,Q} \mid w \in W_Q^{\min}\}$. Let $w \in W_Q^{\min}$, and let $X_Q(w)$ be the Zariski closure of $Be_{w,Q}$ in G/Q. Then $X_Q(w)$ with the canonical reduced scheme structure is called the *Schubert variety* in G/Q associated to wW_Q. In particular, we have bijections between $W_Q^{\min}$ and the set of Schubert varieties in G/Q, and between $W_Q^{\max}$ and the set of Schubert varieties in G/Q. We have the extended version of the Bruhat decomposition

(2.6.1)
$$G/Q = \bigcup_{w \in W_Q^{\min}} Be_{w,Q}, \qquad X_Q(y) = \bigcup_{\substack{w \in W_Q^{\min} \\ e_{w,Q} \in X_Q(y)}} Be_{w,Q}, \qquad y \in W_Q^{\min},$$

(2.6.2)
$$\dim X(w) = l(w).$$

We use the notation $X(w)$ for the Schubert variety in G/Q indexed by $w \in W$ when $Q = B$, or when the parabolic subgroup is understood from context. Note that $G/B = X(w_0)$, w_0 being the element of largest length in W.

2.7. The Bruhat–Chevalley order

There is a partial order on W, known as the Bruhat–Chevalley order (or just Bruhat order), determined by the structure of the Schubert varieties in (2.6.1), namely, for $w_1, w_2 \in W$, $w_1 \geq w_2 \iff X(w_1) \supseteq X(w_2)$. This partial order plays an important role in the theory of Schubert varieties, in particular with their singularities. Chevalley has shown (cf. [36]) that this order can be stated equivalently in terms of the Weyl group, namely $w_1 \geq w_2 \in W$ if every (equivalently, any) reduced expression for w_1 contains a subexpression that is a reduced expression for w_2. For example, $s_1 s_2 s_3 s_2 > s_1 s_3 s_2 > s_3$.

Another test for Bruhat–Chevalley order is given by the following: For $w \in W$, let $X(w^{(d)})$ denote the projection of $X(w)$ under $G/B \to G/P_d$, $1 \leq d \leq l$; then for $u, v \in W$, $u \leq v$ if and only if $X(u^{(d)}) \subseteq X(v^{(d)})$ for all $1 \leq d \leq l$. In Chapter 5, we give specific algorithms for Bruhat–Chevalley relations for the Weyl groups of types A, B, C, D.

2.8. Line bundles on G/Q

For the study of G/B, there is no loss in generality in assuming that G is simply connected; in particular, as remarked in §2.2, we have that the character group $X(T)$ coincides with the weight lattice Λ. Henceforth, we shall suppose that G is simply connected. Let $X = G/B$, and consider the canonical projection $\pi : G \to X$. Then $\pi : G \to X$ is a principal B-bundle with B as both the structure group and fiber. Let $\mathbb{G}_m$ denote the multiplicative group in the field K and let $\lambda \in X(T)$. Then λ defines a character $\lambda_B : B \to \mathbb{G}_m$ obtained by composing the natural map $B \to T$ with $\lambda : T \to \mathbb{G}_m$ (recall that B is the semidirect product of T and U (= the unipotent part of B)). We shall denote λ_B also by just λ. Now

$\lambda : B \to \mathbb{G}_m$ gives rise to an action of B on $K(= \mathbb{G}_a)$, namely $b \cdot k = \lambda(b)k$, $b \in B$, $k \in K$. Set $E = G \times K/ \sim$, where $\sim$ is the equivalence relation defined by $(gb, b \cdot k) \sim (g, k)$, $g \in G$, $b \in B$, $k \in K$. Then E is the total space of a line bundle, say $L(\lambda)$, over X. Thus we obtain a map

$$L : X(T) \to \text{Pic}(G/B), \quad \lambda \mapsto L(\lambda),$$

where $\text{Pic}(G/B)$ is the Picard group of G/B, namely the group of isomorphism classes of line bundles on G/B. We recall (cf. [35]) that the above map is in fact an isomorphism of groups since G is simply connected.

On the other hand, consider the prime divisors $X(w_0 s_i)$, $1 \le i \le l$, on G/B. Let $L_i = \mathcal{O}_{G/B}(X(w_0 s_i))$ be the line bundle defined by $X(w_0 s_i)$, $1 \le i \le l$.[1] The Picard group $\text{Pic}(G/B)$ is a free abelian group generated by the L_i's, and under the isomorphism $L : X(T) \cong \text{Pic}(G/B)$, we have $L(\omega_i) = L_i$, $1 \le i \le l$ [35]. Thus for $\lambda = \sum_{i=1}^{l} \langle \lambda, \alpha_i \rangle \omega_i$, we have $L(\lambda) = \otimes_{i=1}^{l} L_i^{\otimes \langle \lambda, \alpha_i \rangle}$. We shall write $L(\lambda) \ge 0$ if $\lambda \ge 0$, i.e., if λ is dominant.

As above, let E denote the total space of the line bundle $L(\lambda)$, over $X(= G/B)$. Let $\sigma : E \to X$ be the canonical map $\sigma(g, c) = gB$. Let

$$M_\lambda = \{f \in k[G] \mid f(gb) = \lambda(b)f(g), g \in G, b \in B\}.$$

Then M_λ can be identified with the *space of sections* $H^0(G/B, L(\lambda)) := \{s : X \to E \mid \sigma \circ s = \text{id}_X\}$ as follows (see [76] for further details). Let $f \in M_\lambda$. To f, we associate a section $s : X \to E$ by setting $s(gB) = (g, f(g))$. To see that s is well-defined, consider $g' = gb$, $b \in B$. Then $(g', f(g')) = (gb, f(gb)) = (gb, \lambda(b)f(g)) = (gb, b \cdot f(g)) \sim (g, f(g))$. From this, it follows that s is well defined. Conversely, given $s \in H^0(G/B, L(\lambda))$, consider $gB \in G/B$. Let $s(gB) = (g', f(g'))$, where $g' = gb$ for some $b \in B$ (note that $g'B = gB$, since $\sigma \circ s = \text{id}_X$). Now the point $(g', f(g'))$ may also be represented by $(g, \lambda(b)^{-1}f(gb))$ (since $(g', f(g')) = (gb, f(gb)) \sim (g, \lambda(b)^{-1}f(gb))$). Thus given $g \in G$, there exists a unique representative of the form $(g, f(g))$ for $s(gB)$. This defines a function $f : G \to K$. Further, this f has the property that for $b \in B$, $f(g) = \lambda(b)^{-1}f(gb)$, i.e., $f(gb) = \lambda(b)f(g)$, $b \in B$, $g \in G$. Thus we obtain an identification

$$M_\lambda = H^0(G/B, L(\lambda)).$$

It can be easily checked that the above identification preserves the respective G-module structures.

The structure of the line bundle $L(\lambda)$ can often be described in terms of the weight λ. For example, Chevalley has proven the following facts [35] (see also [76]):

1. $H^0(G/B, L(\lambda)) \ne 0$ if and only if λ is dominant.

---

[1]See [63], Ch. II, §6 for the discussion of divisors and line bundles on an algebraic variety.

2. $L(\lambda)$ is ample if and only if $L(\lambda)$ is very ample if and only if λ is dominant and regular.

(A line bundle L on an algebraic variety X is *very ample* if there exists an immersion $i\colon X \hookrightarrow \mathbb{P}^N$ such that $i^*(\mathcal{O}_{\mathbb{P}^N}(1)) = L$. A line bundle L on X is *ample* if L^m is very ample for some positive integer $m \geq 1$. See [63] for details.)

2.8.1. Summary. Let λ be a dominant weight and denote $H^0(G/B, L(\lambda))$ by $F(\lambda)$. We have (see [76] for example), $F(\lambda) \cong V_K(\lambda)^*$, where $V_K(\lambda)$ is the Weyl module corresponding to λ (see §2.11.10 below for the definition of Weyl modules). Now using the fact that the multiplicity of the weight $i(\lambda)$ in $F(\lambda)$ equals one, we obtain that $F(\lambda)$ is G-indecomposable. The indecomposability of $F(\lambda)$, together with the complete reducibility of G in characteristic 0, implies that if $\mathrm{char}K = 0$, then $F(\lambda)$ is irreducible.

We further have the following facts (see [76] for details):
1. There exists a bijection between Λ^+ and the set of isomorphism classes of finite-dimensional irreducible G-modules.
2. $H^0(G/B, L(\lambda)) = \{f \in K[G] \mid f(gb) = \lambda(b)f(g), g \in G, b \in B\}$.
3. $H^0(G/B, L(\lambda)) \neq 0$ if and only if $\lambda \in \Lambda^+$.
4. $L(\lambda)$ is ample $\Longleftrightarrow$ $L(\lambda)$ is very ample $\Longleftrightarrow$ λ is dominant and regular.
5. There exists a unique B-stable line in $H^0(G/B, L(\lambda))$ if λ is dominant, and the G-submodule generated by this line is the irreducible G-module with highest weight $i(\lambda)$, $i(= -w_0)$ being the *Weyl involution*.
6. Let λ be dominant. Then $H^0(G/B, L(\lambda))$ is an indecomposable G-module.
7. In characteristic 0, $H^0(G/B, L(\lambda))$ is G-irreducible with highest weight $i(\lambda)$ and conversely. Thus in characteristic 0, $\{H^0(G/B, L(\lambda)), \lambda \in \Lambda^+\}$ gives (up to isomorphism) all finite-dimensional irreducible G-modules.

2.9. Geometric properties of Schubert varieties

Chevalley showed that Schubert varieties are nonsingular in codimension 1, i.e., the singular locus has codimension at least 2. In types B_2 and C_2 examples can be found where the codimension of the singular locus is exactly 2 (see Chapter 10.) In $SL(4)$ for example, $X(4231)$ is singular in codimension 3, since its singular locus is $X(2143)$. Note that dim $X(4231)$ is 5 while dim $X(2143)$ is 2.

Demazure (cf.[37]) first considered the normality of Schubert varieties when K is of characteristic 0. Seshadri ([147]) proved the normality of Schubert varieties in arbitrary characteristic in the process of fixing a gap in Demazure's above cited paper. The normality of Schubert varieties in arbitrary characteristic was also proved by Andersen (cf. [4]) and Ramanan–Ramanathan (cf. [138]). The normality of Schubert varieties in characteristic 0 also follows from Joseph's paper "On Demazure character formula" (cf. [**77**]); this paper may be considered as providing the first proof of the normality of Schubert varieties in characteristic 0. A very short proof (of 2 pages) of the normality of Schubert varieties was given by Mehta–Srinivas (cf. [**128**]).

Recall that for a variety, the properties of being nonsingular in codimension 1 and Cohen–Macaulay together imply normality (cf. Serre criterion, see [**63**, Theorem 8.22A] for example). Arithmetic normality of a Schubert variety X (that is the normality of the cone $\widehat{X}$) was obtained as a consequence of the standard monomial theory [**105**], [**110**] for G classical. Arithmetic normality, arithmetic Cohen–Macaulayness and rational resolution of singularities for Schubert varieties for any semisimple G were obtained as consequences of Frobenius splitting of Schubert varieties [**138, 139**]; see also [**71**], [**72**] for results relating to Frobenius splitting. Recently Littelmann (cf. [**121**]) has given a "standard monomial theoretic" proof both for the arithmetic normality and arithmetic Cohen–Macaulayness of Schubert varieties for any semisimple G. Further, Kumar and Littelmann (cf. [**92**]) have given a proof for Frobenius splitting of Schubert varieties using quantum groups.

2.10. Equations defining a Schubert variety

Given a projective variety X embedded in a projective space $\mathbb{P}^n$, let $I(X)$ denote the homogeneous ideal defining X for this embedding (cf. [**63**]). By Hilbert's basis theorem, we can find a finite set of generators $f_1, \ldots, f_r$ for the ideal $I(X)$. We shall refer to f_i for $i = 1, \ldots, r$ as the *equations* defining X as a closed subvariety of $\mathbb{P}^n$.

Now take $X = G/Q$. Let L be any ample line bundle on G/Q. From above, L is very ample. Hence L induces a projective embedding $G/Q \hookrightarrow \mathrm{Proj}(H^0(G/Q, L))$ [**63**]; here, $\mathrm{Proj}(H^0(G/Q, L))(= \mathbb{P}(H^0(G/Q, L)^*))$ denotes $\mathrm{Proj}(C)$, where C is the symmetric algebra of $H^0(G/Q, L)$, and $H^0(G/Q, L)^*$ denotes the dual. In fact, identifying $H^0(G/Q, L)^*$ with the Weyl module $V_K(\lambda)$ (cf. §2.11.10 below), the point $e_{w,Q}$ in G/Q gets identified with the point in $\mathbb{P}(V_K(\lambda))$ representing the one-dimensional span of an extremal weight vector in $V_K(\lambda)$ of weight $w(\lambda)$ (see §2.11.9 below for the discussion on extremal weight vectors). Let

$$A = \bigoplus_{n \in \mathbb{Z}^+} H^0(G/Q, L^n)$$

$$A_Q(w) = \bigoplus_{n \in \mathbb{Z}^+} H^0(X_Q(w), L^n).$$

From [**140**], we know that the natural map

$$S^n(H^0(G/Q, L)) \to H^0(G/Q, L^n)$$

is surjective, and the kernel of the surjective map $\oplus_{n\in\mathbb{Z}^+} S^n(H^0(G/Q, L)) \to$ $\oplus_{n\in\mathbb{Z}^+} H^0(G/Q, L^n)$ is generated as an ideal by elements of degree 2. Further, the restriction map $A \to A_Q(w)$ is surjective, and its kernel is generated as an ideal by elements of degree 1. Thus we obtain that $\oplus_{n\in\mathbb{Z}^+} H^0(G/Q, L^n)$ is the homogeneous coordinate ring of G/Q for the embedding $G/Q \hookrightarrow \mathrm{Proj}(H^0(G/Q, L))$ and that the homogeneous ideal of G/Q is generated in degree 2, i.e., $I(G/Q)$ is generated as an ideal by the kernel of the surjective map $S^2(H^0(G/Q, L)) \to H^0(G/Q, L^2)$. Further, $\oplus_{n\in\mathbb{Z}^+} H^0(X_Q(w), L^n)$ is the homogeneous coordinate ring of $X(w)$ and the ideal defining the Schubert variety $X_Q(w)$ in G/Q is generated as an ideal by the kernel of the surjective map $H^0(G/Q, L) \to H^0(X_Q(w), L)$. Thus any Schubert variety $X_Q(w)$ in G/Q is scheme theoretically (even at the cone level) the intersection of G/Q with all the hyperplanes in $\mathrm{Proj}(H^0(G/Q, L))$ containing $X_Q(w)$.

There are similar results for multi-cones over Schubert varieties [81], [71], [72]. For a maximal parabolic subgroup P_i, we have that $\mathrm{Pic}\,(G/P_i)$ $(\simeq \mathbb{Z})$; further, the ample generator of $\mathrm{Pic}\,(G/P_i)$ $(\simeq \mathbb{Z})$ is in fact $L(\omega_i)$. Let us denote $L(\omega_i)$ by just L_i. For any parabolic subgroup Q, let us denote $S \setminus S_Q$ by $\{\alpha_1, \ldots, \alpha_t\}$, for some t. Let

$$C = \bigoplus_{\underline{a}} H^0\left(G/Q, \bigotimes_{i=1}^{t} L_i^{a_i}\right)$$

$$C_Q(w) = \bigoplus_{\underline{a}} H^0\left(X_Q(w), \bigotimes_{i=1}^{t} L_i^{a_i}\right),$$

where $\underline{a} = (a_1, \ldots, a_t) \in \mathbb{Z}_+^t$. It is shown in [81] that the natural map

$$\bigoplus_{\underline{a}} S^{a_1}(H^0(G/Q, L_1)) \otimes \cdots \otimes S^{a_t}(H^0(G/Q, L_t)) \to C$$

is surjective, and its kernel is generated as an ideal by elements of total degree 2. Further, the restriction map $C \to C_Q(w)$ is surjective, and its kernel is generated as an ideal by elements of total degree 1.

2.11. Representations of semisimple algebraic groups

In this section we review the representation theory of semisimple algebraic groups. Let $G, B, T, \ldots$, be as above. For $\alpha \in R$, let U_α be the root subgroup of G associated to α. Let $T_\alpha = (\ker \alpha)^\circ$ (the connected component of $\ker\alpha$), and $Z_\alpha = Z_G(T_\alpha)$, the centralizer of T_α in G. Note that Z_α is the subgroup of G generated by U_α and $U_{-\alpha}$. Therefore, Z_α is an algebraic group of semisimple rank 1, i.e., $\mathrm{rk}\,(Z_\alpha/R(Z_\alpha))$ is 1 here $R(Z_\alpha)$ denotes the radical of Z_α (cf. [20, 68]).

2.11.1. Action of W on $X(T)$. Let $X(T)$ $(= \mathrm{Hom}_{alg.gp.}(T, \mathbb{G}_m))$ be the character group of T, as above. Consider the action of W on T, namely, $w \cdot t = ntn^{-1}$, $w = nT \in W$, $t \in T$. Observe that ntn^{-1} depends only on the coset nT and not on the representative chosen. This action induces an action of W on $X(T)$. Namely, if $\chi \in X(T)$, then we define $w \circ \chi$, $w \in W$, as the character given by

$$w \circ \chi(t) = \chi(n^{-1}tn),$$

where n is a representative in $N(T)$ of w. Note that $w \circ \chi(t)$ has to be defined as $\chi(n^{-1}tn)$, rather than $\chi(ntn^{-1})$, in order to have the multiplicative property for χ as a homomorphism.

2.11.2. Weights in a G-module. Let V be a G-module, i.e., we have a morphism of algebraic groups $\rho : G \to \mathrm{Aut}(V)$, $\mathrm{Aut}(V)$ being the group of linear automorphisms of V. Now V breaks up as a direct sum $V = \sum_\chi V_\chi$, where $\chi \in X(T)$, and $V_\chi = \{v \in V \mid t \cdot v = \chi(t) \cdot v, \ t \in T\}$. Let $\Phi_\rho = \{\chi \mid V_\chi \neq (0)\}$ be the set of weights of V for the representation ρ. Let $\chi \in \Phi_\rho$, and $v \in V_\chi$. Let $w \in W$, say $w = nT$. Then $t \cdot nv = n(n^{-1}tn) \cdot v = \chi(n^{-1}tn)nv = w \circ \chi(t)nv$ since $n^{-1}tn \in T$. Thus we obtain $nv \in V_{w \circ \chi}$. In particular, $w \circ \chi \in \Phi_\rho$. Thus, W permutes the weights of ρ.

2.11.3. Lemma. [68] *Let $\alpha \in R$. Then $\rho(U_\alpha)$ maps a weight space V_α into $\sum_{k \in \mathbb{Z}^+} V_{\lambda + k\alpha}$.*

As a consequence, we obtain that $\rho(Z_\alpha)$ stabilizes $\sum_{k \in \mathbb{Z}} V_{\lambda + k\alpha}$. In particular, we have $s_\alpha(\lambda) = \lambda + k\alpha$, for some $k \in \mathbb{Z}$. Thus $\langle \lambda, \alpha \rangle \in \mathbb{Z}$ for all $\alpha \in R$. Hence we obtain

2.11.4. Proposition. *A weight χ of ρ is in fact an abstract weight, i.e., $\langle \chi, \alpha \rangle$ is an integer, $\alpha \in R$.*

2.11.5. Theorem. *(Borel's Fixed Point Theorem) If a connected solvable group H acts on a (nonempty) complete variety X, then the fixed point set X^H is nonempty.*

See [20] or [68] for a proof.

Let V be a G-module. Then Borel's fixed point theorem applied using the action of a Borel subgroup B on $X = \mathbb{P}(V)$ implies that there exists a nonzero vector $v \in V$, and a character $\lambda : B \to \mathbb{G}_m$ such that $b \cdot v = \lambda(b)v$, for all $b \in B$. Let V_1 be the span of $G \cdot v$. Then V_1 is G-cyclic, generated as a G-module by v. Let U^- denote the unipotent part of B^-, the Borel subgroup opposite to B. Now since $U^- \cdot B$ is dense in G, we find that V_1 is the K-span of $U^- B \cdot v$, i.e., the K-span of $U^- v$, since $B \cdot v = Kv$. Hence, in view of Lemma 2.11.3, we obtain that the weights in V_1 are of the form $\lambda - \sum c_\alpha \alpha$, $\alpha \in R^+$, $c_\alpha \in \mathbb{Z}^+$. Thus for any weight χ of V_1, say $\chi = \lambda - \sum c_\alpha \alpha$, we have $\lambda \geq \chi$. Here, the partial order $\geq$ on the weight lattice Λ is given by $\theta \geq \mu$ if $\theta - \mu$ is a nonnegative integral linear combination of the positive roots. For this reason, λ is called a highest weight in V, and the vector v is said to be a *highest-weight vector* with highest weight λ.

Further, for $w \in W$, $w \cdot \lambda$ is a weight in $\rho_1 : G \to GL(V_1)$, and hence $w \cdot \lambda \leq \lambda$. This implies that λ is dominant, i.e., $\lambda \in \Lambda^+$. In particular we obtain

2.11.6. Proposition. *Let V be a G-module.*

1. *There exist highest-weight vectors.*
2. *Let v be a highest-weight vector in a G-module V with weight λ. Then $\lambda \in \Lambda^+$.*
3. *If in addition V is G-cyclic, then any other weight is of the form $\lambda - \sum_{\alpha \in R^+} c_\alpha \alpha$, $c_\alpha \in \mathbb{Z}^+$; furthermore, the multiplicity of the weight λ is 1.*

2.11.7. Definition. Let V be a G-module (not necessarily finite-dimensional). A nonzero vector $v \in V$ is said to be a *highest-weight vector of weight λ* if B acts on v via λ, i.e., $b \cdot v = \lambda(b)v$ for all $b \in B$, where $\lambda : B \to \mathbb{G}_m$ is a character of B.

Thus, the isomorphism class of a finite-dimensional irreducible G-module determines a unique element of Λ^+. Conversely, given $\lambda \in \Lambda^+$, there exists a unique (up to an isomorphism) irreducible G-module $V(\lambda)$ with highest weight λ (see [**68**] for a proof). Hence we obtain

2.11.8. Theorem. *We have a bijection between Λ^+ and the set of the isomorphism classes of finite-dimensional irreducible G-modules.*

2.11.9. Extremal weight vectors. For a dominant weight λ, let $V(\lambda)$ be the irreducible $G_\mathbb{C}$-module (over $\mathbb{C}$) with highest weight λ (here, $G_\mathbb{C}$ is the group over $\mathbb{C}$ with the same root data as G). Writing $V(\lambda)$ as a direct sum of T weight spaces, we have (see [**70**] for example) that the multiplicity of the weight $w(\lambda)$, $w \in W$ is 1. One usually refers to $w(\lambda)$ as the *extremal weight* in $V(\lambda)$ corresponding to w; a generator of the one-dimensional weight space (of weight $w(\lambda)$) is referred to as the *extremal weight vector* (which is unique up to scalars) in $V(\lambda)$ of weight $w(\lambda)$.

The extremal weight vectors in $V(\lambda)$ play an important role in the study of G/B. In particular, they will be frequently used in the subsequent chapters.

2.11.10. Weyl and Demazure modules. Let $\mathfrak{g}_\mathbb{C} = \text{Lie } G_\mathbb{C}$, and let $U(\mathfrak{g}_\mathbb{C})$ be the universal enveloping algebra of $\mathfrak{g}_\mathbb{C}$. Let $U^+(\mathfrak{g}_\mathbb{C})$ be the sub-algebra of $U(\mathfrak{g}_\mathbb{C})$ generated by $\{X_\alpha, \alpha \in S\}$, and $U_\mathbb{Z}^+(\mathfrak{g}_\mathbb{C})$ be the Kostant $\mathbb{Z}$-form of $U^+(\mathfrak{g}_\mathbb{C})$) (recall (cf. [**84**], p. 95-97)) that $U_\mathbb{Z}^+(\mathfrak{g}_\mathbb{C})$ is the $\mathbb{Z}$-subalgebra of $U^+(\mathfrak{g}_\mathbb{C})$ generated by $\{\frac{X_\alpha^n}{n!}, \alpha \in R^+, n \in \mathbb{N}\}$).

Let λ be a dominant weight. Fix a highest-weight vector u_λ in $V(\lambda)$. For $w \in W$, fix a representative n_w for w in $N_T(G)$, the normalizer of T in G, and set $u_{w,\lambda} = n_w \cdot u_\lambda$. Having fixed λ, we shall denote u_λ by just u, resp. $u_{w,\lambda}$ by u_w. Set $V_{w,\mathbb{Z}}(\lambda) = U_{\mathbb{Z}}^+(\mathfrak{g}_{\mathbb{C}})u_w$. For any field K, let $V_{w,\lambda} = V_{w,\mathbb{Z}}(\lambda) \otimes K$, $w \in W$. Then $V_K(\lambda) := V_{w_0,\lambda}$ ($= V_{w_0,\mathbb{Z}}(\lambda) \otimes K$) is the *Weyl module* corresponding to λ, and for $w \in W$, $V_{w,\lambda}$ is the *Demazure module* corresponding to w and λ. The vectors $u_{w,\lambda}$ for $w \in W$ are also called the extremal weight vectors in $V_K(\lambda)$.

Note, the Bruhat–Chevalley order also appears in the context of Demazure modules as follows: Given a dominant weight λ, let P_λ be the stabilizer of $K \cdot u$, where u as above, is a highest-weight vector u in $V(\lambda)$; note that the Weyl group W_{P_λ} of P_λ is the subgroup of W generated by $\{s_\alpha, \alpha \in S, s_\alpha(\lambda) = \lambda\}$.

2.11.11. Theorem. *For $\tau, w \in W^{P_\lambda}$, $u_\tau \in V_{w,\lambda} \Longleftrightarrow w \geq \tau$.*

For a proof, see [**11**], Theorem 2.9 when Char K is 0, and [**135**, Lemma 1.2], for arbitrary characteristics. Also, we have (see [**76**] for example)

$$(2.11.12) \qquad H^0(G/B, L(\lambda)) \cong V_K(\lambda)^*, \ H^0(X(w), L(\lambda)) \cong V_{w,\lambda}^*,$$

where $L(\lambda)$ is the line bundle on G/B associated to λ.

The vectors u_w, $w \in W$ are called the *extremal weight vectors* in the Weyl module $V_K(\lambda)$ and they again play an important role in the study of G/B.

2.11.13. The vectors $p_{w,\lambda}$, $w \in W$. As above, let λ be a dominant weight. In view of the above identification, we have that $-\lambda$ is the lowest weight in $H^0(G/B, L(\lambda))$; further, the corresponding weight space is one-dimensional. Fix a generator e_λ for the weight space of weight $-\lambda$. For $w \in W$, let $p_{w,\lambda} = n_w e_\lambda, n_w$ being as above. Note that $p_{w,\lambda}$ is a weight vector of weight $-w(\lambda)$. Having fixed λ, we shall denote e_λ by just e, resp. $p_{w,\lambda}$ by p_w. The vectors p_w, $w \in W$, are the extremal weight vectors in $H^0(G/B, L(\lambda))$ and these again play an important role in the study of G/B.

Let λ be a dominant weight, say $\lambda = \sum a_i \omega_i$. Let $Q = P_\lambda$. Let $J = \{i, 1 \leq i \leq l \mid a_i \neq 0\}$; then, Q is the parabolic subgroup with $S \setminus J$ as the associated set of simple roots. Consider the projective embedding

$$G/Q \hookrightarrow \mathbb{P}(V_K(\lambda)) \ (= \mathrm{Proj}(H^0(G/Q, L(\lambda))))$$

induced by the very ample line bundle $L(\lambda)$ on G/Q. The surjectivity of $H^0(G/Q, L(\lambda)) \to H^0(X_Q(w), L(\lambda))$, $w \in W_Q^{\min}$, implies that $\mathbb{P}(V_{w,\lambda})$ is the smallest linear subspace of $\mathbb{P}(V_K(\lambda))$ that contains $X_Q(w)$.

2.11.14. Lemma. *Let notation be as above. For $\tau, w \in W^{P_\lambda}$, we have $p_\tau \mid_{X_Q(w)} \neq 0 \Longleftrightarrow w \geq \tau$.*

PROOF. This result is a consequence of Theorem 2.11.11. There exists a $\mathfrak{g}$-invariant, nondegenerate bilinear form on $V_K(\lambda) \times V_K(\lambda)^*$ (cf. [24]) which we shall denote by $\mathbf{B}(\cdot, \cdot)$; here, the $\mathfrak{g}$-invariance of $\mathbf{B}(,)$ is equivalent to $\mathbf{B}(Hx, x') + \mathbf{B}(x, Hx') = 0$, $H \in \mathfrak{g}$, $x \in V_K(\lambda), x' \in V_K(\lambda)^*$. Now let x, x' be two weight vectors of weight χ, χ' in $V_K(\lambda)$, $V_K(\lambda)^*$ respectively. Taking for H an element of the Cartan subalgebra $\mathfrak{h}$ of $\mathfrak{g}$, we have the following string of equalities

$$0 = \chi(H)\mathbf{B}(x, x') + \chi'(H)\mathbf{B}(x, x')$$
$$= (\chi(H) + \chi'(H))\mathbf{B}(x, x')$$
$$= (\chi + \chi')(H)\mathbf{B}(x, x').$$

Suppose now that $\chi + \chi' \neq 0$. Then choosing H such that $(\chi + \chi')(H) \neq 0$, we obtain $\mathbf{B}(x, x') = 0$. Now choosing $x = u_\phi$, $x' = p_\tau$, we have $\mathbf{B}(u_\phi, p_\tau) = 0$, if $\phi \neq \tau$. Further, if $\phi = \tau$, then $\mathbf{B}(u_\tau, p_\tau) \neq 0$ since $\mathbf{B}(\cdot, \cdot)$ is nondegenerate, and $p_\tau \neq 0$, there exists a $y \in V_K(\lambda)$ such that $\mathbf{B}(y, p_\tau) \neq 0$. Such a y is a weight vector of weight $-\text{weight}(p_\tau) = \text{weight}(u_\tau)$, i.e., y is a scalar multiple of u_τ. Hence for $w \in W_Q^{\min}$, we have $p_\tau \mid_{X_Q(w)} \neq 0 \Longleftrightarrow u_\tau \in V_{w,\lambda}$, i.e., if and only if $w \geq \tau$ in view of Theorem 2.11.11. $\square$

2.11.15. Minuscule fundamental weights. A fundamental weight ω is said to be *minuscule* if $\langle \omega, \beta \rangle \leq 1$, for all $\beta \in R^+$, or equivalently, all weights in the Weyl module $V_K(\omega)$ are extremal. Let P be the maximal parabolic subgroup associated to ω. If ω is minuscule, then P is said to be a maximal parabolic subgroup of minuscule type or just a *minuscule parabolic subgroup*. In Chapter 4, we shall give a geometric definition of a minuscule fundamental weight (cf. §4.8.1). From the definition, it follows that the extremal weight vectors u_w, $w \in W$ in $V_K(\omega)$ (cf. §2.11.10) form a basis for $V_K(\omega)$. Similarly, the extremal weight vectors p_w, $w \in W$ in $H^0(G/P, L_\omega)$ (cf. §2.11.13) form a basis for $H^0(G/P, L_\omega)$.

The minuscule fundamental weights are characterized by the property that in the expression for the co-root $\beta^\vee$, β being the highest root, in terms of the simple co-roots, the co-root associated to ω occurs with coefficient 1. Hence indexing the simple roots as on Page 207, we have the following list of minuscule fundamental weights.

Type **A:** Every fundamental weight is minuscule.

Type **B:** ω_n is minuscule.

Type **C:** ω_1 is minuscule.

Type **D:** $\omega_1, \omega_{n-1}, \omega_n$ are minuscule.

Type **E$_6$:** ω_1, ω_6 are minuscule.

Type **E$_7$:** ω_7 is minuscule.

In types **E$_8$**, **F$_4$**, and **G$_2$**, there are no minuscule fundamental weights!

CHAPTER 3

Specifics for the Classical Groups

The Grassmannian variety $G_{d,n}$ and the flag variety $SL(n)/B$ are important varieties in algebraic geometry and algebraic groups. In this chapter, we spell out the generalities considered in the previous chapter for the semisimple group $SL(n)$. We first review the results for the Grassmannians; we then review the results for $SL(n)/B$. For more details, one may refer to [**49, 57**]. Important details for the other classical groups $Sp(2n), SO(2n+1)$, and $SO(2n)$ are also spelled out.

3.1. The Grassmannian variety $G_{d,n}$

3.1.1. The Plücker embedding. Let us fix integers d, n such that $1 \leq d < n$ and let $V = K^n$. The *Grassmannian* $G_{d,n}$ is the set of all d-dimensional subspaces $U \subset V$. Let U be an element of $G_{d,n}$, and $a_1, \ldots, a_d$ a basis of U, where a_j is a vector of the form

$$a_j = \begin{pmatrix} a_{1j} \\ a_{2j} \\ \cdots \\ a_{nj} \end{pmatrix}, \text{ with } a_{ij} \in K, \text{for } 1 \leq i \leq n, \ 1 \leq j \leq d.$$

Thus, the basis $a_1, \ldots, a_d$ gives rise to an $n \times d$ matrix $A = (a_{ij})$ of rank d, whose columns are the vectors $a_1, \ldots, a_d$. Similarly, if another basis $a'_1, \ldots, a'_d$ for U gives rise to another $n \times d$ matrix A', then there exists a change of basis matrix $C \in GL_d(K)$, such that $A' = AC$. Conversely, given two $n \times d$ matrices A and A' of rank d, such that $A' = AC$, for some matrix $C \in GL_d(K)$, the columns of both A and A' are d linearly independent vectors in V, generating the same d-dimensional subspace of V. Identifying an $n \times d$ matrix with a point in the affine space $\mathbb{A}^{nd}$, we see that $G_{d,n}$ can be viewed as $(\mathbb{A}^{nd} \setminus Z)/ \sim$, where Z is the set of $n \times d$ matrices of rank less than d, and the equivalence relation $\sim$ is defined by

$$A \sim A' \quad \text{if} \quad \text{there exists } C \in GL_d(K) \text{such that } A' = AC.$$

Note the similarity with the definition of the projective space $\mathbb{P}^n$; in fact it is easily seen that $\mathbb{P}^{n-1} = G_{1,n}$.

Define the set

$$I_{d,n} = \{\underline{i} = (i_1, \ldots, i_d) \in \mathbb{Z} | 1 \leq i_1 < \cdots < i_d \leq n\} \,.$$

Then $I_{d,n}$ has $N = \binom{n}{d}$ elements, and the coordinates of the affine space $\wedge^d V = K^N$ will be indexed by the set $I_{d,n}$. Let

$$X = \underbrace{V \oplus \cdots \oplus V}_{d \text{ times}} = K^{nd}.$$

The exterior product map

$$\wedge^d : X \to \wedge^d V$$

sends the element $(a_1, \ldots, a_d)$ to the point $a_1 \wedge \cdots \wedge a_d$ in K^N, whose $\underline{i}$-th coordinate is given by the $d \times d$ minor of the $n \times d$ matrix $A = (a_{ij})$, with row indices $i_1, \ldots, i_d$, where the columns of the matrix $A = (a_{ij})$ are the vectors $a_1, \ldots, a_d$. Note that $a_1 \wedge \cdots \wedge a_d = 0$ if and only if the matrix A belongs to the set Z of matrices of rank $< d$. Moreover, it is easily seen that given $(a_1, \ldots, a_d)$ and $(a'_1, \ldots, a'_d)$ in X such that the associated $n \times d$ matrices A and A' are equivalent, i.e., there exists a matrix $C \in GL_d(k)$ such that $A' = AC$, then

$$a'_1 \wedge \cdots \wedge a'_d = \det(C) a_1 \wedge \cdots \wedge a_d.$$

This shows that the map $\wedge^d$ induces a well-defined map

$$p : G_{d,n} \to \mathbb{P}(\wedge^d V) = \mathbb{P}^{N-1}$$

called the *Plücker map*. It is a well-known fact (see [**57**] for example) that the Plücker map is injective and defines a projective variety structure on $G_{d,n}$. For $\underline{i} \in I_{d,n}$, the $\underline{i}$-th component of p is denoted by $p_{\underline{i}}$, or by $p_{i_1, \ldots, i_d}$, where $\underline{i} = (i_1, \ldots, i_d)$; the $p_{\underline{i}}$'s, with $\underline{i} \in I_{d,n}$, are called the *Plücker coordinates*. If a point U in $G_{d,n}$ is represented by the $n \times d$ matrix A, then $p_{i_1, \ldots, i_d}(U) = \det(A_{i_1, \ldots, i_d})$, where $A_{i_1, \ldots, i_d}$ denotes the matrix whose rows are the rows of A with indices $i_1, \ldots, i_d$, in this order. Note that the Plücker coordinates $p_{\underline{i}}$, $\underline{i} \in I_{d,n}$ form a basis of $(\wedge^d K^n)^*$.

For each $\underline{i} \in I_{d,n}$ consider the point $e_{\underline{i}}$ of $G_{d,n}$ represented by the $n \times d$ matrix whose entries are all 0, except the ones in the i_j-th row and j-th column, for each $1 \le j \le d$, which are equal to 1. Clearly, for $\underline{i}, \underline{j} \in I_{d,n}$,

$$(3.1.2) \qquad p_{\underline{i}}(e_{\underline{j}}) = \begin{cases} 1, & \text{if } \underline{i} = \underline{j}; \\ 0, & \text{otherwise.} \end{cases}$$

Therefore, the $e_{\underline{j}}$'s form a dual basis to the Plücker coordinates.

3.1.3. Schubert varieties of $G_{d,n}$. Let $e_1, \ldots, e_n$ be the standard basis for V. For $1 \le i \le n$, let V_i be the subspace of V spanned by $\{e_1, \ldots, e_i\}$. For each $\underline{i} \in I_{d,n}$, the *Schubert variety in $G_{d,n}$ associated to $\underline{i}$* is defined to be

$$(3.1.4) \qquad X_{\underline{i}} = \{U \in G_{d,n} \mid \dim(U \cap V_{i_t}) \ge t, \ 1 \le t \le d\}.$$

Given $\underline{i} = (i_1, \ldots, i_d), \underline{j} = (j_1, \ldots, j_d) \in I_{d,n}$, define a partial order $\geq$ on $I_{d,n}$ by

$$(3.1.5) \qquad \underline{i} \geq \underline{j} \Leftrightarrow i_t \geq j_t, \text{for all } 1 \leq t \leq d.$$

If $X_{\underline{i}}$, $X_{\underline{j}}$ are the associated Schubert varieties in $G_{d,n}$, we have $\underline{i} \geq \underline{j} \iff X_{\underline{i}} \supseteq X_{\underline{j}}$. In other words, the partial order $\geq$ on $I_{d,n}$ is induced by the Bruhat–Chevalley order on the set of Schubert varieties (see [44, 108, 137] for details).

Let $G = SL(n)$, and let T be the maximal torus in G consisting of diagonal matrices. Then W may be identified with the symmetric group S_n. Let B be the Borel subgroup of G consisting of upper triangular matrices. For the action of G on $\mathbb{P}(\wedge^d V)$ the T-fixed points are precisely the T-eigenvectors in $\wedge^d V$. We have the following decomposition

$$\wedge^d V = \bigoplus_{\underline{i} \in I_{d,n}} K e_{\underline{i}}, \quad \text{as } T\text{-modules}.$$

Thus the T-fixed points in $\mathbb{P}(\wedge^d V)$ are precisely $[e_{\underline{i}}]$, $\underline{i} \in I_{d,n}$, and these points, obviously, belong to $G_{d,n}$. The Schubert variety $X_{\underline{i}}$ associated to $\underline{i}$ can also be viewed as the Zariski closure of the B-orbit $B[e_{\underline{i}}]$ through the T-fixed point $[e_{\underline{i}}]$ with the canonical reduced scheme structure.

3.1.6. Identification of G/P_d with $G_{d,n}$. One can identify G/P_d with $G_{d,n}$ as follows. Let P_d be the maximal parabolic subgroup of G with $S \setminus \{\alpha_d\}$ as the associated set of simple roots. Then

$$P_d = \left\{ A \in G \,\middle|\, A = \begin{pmatrix} * & * \\ 0_{(n-d) \times d} & * \end{pmatrix} \right\},$$
$$W_{P_d} = S_d \times S_{n-d}.$$

Hence the minimal length coset representatives of W/W_P are

$$W_{P_d}^{\min} = \{(w_1 \ldots w_n) \in S_n \mid w_1 < \cdots < w_d, \quad w_{d+1} < \cdots < w_n\}.$$

Thus $W_{P_d}^{\min}$ may be identified with

$$I_{d,n} := \{\underline{i} = (i_1, \ldots, i_d) \mid 1 \leq i_1 < \cdots < i_d \leq n\}$$

by sending $(w_1, \ldots, w_n)$ to $(w_1, \ldots, w_d)$. For the natural action of G on $\mathbb{P}(\wedge^d K^n)$, the isotropy group at $[e_1 \wedge \cdots \wedge e_d]$ is P_d while the orbit through $[e_1 \wedge \cdots \wedge e_d]$ is $G_{d,n}$. Thus we obtain a surjective map $\pi : G \to G_{d,n}, g \mapsto g \cdot a$, where $a = [e_1 \wedge \cdots \wedge e_d]$. Further, the differential $(d\pi)_e : T(G)_e \to T(G_{d,n})_a$ ($=$ the tangent space to $G_{d,n}$ at a) is easily seen to be surjective. Hence we obtain an identification of $G_{d,n}$ with the orbit space G/P_d scheme-theoretically (cf. [20], Proposition 6.7). With this identification, we have $H^0(G/P_d, L(\omega_d)) = (\wedge^d K^n)^*$ and the Plücker coordinates $p_{\underline{i}}$, $\underline{i} \in I_{d,n}$ are simply the extremal weight vectors in $H^0(G/P_d, L(\omega_d))$ (cf. §2.11.9).

Under the identification of $I_{d,n}$ with $W_{P_d}^{\min}$, a d-tuple $\underline{i} \in I_{d,n}$ gets identified with the element $(i_1, \ldots, i_d, j_1, \ldots, j_{n-d}) \in S_n$, where $\{j_1, \ldots, j_{n-d}\}$ is the complement of $\{i_1, \ldots, i_d\}$ in $\{1, \ldots, n\}$ arranged in increasing order. In the sequel, we shall denote an element $(w_1 \ldots w_n) \in W_Q^{\min}$ by just $(w_1 \ldots w_d)$.

We note the following facts for Schubert varieties $X_{\underline{i}} = X_{P_d}(\underline{i})$ in the Grassmannian (see [**57, 131**] for example):

1. **Bruhat Decomposition**: $X_{\underline{i}} = \bigcup\limits_{\underline{j} \le \underline{i}} B e_{\underline{j}}$.

2. **Dimension**: $\dim X_{\underline{i}} = \sum\limits_{1 \le t \le d} i_t - t$.

3. **Young diagram representation**: To $(i_1 \cdots i_d) \in I_{d,n}$, we associate the partition $\mathbf{i} := (\mathbf{i}_1, \ldots, \mathbf{i}_d)$, where $\mathbf{i}_t = i_{d-t+1} - (d - t + 1)$, $1 \le t \le d$. For example, if $d = 3$ and $(i_1 i_2 i_3) = (147)$, then $\mathbf{i} = (4, 2, 0)$:

For a partition $\mathbf{i} = (\mathbf{i}_1, \ldots, \mathbf{i}_d)$, where $\mathbf{i}_d \le n - d$, we shall denote by $X_{\mathbf{i}}$ the Schubert variety corresponding to $(\mathbf{i}_1, \ldots, \mathbf{i}_d)$. Then $\dim X_{\mathbf{i}} = |\mathbf{i}| = \mathbf{i}_1 + \cdots + \mathbf{i}_d$.

4. **Defining equations**: In view of Equation (3.1.2) and the Bruhat decomposition, we have

$$p_{\underline{j}}\big|_{X_{\underline{i}}} \ne 0 \iff \underline{i} \ge \underline{j}.$$

Let $I(\underline{i})$ be the homogeneous ideal in $K[x_{ij}, 1 \le i \le n, 1 \le j \le d]$ defining $X_{\underline{i}}$ for the Plücker embedding and $J(\underline{i})$ be the ideal (in the same polynomial ring) generated by $\{p_{\underline{j}} \mid \underline{i} \not\ge \underline{j}\}$. Clearly $J(\underline{i}) \subseteq I(\underline{i})$. We have in fact $J(\underline{i}) = I(\underline{i})$ [**57, 131**].

3.2. The special linear group $SL(n)$

Let $G = SL(n)$, the special linear group of rank $n - 1$. As above, let T be the maximal torus consisting of all the diagonal matrices in G, and B the Borel subgroup consisting of all the upper triangular matrices in G.

Weyl group of type A_{n-1}. We have $N_G(T)$ is the set of all matrices in G with exactly one nonzero entry in each row and each column. Therefore, $W = N_G(T)/T$ can be identified with S_n, the symmetric group on n letters. We will denote a permutation $w \in S_n$ by its one-line notation $(w_1 w_2 \ldots w_n)$ if 1 maps to w_1, 2 maps to w_2 etc.

Root system of type A_{n-1}. We denote the simple roots by $\alpha_i = \epsilon_i - \epsilon_{i+1}$ for $1 \leq i \leq n-1$. Note that $\epsilon_i - \epsilon_{i+1}$ is the character sending $\mathrm{diag}(t_1, \ldots, t_n)$ to $t_i t_{i+1}^{-1}$ as seen in Chapter 2. Then $R = \{\epsilon_i - \epsilon_j \mid 1 \leq i, j \leq n, \ i \neq j\}$, and the reflection $s_{\epsilon_i - \epsilon_j}$ may be identified with the transposition (i, j) in S_n.

Let us denote the simple reflections in S_n by $\{s_i, \ 1 \leq i \leq n-1\}$, where s_i is the transposition $(i, i+1)$.

The Dynkin diagram of type A_{n-1} is

$$
\begin{array}{ccccccc}
\bullet & \!\!\!-\!\!\!-\!\!\! & \bullet & \!\!\!-\!\!\!-\!\!\! & \bullet & \cdots\cdots & \bullet & \!\!\!-\!\!\!-\!\!\! & \bullet \\
1 & & 2 & & 3 & & n-1 & & n
\end{array}
$$

Chevalley basis. The Lie algebra $sl(n) = \mathrm{Lie}(SL(n))$ consists of all matrices with trace 0. Given a root $\beta = \epsilon_j - \epsilon_k, 1 \leq j, k \leq n$, the element X_β of the Chevalley basis of $sl(n)$ is given by $X_\beta = E_{jk}$, where E_{jk} is the elementary matrix with 1 at the (j, k)-th place, and 0's elsewhere.

Bruhat–Chevalley order. This partial order has a simpler characterization that is much more efficient for computations than using reduced words as in §2.7. For $w_1, w_2 \in W$, we have

$$X(w_1) \subset X(w_2) \iff \pi_d(X(w_1)) \subset \pi_d(X(w_2)), \ \forall 1 \leq d \leq n-1,$$

where π_d is the canonical projection $G/B \to G/P_d$. Hence we obtain that for $(a_1 \ldots a_n), (b_1 \ldots b_n) \in S_n$,

(3.2.5)
$$(a_1 \ldots a_n) \geq (b_1 \ldots b_n) \iff \{a_1 \ldots a_d\} \uparrow \geq \{b_1 \ldots b_d\} \uparrow, \forall 1 \leq d \leq n-1.$$

Here, for a d-tuple $(a_1 \ldots a_d)$ of distinct integers, $\{a_1 \ldots a_d\} \uparrow$ denotes the ordered d-tuple obtained from $\{a_1, \ldots, a_d\}$ by arranging its elements in ascending order. Two sequences of increasing integers are then compared entry by entry as in the partial order on $I_{d,n}$ given by (3.1.5).

Björner and Brenti [**16**] have shown that in fact one only needs to check those d for which d is a descent in $(a_1 \ldots a_n)$.

3.2.6. The flag variety $SL(n)/B$. Let $V = K^n$. A sequence $(0) = V_0 \subset V_1 \subset \cdots \subset V_n = V$, such that $\dim V_i = i$, is called a *full flag* in V. Let $\mathcal{F}(V)$ denote the set of all full flags in V. Let $\{e_i, 1 \leq i \leq n\}$ be the standard basis of K^n. The flag $F_0 = (V_0 \subset \cdots \subset V_i \subset \cdots \subset V_n)$, where V_i is the span of $\{e_1, \ldots, e_i\}$ for $1 \leq i \leq n$, is called the *standard flag*. Any flag $F = (V_0 \subset \cdots \subset V_i \subset \cdots \subset V_n)$ can be represented by an invertible $n \times n$ matrix where the first column spans V_1, the first and second columns together span V_2, etc.

We have a natural action of G on $\mathcal{F}(V)$ by multiplication of matrices. This action is transitive. The isotropy group at F_0 is precisely B. Thus, through the identification of $\mathcal{F}(V)$ with G/B, $\mathcal{F}(V)$ acquires the structure of a projective variety. We have a canonical closed embedding

$$\theta : \mathcal{F}(V) \hookrightarrow G_{1,n} \times G_{2,n} \times \cdots \times G_{n-1,n}.$$

Under this embedding we have

$$\mathcal{F}(V) = \{(U_1, \ldots, U_{n-1}) \in \prod_{i=1}^{n-1} G_{i,n} \mid U_1 \subset \cdots \subset U_{n-1}\}.$$

It is not difficult to see that the incidence relations $U_i \subset U_{i+1}$, $1 \leq i \leq n-2$ identify $\mathcal{F}(V)$ as a closed subset of $\prod_{i=1}^{n-1} G_{i,n}$ (for example, see [**49, 68**] for details).

3.2.7. The partially ordered set $I_{a_1,\ldots,a_k}$. Let Q be a parabolic subgroup in $SL(n)$ containing B. Let $1 \leq a_1 < \cdots < a_k \leq n$, such that $S_Q = S \setminus \{\alpha_{a_1}, \ldots, \alpha_{a_k}\}$. Then $Q = P_{a_1} \cap \cdots \cap P_{a_k}$, and $W_Q = \mathcal{S}_{a_1} \times \mathcal{S}_{a_2-a_1} \times \cdots \times \mathcal{S}_{n-a_k}$. Let

$$I_{a_1,\ldots,a_k} = \{(\underline{i}_1, \ldots, \underline{i}_k) \in I_{a_1,n} \times \cdots \times I_{a_k,n} \mid \underline{i}_t \subset \underline{i}_{t+1} \text{ for all } 1 \leq t \leq k-1\}.$$

Then it is easily seen that $W_Q^{\min}$ may be identified with $I_{a_1,\ldots,a_k}$.

The partial order on the set of Schubert varieties in G/Q (given by inclusion) induces a partial order $\geq$ on $I_{a_1,\ldots,a_k}$, namely, for $\mathbf{i} = (\underline{i}_1, \ldots, \underline{i}_k)$, $\mathbf{j} = (\underline{j}_1, \ldots, \underline{j}_k) \in I_{a_1,\ldots,a_k}$, $\mathbf{i} \geq \mathbf{j} \iff \underline{i}_t \geq \underline{j}_t$ for all $1 \leq t \leq k$.

3.2.8. The minimal and maximal representatives as permutations. Let $w \in W$, and let $w_Q^{\min}$ be the element in $W_Q^{\min}$ that represents the coset wW_Q. Under the identification of $W_Q^{\min}$ with $I_{a_1,\ldots,a_k}$, let $\mathbf{i} = (\underline{i}_1, \ldots, \underline{i}_k)$ be the element in $I_{a_1,\ldots,a_k}$ that corresponds to $w_Q^{\min}$. As a permutation, the element $w_Q^{\min}$ is given by $\underline{i}_1$, followed by $\underline{i}_2 \setminus \underline{i}_1$ arranged in ascending order, and so on, ending with $\{1, \ldots, n\} \setminus \underline{i}_k$ arranged in ascending order.

Similarly, if $w_Q^{\max}$ is the element in $W_Q^{\max}$ that represents the coset wW_Q, then as a permutation, the element $w_Q^{\max}$ is given by $\underline{i}_1$ arranged in descending order, followed by $\underline{i}_2 \setminus \underline{i}_1$ arranged in descending order, etc.

3.2.9. Equations defining Schubert varieties in the flag variety. Let $X(w) \subset G/B$. Let $w^{(d)}$ be the d-tuple corresponding to the Schubert variety that is the image of $X(w)$ under the projection $G/B \to G/P_d$, $1 \leq d \leq l$, i.e., if $w = (w_1, \ldots, w_n)$ (in one-line notation), then $w^{(d)} = (w_1, \ldots, w_d)$ (arranged in ascending order).

3.2.10. Theorem. [**110, 133**] *The ideal sheaf of $X(w)$ in G/B is generated by*

$$\bigcup_{1 \leq d \leq n-1} \{p_{\underline{i}} \mid \underline{i} \in I_{d,n}, \ w^{(d)} \not\geq \underline{i}\}.$$

More generally we have

3.2.11. Theorem. [110, 133] *Let Q be a parabolic subgroup and let $w \in W^Q$. Let $\{\alpha_{a_1}, \ldots, \alpha_{a_k}\} = S \setminus S_Q$. Then the ideal sheaf of $X_Q(w)$ in G/Q is generated by*

$$\bigcup_{1 \leq d \leq k} \{p_{\underline{i}} \mid \underline{i} \in I_{a_d, n}, \ w^{(d)} \not\geq \underline{i}\}.$$

3.3. The symplectic group $Sp(2n)$

Let $V = K^{2n}$ together with a nondegenerate, skew-symmetric bilinear form $(\cdot, \cdot)$. Let $H = SL(V)$ and $G = Sp(V) = \{A \in SL(V) \mid A$ leaves the form $(\cdot, \cdot)$ invariant $\}$. Taking the matrix of the form (with respect to the standard basis $\{e_1, \ldots, e_{2n}\}$ of V) to be

$$E = \begin{pmatrix} 0 & J \\ -J & 0 \end{pmatrix}$$

where J is the anti-diagonal $(1, \ldots, 1)$ of size $n \times n$, we may realize $Sp(V)$ as the fixed point set of a certain involution σ on $SL(V)$, namely $G = H^\sigma$, where $\sigma : H \longrightarrow H$ is given by $\sigma(A) = E(^tA)^{-1}E^{-1}$. Thus

$$\begin{aligned}
G = Sp(2n) &= \{A \in SL(2n) \mid {}^tAEA = E\} \\
&= \{A \in SL(2n) \mid E^{-1}(^tA)^{-1}E = A\} \\
&= \{A \in SL(2n) \mid E(^tA)^{-1}E^{-1} = A\} \\
&= H^\sigma.
\end{aligned}$$

Note that $E^{-1} = -E$. Denoting by T_H the maximal torus in H consisting of diagonal matrices, resp. by B_H the Borel subgroup in H consisting of upper triangular matrices, we see easily that T_H, B_H are stable under σ. We set $T_G = T_H{}^\sigma, B_G = B_H{}^\sigma$. Then it can be seen easily that T_G is a maximal torus in G and B_G is a Borel subgroup in G. We note the following specific facts for this group.

Weyl group of type C_n. Denoting by W_G the Weyl group of G, we have

$$W_G = \{(a_1 \cdots a_{2n}) \in S_{2n} \mid a_i = 2n + 1 - a_{2n+1-i}, \ 1 \leq i \leq 2n\}.$$

Thus $w = (a_1 \cdots a_{2n}) \in W_G$ is known once $(a_1 \cdots a_n)$ is known. We shall denote an element $(a_1 \cdots a_{2n})$ in W_G by just $(a_1 \cdots a_n)$. For example, $(4231) \in S_4$ represents $(42) \in C_2$.

Root system of type C_n. Denoting by R_G the set of roots of G with respect to T_G and by R_G^+ the set of positive roots with respect to B_G, we have

$$(3.3.2) \qquad R_G = \{\pm(\varepsilon_i \pm \varepsilon_j),\ 1 \le i < j \le n\} \cup \{\pm 2\varepsilon_i,\ i = 1, \dots, n\}$$

$$(3.3.3) \qquad R_G^+ = \{(\varepsilon_i \pm \varepsilon_j),\ 1 \le i < j \le n\} \cup \{2\varepsilon_i,\ i = 1, \dots, n\}.$$

The simple roots in R_G^+ are given by

$$(3.3.4) \qquad \{\alpha_i = \varepsilon_i - \varepsilon_{i+1},\ 1 \le i \le n-1\} \cup \{\alpha_n = 2\varepsilon_n\}.$$

Let us denote the simple reflections in W_G by $\{s_i,\ 1 \le i \le n\}$, namely, $s_i =$ reflection with respect to $\varepsilon_i - \varepsilon_{i+1}$, $1 \le i \le n-1$, and $s_n =$ reflection with respect to $2\varepsilon_n$. Then we have

$$s_i = \begin{cases} r_i r_{2n-i}, & \text{if } 1 \le i \le n-1 \\ r_n, & \text{if } i = n, \end{cases}$$

where r_i denotes the transposition $(i, i+1)$ in S_{2n}, $1 \le i \le 2n - 1$. Continuing the example above, $(42) = s_1 s_2 s_1$.

The Dynkin diagram for C_n is

Maximal parabolics. For $1 \le d \le n$, we let P_d be the maximal parabolic subgroup of G with $S \setminus \{\alpha_d\}$ as the associated set of simple roots. Then it can be seen easily that $W_G^{P_d}$, the set of minimal representatives of W_G/W_{P_d} can be identified with

$$\left\{ (a_1 \cdots a_d) \,\middle|\, \begin{array}{ll} (1) & 1 \le a_1 < a_2 < \cdots < a_d \le 2n \\ (2) & \text{for } 1 \le i \le 2n, \text{ if } i \in \{a_1, \dots, a_d\} \\ & \text{then } 2n + 1 - i \notin \{a_1, \dots, a_d\} \end{array} \right\}.$$

Bruhat–Chevalley order. For $w_1 = (a_1 \cdots a_{2n})$ and $w_2 = (b_1 \cdots b_{2n})$ in W_G we have $w_2 \ge w_1 \Leftrightarrow \{b_1, \dots, b_d\} \uparrow \ge \{a_1, \dots, a_d\} \uparrow$ for each $1 \le d \le n$ (cf. [137]). Here $\{a_1, \dots, a_d\}\uparrow$, $\{b_1, \dots, b_d\}\uparrow$ are the corresponding d-tuples arranged in ascending order compared entry by entry as in (3.2.5). Hence for $w \in W_G$, denoting by $w^{(d)}$ the element in $W_G^{P_d}$ that represents the coset wW_{P_d}, we have for $w_1, w_2 \in W_G$ and $1 \le d \le n$,

$$w_2^{(d)} \ge w_1^{(d)},\ 1 \le d \le n \iff \{b_1, \dots, b_d\} \uparrow \ge \{a_1, \dots, a_d\} \uparrow.$$

Further, $w_2 \ge w_1 \iff w_2^{(d)} \ge w_1^{(d)}$, $1 \le d \le n$. But now, the latter condition is equivalent to $w_2 \ge w_1$ in W_H. Thus we obtain that the partial order on W_G is induced by the partial order on W_H (cf. [137]). In particular, for $w_1 = (a_1 \cdots a_d)$, $w_2 = (b_1 \cdots b_d)$, $w_1, w_2 \in W_G^{P_d}$, we have $w_2 \geq w_1 \Leftrightarrow \{b_1, \dots, b_d\} \uparrow \geq \{a_1, \dots, a_d\} \uparrow$.

In the sequel, we shall denote an element $(a_1 \cdots a_n)$ in $W_G^{P_d}$ by just $(a_1 \cdots a_d)$.

Chevalley basis. For $1 \leq i \leq 2n$, set $i' = 2n + 1 - i$. The involution $\sigma :$ $SL(2n) \to SL(2n)$, $A \mapsto E({}^tA)^{-1}E^{-1}$, induces an involution $\sigma : sl(2n) \to$ $sl(2n)$, $A \mapsto -E({}^tA)E^{-1}(= E({}^tA)E$, since $E^{-1} = -E)$. In particular, we have, for $1 \leq i, j \leq 2n$

$$\sigma(E_{ij}) = \begin{cases} -E_{j'i'}, & \text{if } i, j \text{ are both } \leq n \text{ or both } > n \\ E_{j'i'}, & \text{if one of } \{i, j\} \text{ is } \leq n \text{ and the other } > n. \end{cases}$$

where E_{ij} is the elementary matrix with 1 at the (i, j)-th place and 0 elsewhere. Further

$$\text{Lie } Sp(2n) = \{A \in sl(2n) \mid E({}^tA)E = A\}.$$

The Chevalley basis $\{H_{\alpha_i} : \alpha_i \in S\} \cup \{X_\alpha : \alpha \in R\}$ for Lie $Sp(2n)$ may be given as follows:

$$H_{\epsilon_i - \epsilon_{i+1}} = E_{ii} - E_{i+1,i+1} + E_{(i+1)',(i+1)'} - E_{i'i'}$$
$$H_{2\epsilon_n} = E_{nn} - E_{n'n'}$$
$$X_{\epsilon_j - \epsilon_k} = E_{jk} - E_{k'j'}$$
$$X_{\epsilon_j + \epsilon_k} = E_{jk'} + E_{kj'}$$
$$X_{2\epsilon_m} = E_{mm'}$$
$$X_{-(\epsilon_j - \epsilon_k)} = E_{kj} - E_{j'k'}$$
$$X_{-(\epsilon_j + \epsilon_k)} = E_{k'j} + E_{j'k}$$
$$X_{-2\epsilon_m} = E_{m'm}.$$

3.4. The odd orthogonal group SO$(2n+1)$

Let $V = K^{2n+1}$ together with a nondegenerate symmetric bilinear form $(\cdot, \cdot)$. Taking the matrix of the form $(\cdot, \cdot)$, with respect to the standard basis $\{e_1, \dots, e_{2n+1}\}$ of V, to be the $2n + 1 \times 2n + 1$ anti-diagonal matrix with 1's all along the anti-diagonal except at the $n+1 \times n+1$-th place where the entry is 2. We will denote this matrix by E. Note that the associated quadratic form Q on V is given by $Q(\sum_{i=1}^{2n+1} x_i e_i) = x_{n+1}^2 + \sum_{i=1}^n x_i x_{2n+2-i}$. We may realize $G = SO(V)$ as the fixed point set $SL(V)^\sigma$, where $\sigma :$ $SL(V) \to SL(V)$ is given by $\sigma(A) = E^{-1}({}^tA)^{-1}E$. Set $H = SL(V)$.

Denoting by T_H the maximal torus in H consisting of diagonal matrices, resp. B_H the Borel subgroup in H consisting of upper triangular matrices, we see easily that T_H and B_H are stable under σ. We set $T_G = T_H^\sigma$, $B_G = B_H^\sigma$. Then it can be seen easily that T_G is a maximal torus in G and B_G is a Borel subgroup in G.

We note that the following basic facts hold.

Weyl group of type B_n. Denoting by W_G the Weyl group of G, we have

$$W_G = \{(a_1 \ldots a_{2n+1}) \in S_{2n+1} \mid a_i = 2n + 2 - a_{2n+2-i},\ 1 \leq i \leq 2n + 1\}.$$

Thus $w = (a_1 \ldots a_{2n+1}) \in W_G$ is known once $(a_1 \ldots a_n)$ is known. Note that $a_{n+1} = n + 1$, for all $w \in W_G$. In the sequel, we shall denote an element $(a_1 \ldots a_{2n+1})$ in W_G by just $(a_1 \ldots a_n)$. This group is isomorphic to the Weyl group of type C_n.

Root system of type B. Denoting by R_G the set of roots of G with respect to T_G and by R_G^+ the set of positive roots with respect to B_G), we have

$$(3.4.3) \qquad R_G = \{\pm(\varepsilon_i \pm \varepsilon_j),\ 1 \leq i < j \leq n\} \cup \{\pm\varepsilon_i,\ i = 1, \ldots, n\},$$

$$(3.4.4) \qquad R_G^+ = \{(\varepsilon_i \pm \varepsilon_j),\ 1 \leq i < j \leq n\} \cup \{\varepsilon_i,\ i = 1, \ldots, n\}.$$

The simple roots in R_G^+ are given by

$$\{\alpha_i = \varepsilon_i - \varepsilon_{i+1},\ 1 \leq i \leq n - 1\} \cup \{\alpha_n = \varepsilon_n\}.$$

Let us denote the simple reflections in W_G by $\{s_i,\ 1 \leq i \leq n\}$, namely, $s_i =$ the reflection with respect to $\varepsilon_i - \varepsilon_{i+1}$, $1 \leq i \leq n - 1$, and $s_n =$ the reflection with respect to ε_n. Then we have

$$s_i = \begin{cases} r_i r_{2n+1-i}, & 1 \leq i \leq n - 1, \\ r_n r_{n+1} r_n, & i = n \end{cases}$$

where r_i denotes the transposition $(i, i + 1)$ in S_{2n+1}, $1 \leq i \leq 2n$. The Dynkin diagram of B_n is

$$\begin{matrix} \bullet & \bullet & \bullet & \cdots & \bullet \Rightarrow \bullet \\ 1 & 2 & 3 & & n-1 \quad n \end{matrix}$$

Maximal parabolic. For $1 \leq d \leq n$, we let P_d be the maximal parabolic subgroup of G with $S \setminus \{\alpha_d\}$ as the associated set of simple roots. Then it can be seen easily that $W_G^{P_d}$, the set of minimal representatives of W_G/W_{P_d} can be identified with

$$\left\{ (a_1 \cdots a_d) \ \middle| \ \begin{matrix} (1) & 1 \leq a_1 < a_2 < \cdots < a_d \leq 2n + 1 \\ (2) & a_i \neq n + 1, 1 \leq i \leq d \\ (3) & \text{for } 1 \leq i \leq 2n + 1,\ \text{if } i \in \{a_1, \ldots, a_d\} \\ & \text{then } 2n + 2 - i \notin \{a_1, \ldots, a_d\} \end{matrix} \right\}.$$

Bruhat–Chevalley order. For $w_1 = (a_1 \cdots a_{2n+1})$, $w_2 = (b_1 \cdots b_{2n+1})$, $w_1, w_2 \in W_G$, we have $w_2 \geq w_1 \Leftrightarrow$ the d-tuple $\{b_1, \ldots, b_d\} \uparrow \geq \{a_1, \ldots, a_d\} \uparrow$ for all $1 \leq d \leq n$ (cf. [**137**]). Here again recall $\{a_1, \ldots, a_i\} \uparrow$ and $\{b_1, \ldots, b_i\} \uparrow$ are the corresponding i-tuples arranged in ascending order (see (3.2.5))). Let $w^{(d)}$ be the element of $W_G^{P_d}$ representing the coset wW_{P_d}. For $w_1, w_2 \in W_G$,

$$w_2^{(d)} \geq w_1^{(d)}, \ 1 \leq d \leq n \iff \{b_1, \ldots, b_i\} \uparrow \geq \{a_1, \ldots, a_i\} \uparrow,$$

for all i, $1 \leq i \leq 2n + 1$. But now the latter condition is equivalent to $w_2 \geq w_1$ in W_H. Thus we obtain that the partial order on W_G is induced by the partial order on W_H. In particular, for $w_1 = (a_1 \cdots a_d)$, $w_2 = (b_1 \cdots b_d)$, $w_1, w_2 \in W_G^{P_d}$, we have

$$w_2 \geq w_1 \Leftrightarrow \{b_1, \ldots, b_d\} \uparrow \geq \{a_1, \ldots, a_d\} \uparrow.$$

In the sequel, we shall denote an element $(a_1 \cdots a_{2n+1})$ in $W_G^{P_d}$ by just $(a_1 \cdots a_d)$.

Chevalley basis. For $1 \leq k \leq 2n+1$, set $k' = 2n+2-k$. The involution $\sigma : SL(2n+1) \to SL(2n+1), A \mapsto E^{-1}({}^tA)^{-1}E$, induces an involution $\sigma : sl(2n+1) \to sl(2n+1), A \mapsto -E^{-1}({}^tA)E$. In particular, we have $\sigma(E_{ij}) = -E_{j'i'}$, $1 \leq i, j \leq 2n+1$, where E_{ij} is the elementary matrix with 1 at the (i, j)-th place and 0 elsewhere. Further

$$\text{Lie } SO(2n+1) = \{A \in sl(2n+1) \mid E^{-1}({}^tA)E = -A\}.$$

The Chevalley basis $\{H_{\alpha_i} : \alpha_i \in S\} \cup \{X_\alpha : \alpha \in R\}$ for Lie $SO(2n+1)$ may be given as follows:

$$H_{\epsilon_i - \epsilon_{i+1}} = E_{ii} - E_{i+1,i+1} + E_{(i+1)',(i+1)'} - E_{i'i'}$$
$$H_{\epsilon_n} = 2(E_{nn} - E_{n'n'})$$
$$X_{\epsilon_j - \epsilon_k} = E_{jk} - E_{k'j'}$$
$$X_{\epsilon_j + \epsilon_k} = E_{jk'} - E_{kj'}$$
$$X_{\epsilon_m} = 2E_{mn+1} - E_{n+1m'}$$
$$X_{-(\epsilon_j - \epsilon_k)} = E_{kj} - E_{j'k'}$$
$$X_{-(\epsilon_j + \epsilon_k)} = E_{k'j} - E_{j'k}$$
$$X_{-\epsilon_m} = E_{n+1m} - 2E_{m'n+1}.$$

3.5. The even orthogonal group SO($2n$)

Let $V = K^{2n}$ together with a nondegenerate symmetric bilinear form $(\cdot, \cdot)$. Taking the matrix of the form $(\cdot, \cdot)$ (with respect to the standard basis $\{e_1, \ldots, e_{2n}\}$ of V) to be E, the anti-diagonal $(1, \ldots, 1)$ of size $2n \times 2n$. We may realize $G = SO(V)$ as the fixed point set $SL(V)^\sigma$, where $\sigma : SL(V) \to SL(V)$ is given by $\sigma(A) = E({}^tA)^{-1}E$. Set $H = SL(V)$.

Denoting by T_H the maximal torus in H consisting of diagonal matrices and by B_H the Borel subgroup in H consisting of upper triangular matrices, we see easily that T_H, B_H are stable under σ. We set $T_G = T_H{}^\sigma$, $B_G = B_H{}^\sigma$. Then it follows that T_G is a maximal torus in G and B_G is a Borel subgroup in G.

We note that the following basic facts hold.

Weyl group of type D_n. Denoting by W_G the Weyl group of G, we have

$$W_G = \left\{ (a_1 \cdots a_{2n}) \in S_{2n} \left|
\begin{array}{ll}
(1) & a_i = 2n + 1 - a_{2n+1-i}, 1 \le i \le 2n \\
(2) & \#\{i, 1 \le i \le n \,|\, a_i > n\} \text{ is even}
\end{array}
\right. \right\}.$$

Thus $w = (a_1 \ldots a_{2n}) \in W_G$ is known once $(a_1 \ldots a_n)$ is known. This group is a subgroup of the Weyl group of type C_n. In the sequel, we shall denote an element $(a_1 \cdots a_{2n})$ in W by just $(a_1 \cdots a_n)$.

Root system of type D. Denoting by R_G the set of roots of G with respect to T_G and by R_G^+ the set of positive roots with respect to B_G, we have

$$R_G = \{\pm(\varepsilon_i \pm \varepsilon_j),\ 1 \le i < j \le n\},$$
$$R_G^+ = \{(\varepsilon_i \pm \varepsilon_j),\ 1 \le i < j \le n\}.$$

The simple roots in R_G^+ are given by

$$\{\alpha_i = \varepsilon_i - \varepsilon_{i+1},\ 1 \le i \le n-1\} \cup \{\alpha_n = \varepsilon_{n-1} + \varepsilon_n\}.$$

Let us denote the simple reflections in W_G by $\{s_i,\ 1 \le i \le n\}$, namely, s_i = the reflection with respect to $\varepsilon_i - \varepsilon_{i+1}$, $1 \le i \le n-1$, and s_n = the reflection with respect to $\varepsilon_{n-1} + \varepsilon_n$. Then we have (cf. [23]),

$$s_i = \begin{cases} r_i r_{2n-i}, & 1 \le i \le n-1, \\ r_n r_{n-1} r_{n+1} r_n, & i = n, \end{cases}$$

where r_i denotes the transposition $(i, i+1)$ in S_{2n}, $1 \le i \le 2n - 1$. The Dynkin diagram of D_n is

Maximal parabolics. For $1 \le d \le n$, we let P_d be the maximal parabolic subgroup of G with $S \setminus \{\alpha_d\}$ as the associated set of simple roots. For $w \in W$, let $w^{(d)}$ denote the element of $W_G^{P_d}$. Then it can be seen easily that $W_G^{P_d}, d \ne n-1$, can be identified with

$$(3.5.4) \qquad \left\{ (a_1 \cdots a_d) \left|
\begin{array}{ll}
(1) & 1 \le a_1 < a_2 < \cdots < a_d \le 2n \\
(2) & \text{for } 1 \le i \le 2n, \text{ if } i \in \{a_1, \ldots, a_d\} \\
& \text{then } 2n + 1 - i \notin \{a_1, \ldots, a_d\}
\end{array}
\right. \right\}.$$

For $d = n - 1$, if $w \in W_G^{P_d}$, then

$$w \equiv wu_i \pmod{W_{P_{n-1}}},\ 0 \le i \le n, i \ne n - 1,$$

where

$$u_i = \begin{cases} s_{\alpha_n}, & \text{if } i = n \\ id, & \text{if } i = 0 \\ s_{\alpha_i} s_{\alpha_{i+1}} \cdots s_{\alpha_{n-2}} s_{\alpha_n}, & \text{if } 1 \le i \le n - 2. \end{cases}$$

Note that the set $\{wu_i,\ 0 \le i \le n, i \ne n - 1\}$ is totally ordered under the Bruhat order; note also that given $w \in W$, there are n different $n-1$-tuples representing the coset wW_{P_d}, namely, the tuples given respectively by the first $n - 1$ entries in wu_i, $0 \le i \le n, i \ne n - 1$. Hence for $d = n - 1$, $W_G^{P_d}$ gets identified with a certain *proper* subset of (3.5.4); in particular, for $w_1 = (a_1 \cdots a_{2n})$, $w_2 = (b_1 \cdots b_{2n})$, $w_1, w_2 \in W_G$, we can have $w_1^{(n-1)} = w_2^{(n-1)}$, with $\{a_1, \dots, a_{n-1}\} \uparrow$ and $\{b_1, \dots, b_{n-1}\} \uparrow$ being different. For $w \in W$, say $w = (a_1 \cdots a_{2n})$, we see easily that

$$w^{(d)} = \{a_1, \dots, a_d\} \uparrow,\ 1 \le d \le n,\ d \ne n - 1$$

and

$$w^{(n-1)} = \text{the least (under } \ge \text{) in the totally ordered set } Y$$

where

$$Y = \{(y_1^{(i)}, \dots, y_{n-1}^{(i)}) \uparrow\ 0 \le i \le n,\ i \ne n - 1\}.$$

$y_1^{(i)}, \dots, y_{n-1}^{(i)}$ being the first $(n - 1)$ entries in wu_i, $0 \le i \le n$, $i \ne n - 1$. Here, the partial order $\ge$ is the usual partial order, namely, $(i_1, \dots, i_{n-1}) \ge (j_1, \dots, j_{n-1})$, if $i_t \ge j_t$, $1 \le t \le n - 1$, where $(i_1, \dots, i_{n-1})$, $(j_1, \dots, j_{n-1})$ are two increasing sequences of $(n - 1)$-tuples.

Bruhat–Chevalley order. For $1 \le i \le 2n$, let $i' = 2n + 1 - i$, and $|i| = \min\{i, i'\}$. We shall denote the Bruhat order on W_G by $\succeq$. Given $w_1 = (a_1 \cdots a_{2n})$, $w_2 = (b_1 \cdots b_{2n})$, $w_1, w_2 \in W_G$, we have $w_2 \succeq w_1$ if and only if the following two conditions hold [**137**].

1. For $1 \le d \le n$, we have $\{b_1, \dots, b_d\} \uparrow\ \ge \{a_1, \dots, a_d\} \uparrow$, for all d.
2. Let $(c_1, \dots, c_d)$, resp. $(e_1, \dots, e_d)$ be the increasing sequence $\{a_1, \dots, a_d\} \uparrow$, resp. $\{b_1, \dots, b_d\} \uparrow$. Suppose for some r, $1 \le r \le d$, and some i, $0 \le i \le d-r$, $\{|c_{i+1}|, \dots, |c_{i+r}|\} = \{|e_{i+1}|, \dots, |e_{i+r}|\} = \{n+1-r, \dots, n\}$ as sets (order does not matter). Then $\#\{j : i+1 \le j \le i + r \text{ and } c_j > n\}$, and $\#\{j : i + 1 \le j \le i + r \text{ and } e_j > n\}$ should both be even or both be odd.

3.5.5. Remark. Thus the Bruhat order $\succeq$ on W_G is *not* induced from the Bruhat order on W_H. In the sequel, we shall have occasion to use both of the partial orders $\succeq$ and $\ge$.

Following the terminology in [**137**], we shall refer to $\{|c_{i+1}|, \dots, |c_{i+r}|\}$ and $\{|e_{i+1}|, \dots, |e_{i+r}|\}$ as *analogous parts* if they satisfy the hypothesis in Condition 2 above, and if $\{c_1, \dots, c_d\}$ and $\{e_1, \dots, e_d\}$ have analogous parts of the same parity, we say they are **D-compatible**.

3.5.6. Remark. (a) Let $(c_1, \dots, c_d)$, $(e_1, \dots, e_d) \in W_G^{P_d}$, where $(c_1, \dots, c_d) \succeq (e_1, \dots, e_d)$. Suppose $(c_1, \dots, c_d)$, $(e_1, \dots, e_d)$ have analogous parts. Then it is easily seen that the condition (2) is equivalent to the condition that $\#\{j,\ 1 \le j \le d \mid c_j > n\}$ and $\#\{j,\ 1 \le j \le d$ and $e_j > n\}$ are both even or both odd.

(b). Given $\theta \in W$, say $\theta = (a_1 \cdots a_{2n})$, denoting by $y_1^{(i)}, \dots, y_{n-1}^{(i)}$ the first $(n-1)$ entries in θu_i, $0 \le i \le n$, $i \ne n - 1$, we have

$$(y_1^{(i)}, \dots, y_{n-1}^{(i)}) = \begin{cases} (x_1, \dots, x_{n-1}), \ 1 \le i \le n, \ i \ne n - 1 \\ (a_1, \dots, a_{n-1}), \ i = 0, \end{cases}$$

where for $1 \le i \le n - 2$, $(x_1, \dots, x_{n-1})$ is the $(n-1)$-tuple obtained from $(a_1, \dots, a_{n-1})$ by replacing a_i by a_i', and for $i = n$, $(x_1, \dots, x_{n-1}) = (a_1, \dots, a_{n-2}, a_n')$. Further, we have $\theta^{(n-1)}$ is the least (under $\ge$) in

$$\{(y_1^{(i)}, \dots, y_{n-1}^{(i)}) \uparrow, 0 \le i \le n, \ i \ne n - 1\}.$$

(c). Given $\theta, w \in W$, say $\theta = (a_1 \cdots a_{2n})$, $w = (b_1 \cdots b_{2n})$, we have (with notations as in (b) above)

$$w^{(n-1)} \succeq \theta^{(n-1)} \Leftrightarrow \{b_1, \dots, b_{n-1}\} \uparrow \succeq \{y_1^{(i)}, \dots, y_{n-1}^{(i)}\} \uparrow$$

for some i, $0 \le i \le n$, $i \ne n - 1$.

Chevalley basis. Recall that for $1 \le k \le 2n$, we set $k' = 2n + 1 - k$. The involution $\sigma : SL(2n) \to SL(2n)$, $A \mapsto E({}^tA)^{-1}E$, induces an involution $\sigma : sl(2n) \to sl(2n)$, $A \mapsto -E({}^tA)E$. In particular, we have, for $1 \le i, j \le 2n$, $\sigma(E_{ij}) = -E_{j'i'}$, where E_{ij} is the elementary matrix with 1 at the (i, j)-th place and 0 elsewhere. Further,

$$\text{Lie } SO(2n) = \{A \in sl(2n) \mid E({}^tA)E = -A\}.$$

The Chevalley basis for Lie $SO(2n)$ may be given as follows:

$$H_{\epsilon_i - \epsilon_{i+1}} = E_{ii} - E_{i+1,i+1} + E_{(i+1)',(i+1)'} - E_{i'i'}$$
$$H_{\epsilon_{n-1} + \epsilon_n} = E_{n-1,n-1} + E_{n,n} - E_{n',n'} - E_{(n-1)',(n-1)'}$$
$$X_{\epsilon_j - \epsilon_k} = E_{jk} - E_{k'j'}$$
$$X_{\epsilon_j + \epsilon_k} = E_{jk'} - E_{kj'}$$
$$X_{-(\epsilon_j - \epsilon_k)} = E_{kj} - E_{j'k'}$$
$$X_{-(\epsilon_j + \epsilon_k)} = E_{k'j} - E_{j'k}.$$

CHAPTER 4

The Tangent Space and Smoothness

In this chapter we introduce the tangent space at a point and review several criteria for smoothness for arbitrary varieties. It is well known that the tangent space at a point of an irreducible variety always has dimension at least as large as the dimension of the variety. Furthermore, a variety is smooth at a point if the dimensions are equal. If one knows equations defining the variety, then the Jacobian criterion can be used to determine the dimension of the tangent space at a point. In this chapter we also recall the notion of multiplicity at a given point on an algebraic variety and relate this to another criterion for smoothness. There are still many other ways for determining smoothness in Schubert varieties which will be discussed in the remaining chapters of this book. A summary of all criteria for smoothness and rational smoothness appears on Page 208.

4.1. The Zariski tangent space

Let x be a point on a variety X. Let $\mathfrak{m}_x$ be the maximal ideal of the local ring $\mathcal{O}_{X,x}$ with residue field $K(x)(=\mathcal{O}_{X,x}/\mathfrak{m}_x)$. Note that $K(x) = K$, since K is algebraically closed. The Zariski tangent space to X at x is defined as
$$T_x(X) = \mathrm{Der}_K(\mathcal{O}_{X,x}, K(x))$$
where $\mathrm{Der}_K(\mathcal{O}_{X,x}, K(x))$ is the set of all K-linear maps $D : \mathcal{O}_{X,x} \to K(x)$ such that $D(ab) = D(a)b + aD(b)$. Here, $K(x)$ is regarded as an $\mathcal{O}_{X,x}$-module. It can be seen easily that $T_x(X)$ is canonically isomorphic to $\mathrm{Hom}_{K\text{-mod}}(\mathfrak{m}_x/\mathfrak{m}_x^2, K)$.

Let X be an affine variety defined by polynomials $f_1, f_2, \ldots, f_k$ in the variables $x_1, \ldots, x_n$. Then the tangent space to a point $p = (p_1, \ldots, p_n) \in X$ is the zero set of the k linear polynomials
$$\sum_{i=1}^{n} \frac{\partial f_j}{\partial x_i}(p)(x_i - p_i).$$

4.2. Smooth and singular points

A point x on a variety X is said to be a *simple* or *smooth* or *nonsingular point of X* if $\mathcal{O}_{X,x}$ is a regular local ring, i.e., the maximal ideal has a set of $n = \dim \mathcal{O}_{X,x}$ generators. A point x that is not simple is called a *multiple* or *non-smooth* or *singular point* of X. The set of all singular points in X, denoted Sing$X$, is called the *singular locus of X*. A variety X is said to be *smooth* if Sing$X = \emptyset$.

4.2.1. Theorem. *Let $x \in X$. Then $\dim_K T_x(X) \geq \dim \mathcal{O}_{X,x}$ ($\dim \mathcal{O}_{X,x}$ is also denoted $\dim_x X$) with equality if and only if x is a simple point of X.*

PROOF. We have

$$\dim_K T_x(X) = \dim_K(\mathfrak{m}_x/\mathfrak{m}_x^2) \geq \dim \mathcal{O}_{X,x},$$

with equality if and only if $\mathcal{O}_{X,x}$ is regular (here, the inequality $\dim_K(\mathfrak{m}_x/\mathfrak{m}_x^2) \geq \dim \mathcal{O}_{X,x}$ is a consequence of Nakayama's lemma; see [**45**] for details). The result now follows. $\square$

4.3. The space $T(w, \tau)$

In this section we describe a basis for the tangent spaces to a Schubert variety. Because $X(w)$ is the union of its B-orbits, points in an orbit will have isomorphic tangent spaces. Therefore, we only need to examine one point in each orbit. The T-fixed points e_τ for $\tau \leq w$ are well suited for this purpose.

The Zariski tangent space of $X(w)$ at any point naturally sits inside the Lie algebra of G as follows. First, the tangent space to G, resp. B, at e_{id} can be identified with Lie G, resp. Lie B. We have Lie $G = \mathfrak{h} \oplus_{\beta \in R} \mathfrak{g}_\beta$ while Lie $B = \text{Lie}T \oplus_{\beta \in R^+} \mathfrak{g}_\beta$. Hence the tangent space to G/B at e_{id} gets identified with $\text{Lie}G/\text{Lie}B \cong \oplus_{\beta \in R^+} \mathfrak{g}_{-\beta}$. Second, we can identify the tangent space at any other T-fixed point in G/B by conjugating the Borel subgroup B. Namely, for $\tau \in W$, fix a lift n_τ of τ in $N_G(T)$. We can identify G/B with $G/\tau B \tau^{-1}$ via the map $gB \mapsto n_\tau g n_\tau^{-1} n_\tau B n_\tau^{-1} (= n_\tau g B n_\tau^{-1})$. Then the tangent space to G/B at e_τ gets identified with $\oplus_{\beta \in \tau(R^+)} \mathfrak{g}_{-\beta}$.

More specifically, let $T(w, \tau)$ be the Zariski tangent space to $X(w)$ at e_τ for $\tau \leq w \in W$. Since $X(w) \subset X(w_0) = G/B$ where w_0 is the longest element in W, we have $T(w, \tau) \subset T(w_0, \tau) = \oplus_{\beta \in \tau(R^+)} \mathfrak{g}_{-\beta}$. For each $\alpha \in R$, fix X_α to be an element in the one-dimensional root space $\mathfrak{g}_\alpha$. Set

$$N(w, \tau) = \{\beta \in \tau(R^+) \mid X_{-\beta} \in T(w, \tau)\}.$$

Since $T(w, \tau)$ is a T-stable subspace of $T(w_0, \tau)$, $\{X_{-\beta}, \beta \in N(w, \tau)\}$ is a basis for $T(w, \tau)$.

The main goal of Chapter 5 is to give an explicit description of $N(w, \tau)$ for the classical groups using standard monomial theory. The general case of constructing $N(w, \tau)$ has recently been done by Carrell and Kuttler using the Peterson map [**34**].

More generally, let Q be a parabolic subgroup. Let $w, \tau \in W^Q$, $w \geq \tau$. Let $T_Q(w, \tau)$ be the tangent space to $X_Q(w)$ at $e_{\tau,Q}$. For $w = w_0$, we have

$$T_Q(w_0, \tau) = \bigoplus_{\beta \in \tau(R^+ \setminus R_Q^+)} \mathfrak{g}_{-\beta}.$$

Let $N_Q(w, \tau) = \{\beta \in \tau(R^+ \setminus R_Q^+) \mid X_{-\beta} \in T_Q(w, \tau)\}$. Then

$$T_Q(w, \tau) = \text{ the span of } \{X_{-\beta}, \ \beta \in N_Q(w, \tau)\}.$$

4.4. A canonical affine neighborhood of a T-fixed point

Let $B^- := w_0 B w_0$ be the Borel subgroup opposite to B, called the *opposite Borel*. Let $\tau \in W$. Let U_τ^- be the unipotent subgroup of G generated by the root subgroups $U_{-\beta}$, $\beta \in \tau(R^+)$; note that U_τ^- is the unipotent part of the Borel subgroup ${}^\tau B^-$, opposite to ${}^\tau B(= \tau B \tau^{-1})$. Then under the canonical map $G \to G/B$, $g \mapsto g e_\tau$, U_τ^- is mapped isomorphically onto its image $U_\tau^- e_\tau$ which is a (dense) open subset of G/B. We have

$$U_{-\beta} \simeq \mathbb{G}_a,$$

$$U_\tau^- \simeq \prod_{\beta \in \tau(R^+)} U_{-\beta}.$$

Hence $U_\tau^- e_\tau$ gets identified with $\mathbb{A}^N$, where $N = \#R^+ = l(w_0)$ via the above identification, and we shall denote the induced coordinate system on $U_\tau^- e_\tau$ by $\{x_{-\beta}, \ \beta \in \tau(R^+)\}$. We shall denote $U_\tau^- e_\tau$ by $\mathcal{O}_\tau$. Thus we obtain that $\mathcal{O}_\tau$ is an affine neighborhood of e_τ in G/B. This neighborhood is canonical in the sense that this is the unique T-stable affine neighborhood of e_τ (see [**30**] for details).

4.4.1. The affine variety $Y(w, \tau)$. For $w \in W$, $w \geq \tau$, let us denote $Y(w, \tau) := \mathcal{O}_\tau \cap X(w)$. It is a nonempty affine open subvariety of $X(w)$, and a closed subvariety of the affine space $\mathcal{O}_\tau$. For $\tau = \text{id}$, $Y(w, \text{id})$ is usually called the *opposite cell in $X(w)$*, though in general it is not a cell.

4.4.2. Equations defining $Y(w, \tau)$ in $\mathcal{O}_\tau$. Let $w \in W$. For $1 \leq i \leq l$ ($= \text{rank} G$), fix a basis D_i for the kernel of the surjective map $H^0(G/B, L_i)$ $\to H^0(X(w), L_i)$ given by restriction, L_i being the ample generator of $\text{Pic}(G/P_i)$. We have (cf. §2.10) that the ideal sheaf of $X(w)$ in G/B is generated by $\{f \in D_i, 1 \leq i \leq l\}$. Let $\tau \in W$ be such that $\tau \leq w$. Let $I(w, \tau)$ be the ideal defining $Y(w, \tau)$ as a subvariety of $\mathcal{O}_\tau$. Let $f|_{\mathcal{O}_\tau}$ be the restriction of f to $\mathcal{O}_\tau$, then we have that

(4.4.3) $I(w, \tau)$ is generated by $\{f|_{\mathcal{O}_\tau}, \ f \in D_i, 1 \leq i \leq l\}$.

For example, take $G = SL(4)$. By Theorem 3.2.11, the ideal defining $Y((3412), \text{id})$ in $\mathcal{O}_{\text{id}}$ is generated by $\{p_4, p_{234}\}$, and the ideal of $Y((2413), \text{id})$ is generated by $\{p_3, p_4, p_{34}, p_{134}, p_{234}\}$.

4.4.4. The affine variety $Y_Q(w, \tau)$. More generally, given a parabolic subgroup Q and $\tau \in W^Q$, let $U_{\tau,Q}^-$ be the subgroup of G generated by the root subgroups $U_{-\beta}$, $\beta \in \tau(R^+ \setminus R^+(Q))$. Then under the canonical map $G \to G/Q$, $g \mapsto ge_{\tau,Q}$, $U_{\tau,Q}^-$ is mapped isomorphically onto its image $U_{\tau,Q}^- e_{\tau,Q}$ which is a (dense) open subset of G/Q. We have

$$U_{\tau,Q}^- \simeq \prod_{\beta \in \tau(R^+ \setminus R^+(Q))} U_{-\beta}.$$

Hence $U_{\tau,Q}^- e_{\tau,Q}$ gets identified with $\mathbb{A}^{N_Q}$, where $N_Q = \#\{R^+ \setminus R^+(Q)\}$ via the above identification, and we shall denote the induced coordinate system on $U_{\tau,Q}^- e_{\tau,Q}$ by $\{x_{-\beta}, \ \beta \in \tau(R^+ \setminus R_Q^+)\}$. In the sequel, we shall denote $U_{\tau,Q}^- e_{\tau,Q}$ by $\mathcal{O}_{\tau,Q}$. Thus we obtain that $\mathcal{O}_{\tau,Q}$ is an affine neighborhood of $e_{\tau,Q}$ in G/Q.

For $w \in W^Q$, $w \geq \tau$, let us denote $Y_Q(w, \tau) := \mathcal{O}_{\tau,Q} \cap X_Q(w)$. It is a nonempty affine open subvariety of $X_Q(w)$, and a closed subvariety of the affine space $\mathcal{O}_{\tau,Q}$. For $\tau = \mathrm{id}$, $Y_Q(w, \mathrm{id})$ is called the *opposite cell in* $X_Q(w)$, though in general it is not a cell.

4.5. Tangent cone and Jacobian criteria for smoothness

In this section, we give two general criteria for smoothness.

Tangent cone criterion. Let $\mathcal{O}_{X,P}$ be the local ring at $P \in X$ with maximal ideal $\mathfrak{m}_P$. The associated graded ring is

$$\mathrm{gr}\mathcal{O}_{X,P} = \sum_{n \geq 0} \mathfrak{m}_P^n / \mathfrak{m}_P^{n+1}.$$

$\mathrm{Spec}(\mathrm{gr}\mathcal{O}_{X,P})$ is the *tangent cone* to X at P. We have

$$(4.5.2) \qquad\qquad \mathrm{gr}\mathcal{O}_{X,P} \xrightarrow{\sim} \mathrm{Sym}(\mathrm{gr}_1\mathcal{O}_{X,P})/I$$

for some ideal I. In particular, X is smooth at P if and only if $\mathrm{gr}\mathcal{O}_{X,P}$ is a polynomial algebra generated by degree one elements, i.e., I is trivial [**62**, Lecture 14].

Jacobian criterion. Let Y be an affine variety in $\mathbb{A}^n$, and let $I(Y)$ be the ideal defining Y in $\mathbb{A}^n$. Let $I(Y)$ be generated by $\{f_1, f_2, \ldots, f_r\}$. Let J be the Jacobian matrix $(\frac{\partial f_i}{\partial x_j})$ and let J_P be the evaluation of J at a point P. Then the dimension of the tangent space to Y at a point P is greater than or equal to the dimension of Y with equality if and only if P is a smooth point. Equivalently, rank $J_P \leq \mathrm{codim}_{\mathbb{A}^n} Y$ with equality if and only if P is a smooth point of Y.

For example, take $X(2413)$ and $X(3412)$. As we have stated in Section 4.4.2, the equations defining $Y((3412), \mathrm{id})$ in $\mathcal{O}_{\mathrm{id}}$ are $\{p_4, p_{234}\}$, and the equations defining $Y((2413), \mathrm{id})$ are $\{p_3, p_4, p_{34}, p_{134}, p_{234}\}$. Identifying $\mathcal{O}_{\mathrm{id}}$ with the group of unipotent lower triangular matrices in SL_4, we have

(4.5.4) $\qquad p_3 = x_{31}$

(4.5.5) $\qquad p_4 = x_{41}$

(4.5.6) $\qquad p_{34} = x_{42}x_{31} - x_{41}x_{32}$

(4.5.7) $\qquad p_{134} = x_{43}x_{32} - x_{42}$

(4.5.8) $\qquad p_{234} = x_{21}(x_{43}x_{32} - x_{42}) - (x_{43}x_{31} - x_{41}).$

Hence, the equations defining $Y((3412), \mathrm{id})$ are $\{x_{41},\ x_{21}(x_{43}x_{32} - x_{42}) - (x_{43}x_{31} - x_{41})\}$. The Jacobian matrix is

(4.5.9)
$$\begin{bmatrix} 0 & 0 & 0 & 1 & 0 & 0 \\ (x_{43}x_{32} - x_{42}) & -x_{43} & x_{21}x_{43} & 1 & -x_{21} & x_{21}x_{32} - x_{31} \end{bmatrix}.$$

Here, the columns are indexed by $x_{21}, x_{31}, x_{32}, x_{41}, x_{42}, x_{43}$. The rank of this matrix at $p = \mathrm{id}$ (i.e., all $x_{ij} = 0$) is $1 < 2 = 6 - l(3412) = \mathrm{codim} Y((3412), \mathrm{id})$. Therefore, $X(3412)$ is singular at e_{id}.

In contrast, the equations defining $Y((2413), \mathrm{id})$ are $\{x_{31},\ x_{41},\ x_{31}x_{42} - x_{41}x_{32},\ x_{43}x_{32} - x_{42},\ x_{21}(x_{43}x_{32} - x_{42}) - (x_{43}x_{31} - x_{41})\}$. The Jacobian matrix is

(4.5.10)
$$\begin{bmatrix} 0 & 1 & 0 & 0 & 0 & 0 \\ 0 & 0 & 0 & 1 & 0 & 0 \\ 0 & x_{42} & -x_{41} & -x_{32} & x_{31} & 0 \\ 0 & 0 & x_{43} & 0 & -1 & x_{32} \\ (x_{43}x_{32} - x_{42}) & -x_{43} & x_{21}x_{43} & 1 & -x_{21} & (x_{32}x_{21} - x_{31}) \end{bmatrix}.$$

This matrix is easily seen to have rank 3 when all variables are set to zero; $3 = 6 - l(2413) = \mathrm{codim} Y((2413), \mathrm{id})$ which implies that $X(2413)$ is nonsingular.

4.6. Discussion of smoothness at a T-fixed point

Let $w, \tau \in W$, $w \geq \tau$. The problem of determining whether or not e_τ is a smooth point of $X(\tau)$ is equivalent to determining whether or not e_τ is a smooth point of $Y(w, \tau)$, since $Y(w, \tau)$ is an open neighborhood of e_τ in $X(w)$. Let D_i, $1 \leq i \leq l$, be as in §4.4.2. In view of the Jacobian criterion, the problem is reduced to computing $\partial f / \partial x_\beta$, $f \in D_i$, $1 \leq i \leq l$, $\beta \in \tau(R^-)$. To carry out this computation, we first observe the following:

1. Let λ be a dominant character of T (or B), and let L_λ be the associated line bundle on G/B. Let V be the G-module $H^0(G/B, L_\lambda)$. Then as seen in Chapter 2, V can be identified as

$$V = \{f : G \to K \mid f(gb) = \lambda(b)f(g),\ b \in B,\ g \in G\}.$$

2. Now V is also a $\mathfrak{g}$-module. Given X in $\mathfrak{g}$, we identify X with the corresponding right invariant vector field D_X on G. Thus, if $v \in V$ corresponds to a function f on G as above, then we have $D_X f = X f$. If $f \in V$, then denoting the restriction of f to U_τ^- also by just f, we note that the evaluations of $\partial f / \partial x_\beta$ and $X_\beta f$, $\beta \in \tau(R^-)$, at e_τ coincide. Further, observe that $X_\beta \in T(w, \tau)$ if and only if $(X_\beta f)(e_\tau) = 0$ for all $f \in k[\mathcal{O}_\tau]$ such that $f|_{Y(w,\tau)} = 0$.

Summarizing, given $w, \tau \in W$, $w \geq \tau$, the problem of determining whether or not e_τ is a smooth point of $X(w)$ ($\subset G/B$) is reduced to constructing bases D_i, $1 \leq i \leq l$, as in §4.4.2 and computing $X_\beta f$, $\beta \in \tau(R^-)$, $f \in D_i$, $1 \leq i \leq l$. In [**110**] (see also [**97**]) such bases have been constructed for G classical and the computations of $X_\beta f$ (for G classical) have been carried out in [**94**]. In Chapter 5, we give a brief review of the bases as constructed in [**110**] and then recall the results on $T(w, \tau)$ (cf. [**97**], [**99**], [**100**], [**135**]).

4.7. Multiplicity at a point P on a variety X

4.7.1. Multiplicity of a local ring. Let A be a finitely generated K-algebra. Furthermore, let A be local with $\mathfrak{m}$ as the unique maximal ideal. For $l \geq 0$, let $\psi_A(l) = \text{length}(A/\mathfrak{m}^l) (= \dim_K(A/\mathfrak{m}^l))$. ψ_A is called the *Hilbert–Samuel* function of A. Recall (see [**45**] for example) the following:

4.7.2. Theorem. *There exists a polynomial $P_A(x) \in \mathbb{Q}[x]$, called the Hilbert–Samuel polynomial of A such that*

1. *$\psi_A(l) = P_A(l)$, $l \gg 0$.*
2. *$\deg P_A(x) = \dim A$.*
3. *The leading coefficient of $P_A(x)$ is of the form $e_A/n!$, where $e_A \in \mathbb{Z}^+$ and $n = \dim A$.*

4.7.3. Definition. With notation as in Theorem 4.7.2, the number e_A is called the *multiplicity* of A.

4.7.4. Definition. For a point P on an algebraic variety X, the multiplicity of X at P is defined to be the number e_A, where $A = \mathcal{O}_{X,P}$, the stalk at P, and is denoted by $\text{mult}_P X$.

We have the following important and useful characterization for a point to be smooth in terms of its multiplicity.

4.7.5. Proposition. [**130**] *Let P be a point on an algebraic variety X. Then P is a smooth point of X if and only if $\text{mult}_P X = 1$.*

4.7.6. Theorem. [**45**] *Let $A = \mathcal{O}_{X,P}$, and let $\text{gr}(A, \mathfrak{m}) = \oplus_{l \geq 0} \mathfrak{m}^l/\mathfrak{m}^{l+1}$. There exists a polynomial $Q_A(x) \in \mathbb{Q}[x]$, such that*

1. *$Q_A(l) = \dim_K \mathfrak{m}^l/\mathfrak{m}^{l+1}$, $l \gg 0$.*
2. *$\deg Q_A(x) = \dim(A) - 1 = n - 1$.*

3. *The leading coefficient of $Q_A(x)$ is of the form $\frac{f_A}{(n-1)!}$, where $f_A \in \mathbb{Z}^+$.*

4.7.7. Remark. With notation as above, by considering the exact sequence

$$0 \to \mathfrak{m}^l/\mathfrak{m}^{l+1} \to A/\mathfrak{m}^{l+1} \to A/\mathfrak{m}^l \to 0,$$

we have for $l \gg 0$,

$$\begin{aligned}
Q_A(l) &= P_A(l+1) - P_A(l) \\
&= \left[e_A/n!(l+1)^n + c_{n-1}(l+1)^{n-1} + \cdots \right] \\
&\quad - \left[e_A/n!(l)^n + c_{n-1}(l)^{n-1} + \cdots \right].
\end{aligned}$$

Hence we obtain

$$e_A = f_A.$$

Example. Let $K[x_1, \ldots, x_n]$ be the polynomial algebra and let $\mathfrak{a}$ be the maximal ideal generated by $\{x_1, \ldots, x_n\}$. Let A be the localization at $\mathfrak{a}$. Let $\mathfrak{m} = \mathfrak{a}A$. We have that $\mathfrak{m}^l/\mathfrak{m}^{l+1}$ is the span of monomials of total degree l and has dimension $\binom{l+n-1}{n-1}$. Hence $Q_A(l) = \binom{l+n-1}{n-1}$ and is a polynomial in l of degree $n-1$; further, the leading coefficient of $Q_A(l)$ is equal to $\frac{1}{(n-1)!}$. Thus $e_A = 1$.

4.7.8. Multiplicity of a graded affine K-algebra. For this subsection we refer the reader to [**45, 63**] for more details. Let B be a graded, finitely generated K-algebra. The function $f_B(n) = \dim_K B_n, n \in \mathbb{Z}^+$ is called the *Hilbert function* of B.

4.7.9. Theorem. *There exists a polynomial $P_B(x) \in \mathbb{Q}[x]$ of degree equal to $\dim(X)$ where $X = \mathrm{Proj}(B)$ such that $f_B(n) = P_B(n)$, for $n \gg 0$. Further, the leading coefficient of $P_B(n)$ is of the form $c_B/r!$ where $c_B \in \mathbb{N}$ and $r = \deg P_B(x)$.*

4.7.10. Definition. Let $X \subset \mathbb{P}^n$ be a projective variety with B as the associated homogeneous coordinate ring. Then c_B is called the *degree* of X and is denoted by $\deg X$.

4.7.11. Remark. Let X, B be as above. Consider $\widehat{X} = \mathrm{Spec} B$, the cone over X; denote its vertex by O. Then

$$(4.7.12) \qquad \mathrm{mult}_O \widehat{X} = \deg X.$$

4.7.13. Remark. Let $P \in X$, X being an algebraic variety, and let $\mathcal{O}_{X,P} = A$ with $\mathfrak{m}$ as the unique maximal ideal of the local ring A. Let C_P be the *tangent cone* at P, namely $C_P = \mathrm{Specgr}((A, \mathfrak{m}))$. Then $\mathrm{mult}_P X = \mathrm{mult}_O C_P = \deg \mathrm{Proj}(\mathrm{gr}(A, \mathfrak{m}))$, where O is the vertex of C_P.

We recall the following result (cf. [**63**, Th. 7.7]) on $\deg X$:

4.7.14. Theorem. *Let X be a variety of dimension ≥ 1 in $\mathbb{P}^n$, and let H be a hypersurface not containing X. Let $Z_1, \ldots, Z_s$ be the irreducible components of $X \cap H$. Then*

$$\deg X \cdot \deg H = \sum_{j=1}^{s} i(X, H; Z_j) \deg Z_j,$$

where $i(X, H; Z_j)$ denotes the intersection multiplicity of Z_j in $X \cap H$ (refer to [63, Ch. I, §7], for the definition of intersection multiplicity), and $\deg H$ is the degree of the homogeneous polynomial defining H.

4.8. Degree of $X(w)$

Chevalley multiplicity. Let $X_Q(w)$ be a Schubert variety in G/Q, Q being a parabolic subgroup. Let $L = L(\lambda)$ be a very ample line bundle on G/Q; note that for a simple root α, $\langle \lambda, \alpha \rangle \neq 0$ if and only if α does not belong to S_Q (such a λ is also called Q-regular). Consider the projective embedding

$$X_Q(w) \hookrightarrow G/Q \hookrightarrow \operatorname{Proj} H^0(G/Q, L).$$

If $w' = ws_\beta$, for some $\beta \in R^+$ such that $l(w') = l(w) - 1$, then $X_Q(w')$ is a Schubert divisor in $X_Q(w)$ (here we assume that $w, w' \in W_Q^{\min}$). The positive integer $m_\lambda(w, w') := \langle \lambda, \beta \rangle$ is called the *Chevalley multiplicity* of $X_Q(w')$ in $X_Q(w)$.

Geometric interpretation of $m_\lambda(w, w')$. Let p_w be the extremal weight vector in $H^0(G/Q, L)$ of weight $-w(\lambda)$ (cf. §2.11.13), and let H_w be the hyperplane in $\operatorname{Proj} H^0(G/Q, L)$ defined by p_w, note that $\deg H_w = 1$. We have that $X_Q(w) \cap H_w$ is the union of all the Schubert divisors in $X_Q(w)$, since $p_w \mid_{X_Q(\tau)} \neq 0 \Leftrightarrow \tau \geq w$, $\tau \in W_Q^{\min}$. Furthermore,

$$m_\lambda(w, w') = i(X_Q(w), H_w; X_Q(w'))$$

where the latter is the intersection multiplicity of $X_Q(w')$ in $X_Q(w) \cap H_w$ [36]. Hence we obtain (in view of Theorem 4.7.14)

4.8.3. Theorem.

$$\deg X_Q(w) = \sum m_\lambda(w, w') \cdot \deg X_Q(w'),$$

where the summation runs over all the Schubert divisors $X_Q(w')$ in $X_Q(w)$.

4.8.4. Remark. Let ω be a fundamental weight with P as the associated maximal parabolic subgroup. It is easily checked that ω is minuscule if and only if for any pair (w, τ) in W^P, where $w \geq \tau$, $l(w) = l(\tau) + 1$ (i.e., $X_P(\tau)$ is a Schubert divisor in $X_P(w)$), $m_\omega(w, \tau) = 1$. As above, let p_w be the extremal weight vector in $H^0(G/Q, L_\omega)$ of weight $-w(\omega)$, and let H_w be the zero set in G/P of p_w. Then, ω is minuscule if and only if for any pair (w, τ) in W^P such that $X_P(\tau)$ is a Schubert divisor in

$X_P(w)$, $X_P(\tau)$ occurs with multiplicity 1 in the intersection $X(w) \cap H_w$, i.e., $i(X_P(w), H_w; X_P(\tau)) = 1$.

Combinatorial interpretation of deg$X_Q(w)$. Consider the Hasse diagram given by the Bruhat–Chevalley order on $[\mathrm{id}, w_0] := \{\tau \in W_Q^{\min} \mid \tau \leq w\}$. Note that $[\mathrm{id}, w]$ is a ranked poset, i.e., all maximal chains have the same length, namely $l(w) + 1$; here, by a chain we mean a totally ordered subset of $[\mathrm{id}, w]$. Consider an edge $\tau' \to \tau$, where $X(\tau')$ is a Schubert divisor in $X(\tau)$, and give it the weight $m(\tau, \tau')$. Then deg$X_Q(w)$ is simply the number of maximal chains with the edges counted with the respective multiplicities. To be very precise, to a maximal chain $\underline{c} : \{w = \phi_0 > \phi_1 > \cdots > \phi_r = \mathrm{id}\}$, where $r = l(w) = \dim(X(w))$, assign the weight $n(\underline{c}) := \prod_{i=1}^{r} m(\phi_{i-1}, \phi_i)$. Then

$$\deg X_Q(w) = \sum_{\underline{c}} n(\underline{c}),$$

where the summation runs over all the maximal chains $\underline{c}$ in $[\mathrm{id}, w]$.

In particular, let $G = GL(n)$ and fix a maximal parabolic subgroup P_d so that G/P_d is the Grassmannian $G_{d,n}$. Then $m(w, w') = 1$ for all w, w' such that $X_Q(w')$ is a Schubert divisor in $X_Q(w)$; note that ω_d is minuscule for all $1 \leq d \leq n-1$, and $0 \leq (\omega_d, \beta^*) \leq 1$, $\forall \beta \in R^+$. Hence we obtain that the degree of a Schubert variety $X_{P_d}(w)$ in $G_{d,n}$ for the Plücker embedding $X_{P_d}(w) \hookrightarrow \mathbb{P}(\wedge^d K^n)$ is simply the number of maximal chains in $[\mathrm{id}, w]$. This result was first proved by Stanley in [**151**].

4.8.6. Example. Let $G = SL(3)$, $\lambda = \omega_1 + \omega_2$, the adjoint representation. The corresponding weighted Hasse graph of $[\mathrm{id}, w]$ is given by

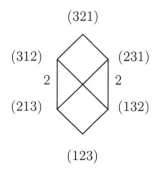

The edges $(213) \to (312)$, $(132) \to (231)$ have weights 2. All other edges have weight 1.

For the embedding $X(w) \hookrightarrow G/B \hookrightarrow \mathrm{Proj}\, H^0(G/B, L(\lambda))$, one can compute $\deg G/B = 6$, $\deg X(312) = \deg X(231) = 3$, $\deg X(213) = \deg X(132) = 1$.

4.8.7. Example. Consider $X(235) \subset G_{3,6}$ (= the Grassmannian of 3-planes in K^6). The corresponding weighted Hasse diagram of $[\text{id}, (235)]$ is given by

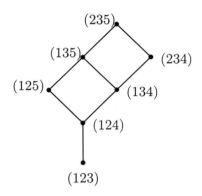

(The weight for any edge is just 1.) For the corresponding Plücker embedding, we have that $\deg X(235) = 3$.

4.9. Summary of smoothness criteria

Let X be a variety and $P \in X$. We may assume that X is affine since smoothness is a local condition. Let $\mathfrak{m}_{X,P}$ be the maximal ideal of the local ring $\mathcal{O}_{X,P}$. Then the following are equivalent:

1. X is smooth at P.
2. $\mathcal{O}_{X,P}$ is a regular local ring.
3. $\text{gr}(\mathcal{O}_{X,P}, \mathfrak{m}_{X,P})$ is a polynomial algebra generated by degree one elements.
4. The dimensions of X and of the tangent space to X at P are equal.
5. The rank of the Jacobian matrix $J_P = \text{codim}_{\mathbb{A}^n} X$.
6. $\text{mult}_P X = 1$.

A summary of all smoothness and rational smoothness criteria for Schubert varieties is given on Page 208.

CHAPTER 5

Root System Description of $T(w, \tau)$

In this chapter we recall the construction due to Lakshmibai–Musili–Seshadri (cf. [105], [110]) of a special basis for the Weyl module associated to a fundamental weight in the case G is a classical group, and we compute the action of the Chevalley basis elements of $\mathfrak{g}$ on this basis. Using these computations and a result of Polo (cf. [135]) relating tangent spaces and Demazure modules, we give a description of the tangent space $T(w, \tau)$ $\tau \leq w$ due to Lakshmibai ([98, 99, 100]) in terms of the root system. We also bring out the relationship between $T(w, \tau)$ and the multiplicities of certain weights in the fundamental representations of G. We have separated out the discussions of the case $\tau = id$ and the case of a general τ, since the results as well as the discussion for $\tau = id$ turn out to be more compact. We have also included a result of Polo (cf. [135]) describing the module structure of the tangent space at the identity. Also included are two criteria for smoothness due to Carrell–Kuttler ([34])

The following proposition will be frequently used in this chapter. It will be proved in Chapter 6, Proposition 6.2.12 using T-stable curves lying inside Schubert varieties.

5.0.1. Proposition. [33] *A lower bound for* $\dim T(w, \tau)$. *Let* $w, \tau \in W$, $\tau \leq w$. *Let* $\alpha \in R^+$ *such that* $\tau s_\alpha \leq w$. *Then* $X_{-\alpha} \in T(w, \tau)$. *In particular,* $\dim T(w, \tau) \geq \#\{\alpha \in R^+ \mid \tau s_\alpha \leq w\}$.

5.1. Polo's results

We recall the notation of Chapter 2. For $1 \leq d \leq l$, l being the rank of G, we shall denote by P_d the maximal parabolic subgroup associated to the simple root α_d, W_d the Weyl group of P_d, and W^d the set of minimal representatives of W/W_d. For $\tau \in W$, let $\tau^{(d)}$ denote the element in W^d representing τW_d. In this chapter we will have occasion to look at $\tau^{(d)}, \tau \in W$ as well as elements θ in W^d; thus, a superscript "(d)" as in $\tau^{(d)}$ would mean that it comes from the Weyl group element τ, while an element $\theta \in W^d$ need not necessarily have any reference to a Weyl group element. Let ω_d, $1 \leq d \leq l$ be the fundamental weights of G. The Weyl modules $V_K(\lambda)$, $\lambda \in \Lambda^+$, and the Demazure modules $V_{w,\lambda}$, $w \in W$ are as in §2.11.10.

Let $w, \tau \in W^Q$, $w \geq \tau$. In Chapter 4 we introduced the set $N_Q(w, \tau) = \{\beta \in \tau(R^+ \setminus R_Q^+) \mid X_{-\beta} \in T_Q(w, \tau)\}$, $T_Q(w, \tau)$ being the tangent space to

$X_Q(w)$ at $e_{\tau,Q}$. Recall (cf. Chapter 4, §4.3)

$$T_Q(w,\tau) = \text{ the span of } \{X_{-\beta},\ \beta \in N_Q(w,\tau)\}.$$

Let us denote $S \setminus S_Q$ by $\{\alpha_1, \ldots, \alpha_t\}$ for some t.

5.1.1. Theorem. [135] *Let $w, \tau \in W^Q$, $w \geq \tau$. Let $\beta \in \tau(R^+ \setminus R_Q^+)$. Then $\beta \in N_Q(w,\tau)$ if and only if $X_{-\beta}q_{\tau^{(i)}} \in V_{w,\omega_i}$, for all $1 \leq i \leq t$, where $q_{\tau^{(i)}}$ is the extremal weight vector in V_{ω_i} of weight $\tau(\omega_i)$ $(= \tau^{(i)}(\omega_i))$.*

PROOF. Let us denote $U_{\tau,Q}^- e_{\tau,Q}$ (cf. §4.4.2) by just $\mathcal{O}_{\tau,Q}$. Let

$$A = \bigoplus_{\underline{a}} H^0(G/Q, \bigotimes_i L_i^{a_i})$$

$$A_w = \bigoplus_{\underline{a}} H^0(X_Q(w), \bigotimes_i L_i^{a_i}),$$

where L_i is the ample generator of $\text{Pic} G/P_i$, $1 \leq i \leq t$, and $\underline{a} = (a_1, \ldots, a_t)$ is a sequence of positive integers. We have from §2.10 that the kernel of the restriction map $A \to A_w$ is generated in degree 1. Hence denoting the ideal of $Y_Q(w,\tau)$ (cf. §4.4.2) in $\mathcal{O}_{\tau,Q}$ by $I_Q(w,\tau)$, we have $I_Q(w,\tau)$ is generated as an ideal in $K[\mathcal{O}_{\tau,Q}]$ by

$$\bigcup_{1 \leq i \leq t} \{f|_{\mathcal{O}_{\tau,Q}} | f \in H^0(G/Q, L_i),\ f|_{X_Q(w)} = 0\}.$$

Now given $\beta \in \tau(R^+ \setminus R_Q^+)$, we have (cf. §4.6) the following sequence of implications that are equivalent:

1. $X_{-\beta} \in T_Q(w,\tau)$.
2. $X_{-\beta}f(e_{\tau,Q}) = 0$, for all $f \in I_Q(w,\tau)$.
3. $X_{-\beta}f(e_{\tau,Q}) = 0$, for all $f \in \ker: H^0(G/Q, L_i) \to H^0(X_Q(w), L_i)$, $1 \leq i \leq t$.
4. $B_i(q_{\tau^{(i)}}, X_{-\beta}f) = 0$, B_i being the $\mathfrak{g}$-invariant, nondegenerate bilinear form on $V_K(\omega_i) \times V_K(\omega_i)^*$.
5. $B_i(X_{-\beta}q_{\tau^{(i)}}, f) = 0$, for all $f \in \ker: H^0(G/Q, L_i) \to H^0(X_Q(w), L_i)$, $1 \leq i \leq t$.
6. $X_{-\beta}q_{\tau^{(i)}} \in V_{w,\omega_i}$.

$\square$

5.1.2. Remark. A second proof of the above theorem (for G classical) will be given in the following section using the special bases for $V_K(\omega_i)$ described in that section.

Given $c_i \in \mathbb{N}$, replacing L_i by $L_i^{\otimes c_i}$, the above proof yields the following, with notation as in Theorem 5.1.1:

5.1.3. Corollary. *We have $\beta \in N(w, \tau)$ if and only if $X_{-\beta} f_{\tau^{(i)}} \in V_{w, c_i \omega_i}$ for all $1 \le i \le t$, where $f_{\tau^{(i)}}$ is the extremal weight vector in $V_{c_i \omega_i}$ of weight $\tau(c_i \omega_i)(= \tau^{(i)}(c_i \omega_i))$.*

Thus we obtain a particular version of Polo's theorem:

5.1.4. Theorem. **[135**, Thm. 3.2] *Let notation be as in Theorem 5.1.1. Let $\beta \in \tau(R^+ \setminus R_Q^+)$. The following are equivalent:*

1. $X_{-\beta} \in T_Q(w, \tau)$.
2. $X_{-\beta} q_{\tau^{(i)}} \in V_{w, \omega_i}$, *for all $1 \le i \le t$.*
3. $X_{-\beta} f_{\tau^{(i)}} \in V_{w, c_i \omega_i}$, $c_i \in \mathbb{N}$ *for all $1 \le i \le t$.*

In particular, taking $Q = B$, we obtain

5.1.5. Corollary. *Let $\beta \in R^+$. Then $\beta \in N(w, \tau)$ if and only if $X_{-\beta} q_{\tau^{(i)}} \in V_{w, \omega_i}$, for all $1 \le i \le l$, l being the rank of G.*

5.1.6. Remark. The result in Theorem 5.1.1, though pretty, does not help much unless we have some precise knowledge of $V_K(\lambda)$, say for example some "nice" basis of $V_K(\lambda)$ consisting of weight vectors.

5.2. Bases $\mathcal{B}_\lambda$, $\mathcal{B}_\lambda^*$ for $V_K(\lambda)$ and $H^0(G/B, L_\lambda)$

For the rest of this chapter, we shall suppose that G is classical. Note that $(\omega_d, \beta^*) \le 2$, for all $\beta \in R^+$ (G being classical).

Let λ be a dominant weight, and let $V_K(\lambda)$ be the associated Weyl module (cf. §2.11.10). In [**97**], dual bases $\mathcal{B}_\lambda$ and $\mathcal{B}_\lambda^*$ for $V_K(\lambda)$ and $H^0(G/B, L_\lambda)$ respectively have been constructed consisting of weight vectors; also, the extremal weight vectors in $V_K(\lambda)$, resp. $H^0(G/B, L_\lambda)$, belong to $\mathcal{B}_\lambda$, resp. $\mathcal{B}_\lambda^*$. Further, $\mathcal{B}_\lambda$ and $\mathcal{B}_\lambda^*$ are Bruhat-order compatible, i.e., $\{u \in \mathcal{B}_\lambda \mid u \in V_{w, \lambda}\}$ is a basis for $V_{w, \lambda}$, and $\{f|_{X(w)} \mid f \in \mathcal{B}_\lambda^*, \, f|_{X(w)} \neq 0\}$ is a basis for $V_{w, \lambda}^*$. If $\lambda = \omega_d$, a fundamental weight, we shall denote $\mathcal{B}_\lambda$ and $\mathcal{B}_\lambda^*$ by $\mathcal{B}_d$ and $\mathcal{B}_d^*$ respectively. For $\lambda = \omega_d$, the bases $\mathcal{B}_d$ and $\mathcal{B}_d^*$ coincide with the bases constructed in [**110**]. We recall below the bases $\mathcal{B}_d$, $\mathcal{B}_d^*$.

5.2.1. Admissible pairs in classical groups. Fix $d, 1 \le d \le l$. In this section we shall denote W^{P_d} by just W^d. Let $\tau, \phi \in W^d$ be such that the Schubert variety $X_{P_d}(\phi)$ is a Schubert divisor in $X_{P_d}(\tau)$. We have, $\phi = s_\beta \tau$, for some $\beta \in R^+$, and $l(\phi) = l(\tau) - 1$. Let $m(\tau, \phi) = (\phi(\omega_d), \beta^*)$ be the Chevalley multiplicity of $X_{P_d}(\phi)$ in $X_{P_d}(\tau)$ (cf. §4.8.1).

5.2.2. Definition. Admissible pairs. A pair of elements $(\tau, \phi) \in W^d$ for $\tau \ge \phi$ is called an *admissible pair* if either $\tau = \phi$, in which case we call (τ, τ) a *trivial admissible pair*, or if there exists a chain $\tau = \tau_0 > \tau_1 > \cdots > \tau_r = \phi$ such that $X_{P_d}(\tau_{i+1})$ is a Schubert divisor in $X_{P_d}(\tau_i)$, and $m(\tau_i, \tau_{i+1}) = 2$, $0 \le i \le r - 1$.

5.2.3. Remark. In Type $\mathbf{A}_n$, any fundamental weight ω_d satisfies $(\omega, \beta^*) \le 1$, $\forall \beta \in R^+$, and hence the trivial pairs (τ, τ), $\tau \in W^d$ are the only admissible pairs in W^d. More generally, if ω_d is a minuscule fundamental weight, i.e., $(\omega, \beta^*) \le 1$, $\forall \beta \in R^+$, then the trivial pairs are the only admissible pairs in W^d.

Example. Let $(.,.)$ be the nondegenerate, skew-symmetric bilinear form on K^6 given by the matrix, with respect to the standard basis,

$$E = \begin{pmatrix} 0 & J \\ -J & 0 \end{pmatrix},$$

where J is the anti-diagonal $(1,1,1)$ of size 3×3. Let $G = Sp(6)$ be the subgroup of $SL(6)$ leaving $(.,.)$ invariant. Then W^3 may be identified with $\{(a,b,c), 1 \le a < b < c \le 6 \,|\, \text{the sum of any two of them} \ne 7\}$ (see [**94**] for details). The admissible pairs in W^3 are given by

$$((236), (135)) \quad ((135), (124)) \quad ((236), (124))$$
$$((356), (246)) \quad ((246), (145)) \quad ((356), (145))$$

5.2.4. Definition. Moving divisors. Let $\tau, \phi \in W^d$ be such that $X_{P_d}(\phi)$ is a Schubert divisor in $X_{P_d}(\tau)$. Let $\phi = s_\beta \tau$, where $\beta \in R^+$. Then $X_{P_d}(\phi)$ is said to be a *moving divisor* in $X_{P_d}(\tau)$, if β is simple. The terminology "moving divisors" is taken after [**80**].

5.2.5. Proposition. *Let $\tau, \phi \in W^d$ be such that $X_{P_d}(\phi)$ is a divisor in $X_{P_d}(\tau)$ with $m(\tau, \phi) = 2$. Then $X_{P_d}(\phi)$ is a moving divisor in $X_{P_d}(\tau)$.*

5.2.6. Proposition. *Let $\tau, \phi \in W^d, \tau \ge \phi$. If there exists one chain $\tau = \tau_0 > \tau_1 > \cdots > \tau_r = \phi$ such that $X_{P_d}(\tau_{i+1})$ is a moving divisor in $X_{P_d}(\tau_i)$, and $m(\tau_i, \tau_{i+1}) = 2$, $0 \le i \le r-1$, then any other chain $\tau = \tau_0' > \tau_1' > \cdots > \tau_r' = \phi$ also has the property that $m(\tau_i', \tau_{i+1}') = 2$, $0 \le i \le r-1$.*

5.2.7. Corollary. *Let $\tau, \phi \in W^d$ be such that (τ, ϕ) is an admissible pair. Then any chain $\tau = \tau_0 > \tau_1 > \cdots > \tau_r = \phi$ such that $X_{P_d}(\tau_{i+1})$ is a divisor in $X_{P_d}(\tau_i)$ has the property that $m(\tau_i, \tau_{i+1}) = 2$, $0 \le i \le r-1$.*

See [**110**] for a proof of Propositions 5.2.5, 5.2.6.

We recall from [**94**] the following three propositions:

5.2.8. Proposition. *Let $G = Sp(2n)$. Let $\tau, \phi \in W^d, \tau \ge \phi$. Then (τ, ϕ) is an admissible pair if and only if either $\tau = \phi$, or τ is obtained from ϕ by a sequence of Type I operations defined as follows:*

> *Type I operation: Say $\phi = (a_1 \cdots a_d)$, and let i, $1 \le i \le n$, be such that $i, (i+1)' \in \{a_1, \ldots, a_d\}$. Let $\tau = (b_1 \cdots b_d)$ be the element of W^d obtained from $(a_1 \cdots a_d)$ by replacing i by $i+1$, and $(i+1)'$ by i'.*

5.2.9. Proposition. *Let $G = SO(2n + 1)$. Let $\tau, \phi \in W^d$, $1 \leq d \leq n - 1, \tau \geq \phi$. Then (τ, ϕ) is an admissible pair if and only if either $\tau = \phi$, or τ is obtained from ϕ by a sequence of operations of Type I or II.*

1. *Type I: Let $\phi = (a_1 \cdots a_d) \in W^d$, $1 \leq d \leq n-1$, and let i, $1 \leq i \leq n$ be such that $i, (i + 1)' \in \{a_1, \ldots, a_d\}$. Let $\tau = (b_1 \cdots b_d)$ be the element of W^d obtained from $(a_1 \cdots a_d)$ by replacing i by $i + 1$, and $(i + 1)'$ by i'.*

2. *Type II: Let $\phi = (a_1 \cdots a_d) \in W^d$, $1 \leq d \leq n - 1$, and let $n \in \{a_1, \ldots, a_d\}$. Let $\tau = (b_1 \cdots b_d)$ be the element of W^d obtained from $(a_1 \cdots a_d)$ by replacing n by n'.*

5.2.10. Proposition. *Let $G = SO(2n)$. Let $\tau, \phi \in W^d$, $1 \leq d \leq n - 2$, $\tau \geq \phi$. Then (τ, ϕ) is an admissible pair if and only if either $\tau = \phi$, or τ is obtained from ϕ by a sequence of operations either Type I or Type II below.*

1. *Type I: Let $\phi = (a_1 \cdots a_d) \in W^d$, $1 \leq d \leq n-2$, and let i, $1 \leq i \leq n$ be such that $i, (i + 1)' \in \{a_1, \ldots, a_d\}$. Let $\tau = (b_1 \cdots b_d)$ be the element of W^d obtained from $(a_1 \cdots a_d)$ by replacing i by $i + 1$, and $(i + 1)'$ by i'.*

2. *Type II: Let $\phi = (a_1 \cdots a_d) \in W^d$, $1 \leq d \leq n - 2$. Further let $n - 1, n \in \{a_1, \ldots, a_d\}$. Let $\tau = (b_1 \cdots b_d)$ be the element of W^d obtained from $(a_1 \cdots a_d)$ by replacing $n - 1$ by n', and n by $(n - 1)'$.*

5.2.11. Proposition. [110] *Let $\tau, \phi \in W^d$ be a nontrivial admissible pair (refer to §5.2.1). Let $\tau = \tau_0 > \tau_1 > \cdots > \tau_r = \phi$ be any chain (so that $X_{P_d}(\tau_{i+1})$ is a Schubert divisor in $X_{P_d}(\tau_i)$), such that $\tau_{i+1} = s_{\beta_i} \tau_i$, $l(\tau_i) = l(\tau_{i+1})$, $0 \leq i \leq r - 1$. Define $v \in V_K(\omega_d)$ as $v = X_{-\beta_0} X_{-\beta_1} \cdots X_{-\beta_{r-1}} q_\phi$, where q_ϕ is the extremal weight vector in $V_K(\omega_d)$ of weight $\phi(\omega_d)$. Then v is independent of the chain chosen and depends only on τ and ϕ. Further, v is a weight vector of weight $\frac{1}{2}(\tau(\omega_d) + \phi(\omega_d))$.*

5.2.12. Definition. The sets $\mathcal{B}$ and $\mathcal{B}_w$. Let $\tau, \phi \in W^d$ be such that (τ, ϕ) is an admissible pair. If $\tau = \phi$, then set $q_{\tau,\tau}$ (or just q_τ) as the extremal weight vector in $V_K(\omega_d)$ of weight $\tau(\omega_d)$ (which is unique up to scalars). If $\tau > \phi$, then set $q_{\tau,\phi}$ as the vector v as given by Proposition 5.2.11. Set $\mathcal{B} = \{q_{\tau,\phi}, (\tau, \phi)$ an admissible pair$\}$. For $w \in W^d$, set $\mathcal{B}_w = \{q_{\tau,\phi} \in \mathcal{B} \mid w \geq \tau\}$.

5.2.13. Definition. The sets $\mathcal{B}^*$ and $\mathcal{B}_w^*$. Define $\mathcal{B}^*$ to be the basis of $H^0(G/P_d, L_{\omega_d})$ $(= V_K(\omega_d)^*)$ dual to $\mathcal{B}$. Let us denote the elements of $\mathcal{B}^*$ by $\{p_{\tau,\phi}, (\tau, \phi)$ an admissible pair$\}$. For $w \in W^d$, set

$$\mathcal{B}_w^* = \{p_{\tau,\phi}|_{X_{P_d}(w)} \mid p_{\tau,\phi} \in \mathcal{B}^*, \ p_{\tau,\phi}|_{X_{P_d}(w)} \neq 0\}.$$

5.2.14. Theorem. [110] *Let notation be as above.*

1. *The set $\mathcal{B}$ is a basis for $V_K(\omega_d)$.*

2. *For $w \in W^d$, the set $\mathcal{B}_w$ is a basis for V_{w,ω_d}.*
3. *For $w \in W^d$, the set $\mathcal{B}_w^*$ is a basis of $H^0(X_{P_d}(w), L(\omega_d))$.*

A second proof of Theorem (5.1.1). We give below a constructive proof of Theorem (5.1.1) using the bases $\mathcal{B}_d, \mathcal{B}_d^*$, $1 \le d \le l$ for G classical.

PROOF. Let us take $Q = B$, the proof being similar for an arbitrary parabolic. For $\tau \in W$, as above, let $\tau^{(d)}$ denote the element in W^d representing τW_d. Let $q_{\tau^{(d)}}$ denote the extremal weight vector in $V_k(\omega_d)$ of weight $\tau^{(d)}(\omega_d)$, and let $p_{\tau^{(d)}}$ denote the extremal weight vector in $H^0(G/B, L_{\omega_d})$ of weight $-\tau^{(d)}(\omega_d)$, $1 \le d \le l$, here l is the rank of G. Let $I(w,\tau)$ be the ideal of $Y(w,\tau)$ as a closed subvariety of $\mathcal{O}_\tau$. We have from §4.4.2 that $I(w,\tau)$ is generated by $\{f|_{\mathcal{O}_\tau}, f \in \mathcal{B}_d^*, 1 \le d \le l \mid f|_{X(w)} = 0\}$. Hence a root β in $\tau(R^+)$ belongs to $N(w,\tau)$ if and only if $f|_{X(w)} \ne 0$ for all $f \in \mathcal{B}_d^*$, $1 \le d \le l$, such that $X_{-\beta}f = cp_{\tau^{(d)}}$, $c \ne 0$ (cf. §4.6). Given $f \in \mathcal{B}_d^*$ and any $1 \le d \le l$, let u_f be the element of $\mathcal{B}_d$ corresponding to f. Then, in view of the duality between $\mathcal{B}_d$ and $\mathcal{B}_d^*$, we have $X_{-\beta}f = cp_{\tau^{(d)}}$, where $c \ne 0$ if and only if u_f appears with a nonzero coefficient in the expression for $X_{-\beta}q_{\tau^{(d)}}$ as a linear combination of the basis vectors in $\mathcal{B}_d$. Thus we obtain that a root β in $\tau(R^+)$ belongs to $N(w,\tau)$ if and only if $f|_{X(w)} \ne 0$ for all u_f appearing with a nonzero coefficient in $X_{-\beta}q_{\tau^{(d)}}$, $1 \le d \le l$, i.e., all such u_f belong to V_{w,ω_d}, for all $1 \le d \le l$; and this last condition is equivalent to the condition that $X_{-\beta}q_{\tau^{(d)}} \in V_{w,\omega_d}$, for all $1 \le d \le l$, in view of Bruhat-order compatibility of $\mathcal{B}_d$.   □

5.2.15. Remark. Given $c_i \in \mathbb{N}$, replacing L_i by $L_i^{\otimes c_i}$, a proof similar to the above yields the equivalence of (1) and (3) in Theorem 5.1.4.

5.2.16. Remark. In the above proof, one may also work with any Bruhat-Chevalley order compatible bases for $V_K(\omega_d), V_K(\omega_d)^*$.

5.3. Description of $T(w, \mathrm{id})$

In this section, we give a root system description of $T(w, \mathrm{id})$ for G classical. As mentioned in the beginning of this chapter, the results below are proved using the basis $\mathcal{B}_d$ for the Weyl module $V(\omega_d)$ for G classical and Theorem 5.1.1.

5.3.1. Theorem. *(cf. [98], [99]). Let $\beta \in R^+$.*

1. *Let G be of type $\mathbf{A}_n$. Then $\beta \in N(w, \mathrm{id}) \iff w \ge s_\beta$.*
2. *Let G be of type $\mathbf{C}_n$.*
   (a) *If $\beta = \epsilon_i - \epsilon_j$, or $2\epsilon_i$, then $\beta \in N(w, \mathrm{id}) \iff w \ge s_\beta$.*
   (b) *If $\beta = \epsilon_i + \epsilon_j$, then $\beta \in N(w, \mathrm{id}) \iff w \ge$ either $s_{\epsilon_i + \epsilon_j}$ or $s_{2\epsilon_i}$.*
3. *Let G be of type $\mathbf{B}_n$.*
   (a) *If $\beta = \epsilon_i - \epsilon_j$, ϵ_n, or $\epsilon_i + \epsilon_n$, then $\beta \in N(w, \mathrm{id}) \iff w \ge s_\beta$.*

(b) *If $\beta = \epsilon_i, i < n$, then $\beta \in N(w, \mathrm{id}) \iff w \geq$ either s_{ϵ_i} or*
$s_{\epsilon_i + \epsilon_n}$.

(c) *If $\beta = \epsilon_i + \epsilon_j, j < n$, then $\beta \in N(w, \mathrm{id}) \iff w \geq$ either $s_{\epsilon_i + \epsilon_j}$*
or $s_{\epsilon_i} s_{\epsilon_j + \epsilon_n}$.

4. *Let G be of type $\mathbf{D}_n$.*

   (a) *If $\beta = \epsilon_k - \epsilon_l$, or $\epsilon_i + \epsilon_j$, $j = n - 1, n$, then $\beta \in N(w, \mathrm{id}) \iff$*
   $w \geq s_\beta$.

   (b) *If $\beta = \epsilon_i + \epsilon_j$, $j < n - 1$, then $\beta \in N(w, \mathrm{id}) \iff w \geq$ either*
   $s_{\epsilon_i + \epsilon_j}$ *or* $s_{\epsilon_i - \epsilon_n} s_{\epsilon_i + \epsilon_n} s_{\epsilon_j + \epsilon_{n-1}}$.

PROOF. We have by Theorem 5.1.1, $\beta \in N(w, \mathrm{id})$ if and only if $X_{-\beta} q_{\mathrm{id}(d)}$ belongs to V_{w, ω_d} for all $1 \leq d \leq l$. We shall now compute $X_{-\beta} q_{\mathrm{id}(d)}$, $\beta \in R^+$, and show that $X_{-\beta} q_{\mathrm{id}(d)}$ belongs to V_{w, ω_d} for all $1 \leq d \leq l$ if and only if β satisfies the conditions in the theorem for the respective types. We do this computation case by case.

Case 1: The special linear group $SL(n)$. Let $G = SL(n)$, and $V = K^n$. We denote the standard basis for K^n by $\{e_1, \ldots, e_n\}$. Given a root $\beta = \epsilon_j - \epsilon_k, 1 \leq j, k \leq n$, the element $X_{-\beta}$ of the Chevalley basis of $\mathfrak{g}$ is given by $X_{-\beta} = E_{kj}$, where E_{kj} is the elementary matrix with 1 at the (k, j)-th place, and 0's elsewhere. For $1 \leq d \leq l (= n - 1)$, we have that $V_K(\omega_d) = \wedge^d V$. Given $w = (a_1 \cdots a_n) \in W$, denoting by $w^{(d)}$ the projection of w under $W \to W/W_d$, we have the extremal weight vector $q_{w^{(d)}} = e_{a_1} \wedge \cdots \wedge e_{a_d}$. Thus $\mathcal{B}_d = \{q_{w^{(d)}}, w \in W\}$. We have

$$X_{-\beta} e_i = E_{kj} e_i = \begin{cases} 0, & \text{if } i \neq j \\ e_k, & \text{if } i = j. \end{cases}$$

Let $\beta = \epsilon_j - \epsilon_k, 1 \leq j < k \leq n$. Then it follows that $X_{-\beta} q_{\mathrm{id}(d)} \neq 0$ if and only if $j \in \{1, \ldots, d\}$, and $k \notin \{1, \ldots, d\}$, i.e., if and only if $j \leq d < k$; further, for $j \leq d < k$, we have

$$X_{-\beta} q_{\mathrm{id}(d)} = \sum_{m=1}^{d} e_1 \wedge \cdots \wedge e_{m-1} \wedge X_{-\beta} e_m \wedge e_{m+1} \wedge \cdots \wedge e_d$$

$$= e_1 \wedge \cdots \wedge e_{j-1} \wedge e_k \wedge e_{j+1} \wedge \cdots \wedge e_d.$$

Hence we obtain

$$X_{-\beta} q_{\mathrm{id}(d)} = q_{s_\beta^{(d)}}.$$

From this it follows that $X_{-\beta} q_{\mathrm{id}(d)}$ belongs to V_{w, ω_d} for all $1 \leq d \leq l$ if and only if $w^{(d)} \geq s_\beta^{(d)}, j \leq d < k$, i.e., if and only if $w \geq s_\beta$, note that for $d < j$, or $d \geq k$, $s_\beta^{(d)} = (1 \cdots d)$.

Case 2: The Symplectic Group $Sp(2n)$. For $1 \leq d \leq n$, we have $\omega_d = \epsilon_1 + \cdots + \epsilon_d$. If $d = 1$, then $V_K(\omega_d) = V(= K^{2n})$. Let us next suppose that $d \geq 2$. Consider the 2-form $f \in \wedge^2 V$ given by

$$f = e_1 \wedge e_{2n} + e_2 \wedge e_{2n-1} + \cdots + e_n \wedge e_{n+1},$$

where $\{e_1, \ldots, e_{2n}\}$ is the standard basis in V. We have

$$V_K(\omega_d) = \{v \in \overset{d}{\bigwedge} V \mid v \wedge f^{n+1-d} = 0\}.$$

The extremal weight vectors $\{q_\tau, \tau \in W^d\}$ for $\tau = (a_1 \cdots a_d)$ are given by

$$q_\tau = e_{a_1} \wedge \cdots \wedge e_{a_d}.$$

Let $\beta \in R^+$ and let X_β be the element of the Chevalley basis given by §3.3. We have (cf. [95]):

1. Let $\beta = \epsilon_j - \epsilon_k, 1 \leq j < k \leq n$. Then

$$X_{-\beta}q_{\mathrm{id}(d)} = \begin{cases} 0, & \text{if } d < j \text{ or } d \geq k \\ \pm q_{s_\beta^{(d)}}, & \text{if } j \leq d < k. \end{cases}$$

2. Let $\beta = 2\epsilon_j, 1 \leq j \leq n$. Then

$$X_{-\beta}q_{\mathrm{id}(d)} = \begin{cases} 0, & \text{if } d < j \\ \pm q_{s_\beta^{(d)}}, & \text{if } j \leq d \leq n. \end{cases}$$

3. Let $\beta = \epsilon_j + \epsilon_k, 1 \leq j < k \leq n$. Then

$$X_{-\beta}q_{\mathrm{id}(d)} = \begin{cases} 0, & \text{if } d < j \\ \pm q_{s_\beta^{(d)}}, & \text{if } j \leq d < k \\ \pm q_{\tau,\phi}, & \text{if } k \leq d \leq n \end{cases}$$

where $\tau = (1 2 \cdots j-1\, j+1 \cdots d\, j')$ and $\phi = (1 2 \cdots k-1\, k+1 \cdots d\, k')$. Note that (τ, ϕ) is an admissible pair. From the above description it follows that if $\beta = \epsilon_i - \epsilon_j$ or $2\epsilon_i$, then $X_{-\beta}q_{\mathrm{id}(d)}$ belongs to $V_{w,\omega_d}, 1 \leq d \leq l$, if and only if $w \geq s_\beta$, and if $\beta = \epsilon_i + \epsilon_j$, then $X_{-\beta}q_{\mathrm{id}(d)}$ belongs to $V_{w,\omega_d}, 1 \leq d \leq l$ if and only if $w \geq$ *either* $s_{\epsilon_i + \epsilon_j}$ *or* $s_{2\epsilon_i}$.

Case 3: The Special orthogonal group $SO(2n+1)$. For $d = n$, $V_K(\omega_d)$ is the spin representation, and the extremal weight vectors, $q_\tau, \tau \in W^d$, form a basis for $V_K(\omega_d)$. For $1 \leq d < n$, we have $V_K(\omega_d) = \wedge^d V$, recall $V = K^{2n+1}$. The extremal weight vectors $\{q_\tau, \tau \in W^d\}$ for $\tau = (a_1 \cdots a_d)$ are given by

$$q_\tau = e_{a_1} \wedge \cdots \wedge e_{a_d}.$$

Let $\beta \in R^+$. We have (cf. [96])

1. Let $\beta = \epsilon_j - \epsilon_k, 1 \leq j < k \leq n$. Then

$$X_{-\beta}q_{\mathrm{id}(d)} = \begin{cases} 0, & \text{if } d < j \text{ or } d \geq k \\ \pm q_{s_\beta^{(d)}}, & \text{if } j \leq d < k. \end{cases}$$

2. Let $\beta = \epsilon_j, 1 \leq j \leq n$.
   (a) If $j = n$, then

$$X_{-\beta}q_{\mathrm{id}(d)} = \begin{cases} 0, & \text{if } d < n \\ \pm q_{s_\beta^{(d)}}, & \text{if } d = n. \end{cases}$$

   (b) If $j < n$, then

$$X_{-\beta}q_{\mathrm{id}(d)} = \begin{cases} 0, & \text{if } d < j \\ \pm q_{s_\beta^{(d)}}, & \text{if } d = n \\ \pm q_{s_{\epsilon_j+\epsilon_n}^{(d)}, s_{\epsilon_j-\epsilon_n}^{(d)}}, & \text{if } j \leq d < n. \end{cases}$$

3. Let $\beta = \epsilon_j + \epsilon_k, 1 \leq j < k \leq n$.
   (a) If $k = n$, then

$$X_{-\beta}q_{\mathrm{id}(d)} = \begin{cases} 0, & \text{if } d < j \\ \pm q_{s_\beta^{(d)}}, & \text{if } j \leq d. \end{cases}$$

   (b) If $k < n$, then

$$X_{-\beta}q_{\mathrm{id}(d)} = \begin{cases} 0, & \text{if } d < j \\ \pm q_{s_\beta^{(d)}}, & \text{if } j \leq d < k \text{ or } d = n \\ \pm(\sum_{i=0}^{n-d} c_i q_{\theta_i, \delta_i} + a q_{\tau, \phi}), & \text{if } k \leq d < n \end{cases}$$

where $c_i = \pm 2$, $i < n - d$, $c_{n-d} = \pm 1 = a$,
$\tau = (1 \cdots j - 1\, j + 1 \cdots d j')$,
$\phi = (1 \cdots k - 1\, k + 1 \cdots d k')$,
$\theta_0 = (1 \cdots j - 1\, j + 1 \cdots k - 1\, k + 1 \cdots d\, d + 1\, k')$,
$\delta_0 = (1 \cdots j - 1\, j + 1 \cdots d(d+1)')$,
$\theta_i = (12 \cdots j - 1\, j + 1 \cdots k - 1\, k + 1 \cdots (d+i+1)(d+i)')$,
$\delta_i = (12 \cdots j - 1\, j + 1 \cdots k - 1\, k + 1 \cdots (d+i)(d+i+1)')$, $1 \leq i < n - d$,
$\theta_{n-d} = (12 \cdots j - 1\, j + 1 \cdots k - 1\, k + 1 \cdots n'(n-1)')$,
$\delta_{n-d} = (12 \cdots j - 1\, j + 1 \cdots k - 1\, k + 1 \cdots n - 1n)$.
From the above description it is not difficult to check the following:

1. If $\beta = \epsilon_i - \epsilon_j$, ϵ_n, or $\epsilon_i + \epsilon_n$, then $X_{-\beta}q_{\mathrm{id}(d)}$ belongs to V_{w,ω_d} for all $1 \leq d \leq l$ if and only if $w \geq s_\beta$.
2. If $\beta = \epsilon_i, i < n$, then $X_{-\beta}q_{\mathrm{id}(d)}$ belongs to V_{w,ω_d}, $1 \leq d \leq l$ if and only if *either* $w \geq s_{\epsilon_i}$ *or* $w \geq s_{\epsilon_i+\epsilon_n}$.

3. If $\beta = \epsilon_i + \epsilon_j, j < n$, then $X_{-\beta}q_{\text{id}(d)}$ belongs to $V_{w,\omega_d}, 1 \le d \le l$ if and only if *either* $w \ge s_{\epsilon_i+\epsilon_j}$ *or* $w \ge s_{\epsilon_i}s_{\epsilon_j+\epsilon_n}$.

Case 4: The special orthogonal group $SO(2n)$. For $d = n - 1, n$, $V_K(\omega_d)$ is the spin representation, and the extremal weight vectors, $q_\tau, \tau \in W^d$, form a basis for $V_K(\omega_d)$. For $1 \le d \le n - 2$, we have $V_K(\omega_d) = \wedge^d V$ (here, $V = K^{2n}$), and the extremal weight vectors $\{q_\tau. \tau \in W^d\}$, say $\tau = (a_1 \cdots a_d)$ are given by

$$q_\tau = e_{a_1} \wedge \cdots \wedge e_{a_d}.$$

Let $\beta \in R^+$. We have (cf. [**106**])

1. Let $\beta = \epsilon_j - \epsilon_k, 1 \le j < k \le n$. Then

$$X_{-\beta}q_{\text{id}(d)} = \begin{cases} 0, & \text{if } d < j \text{ or } d \ge k \\ \pm q_{s_\beta^{(d)}}, & \text{if } j \le d < k. \end{cases}$$

2. Let $\beta = \epsilon_j + \epsilon_k, 1 \le j < k \le n$.
   (a) If $k = n - 1, n$, then

$$X_{-\beta}q_{\text{id}(d)} = \begin{cases} 0, & \text{if } d < j \\ \pm q_{s_\beta^{(d)}}, & \text{if } j \le d. \end{cases}$$

   (b) If $k < n - 1$, then

$$X_{-\beta}q_{\text{id}(d)} = \begin{cases} 0, & \text{if } d < j \\ \pm q_{s_\beta^{(d)}}, & \text{if } j \le d < k \text{ or } d = n - 1, n \\ \pm(\sum_{i=0}^{n-d} c_i q_{\theta_i,\delta_i} + q_{\tau,\phi}), & \text{if } k \le d < n - 1. \end{cases}$$

where $\tau, \phi, \theta_i, \delta_i, 0 \le i \le n - d$ are defined in the same way as in the $SO(2n+1)$ case above, and $c_i = \pm 2$ or ± 1, according as $i <$ or $\ge n-d-1$. From the above description the following could be checked easily:

If $\beta = \epsilon_k - \epsilon_l$, or $\epsilon_i + \epsilon_j$, $j = n - 1, n$, then $X_{-\beta}q_{\text{id}(d)}$ belongs to $V_{w,\omega_d}, 1 \le d \le l$ if and only if $w \ge s_\beta$.

If $\beta = \epsilon_i + \epsilon_j$, $j < n - 1$, then $X_{-\beta}q_{\text{id}(d)}$ belongs to $V_{w,\omega_d}, 1 \le d \le l$ if and only if $w \ge$ either $s_{\epsilon_i+\epsilon_j}$ or $s_{\epsilon_i-\epsilon_n}s_{\epsilon_i+\epsilon_n}s_{\epsilon_j+\epsilon_{n-1}}$. $\square$

5.4. Description of $T(w, \tau)$

In analogy with §5.3, one can construct a basis for $T(w, \tau)$, $w \ge \tau$. Results for classical groups are described below. Types B and D require extra notation so we chose to describe each case separately.

5.4.1. The special linear group $SL(2n)$. Let $\tau = (a_1 \cdots a_n)$; denoting as above by $\tau^{(d)}$ the element in W^{P_d} representing the coset τW_{P_d}, we have the extremal weight vector $q_{\tau^{(d)}} = e_{a_1} \wedge \cdots \wedge e_{a_d}$. Let $\beta \in \tau(R^+)$, say $\beta = \tau(\epsilon_j - \epsilon_k)$, $1 \le j < k \le n$. We have $X_{-\beta} = E_{a_k a_j}$, and it follows easily that for $1 \le d \le l$, $X_{-\beta} q_{\tau^{(d)}} \ne 0$ if and only if $a_j \in \{a_1, \ldots a_d\}$, $a_k \notin \{a_1, \ldots, a_d\}$, i.e., if and only if $j \le d < k$, in which case

$$X_{-\beta} q_{\tau^{(d)}} = e_{a_1} \wedge \cdots \wedge e_{a_{j-1}} \wedge e_{a_k} \wedge e_{a_{j+1}} \wedge \cdots \wedge e_{a_d}.$$

Hence we obtain

$$X_{-\beta} q_{\tau^{(d)}} = q_{(s_\beta \tau)^{(d)}}.$$

This implies in view of Theorem 5.1.1 that for $w \ge \tau$, $\beta \in N(w, \tau)$ if and only if $w^{(d)} \ge (s_\beta \tau)^{(d)}$, for all $1 \le d \le l$, i.e., if and only if $w \ge s_\beta \tau$, note that for $d < j$, or $d \ge k$, $(s_\beta \tau)^{(d)} = \tau^{(d)}$. Hence we obtain

5.4.2. Theorem. *Let $G = SL(n)$, and $w, \tau \in S_n$, $w \ge \tau$. Then $T(w, \tau)$ is spanned by $\{X_{-\beta}, \beta \in \tau(R^+) \mid w \ge s_\beta \tau\}$.*

5.4.3. Remark. We have used Theorem 5.1.1 in the proofs of Theorems 5.3.1(1) and 5.4.2. In [**109**], the two theorems are proved using the Jacobian criterion.

5.4.4. The Symplectic Group $Sp(2n)$.

5.4.5. Theorem. *(cf. [**100**]) Let $w, \tau \in W, w \ge \tau$, and $\beta \in \tau(R^+)$, say $\beta = \tau(\alpha)$, where $\alpha \in R^+$. Note, $\tau s_\alpha = s_\beta \tau$.*

1. *Let $\alpha = \epsilon_j - \epsilon_k, 1 \le j < k \le n$, or $2\epsilon_j, 1 \le j \le n$. Then $\beta \in N(w, \tau)$ if and only if $w \ge s_\beta \tau$.*
2. *Let $\alpha = \epsilon_j + \epsilon_k, 1 \le j < k \le n$.*
   (a) *Let $\tau > \tau s_\alpha$. Then $\beta \in N(w, \tau)$ necessarily.*
   (b) *Let $\tau < \tau s_\alpha$.*
      (i) *Let $\tau > \tau s_{2\epsilon_j}$ or $\tau s_{2\epsilon_k}$. Then $\beta \in N(w, \tau)$ if and only if $w \ge s_\beta \tau$.*
      (ii) *Let $\tau < \tau s_{2\epsilon_j}$ and $\tau s_{2\epsilon_k}$.*
         (A) *If $\tau < \tau s_{\epsilon_j - \epsilon_k}$, then $\beta \in N(w, \tau)$ if and only if $w \ge s_\beta \tau \ (= \tau s_{\epsilon_j + \epsilon_k})$ or $\tau s_{2\epsilon_j}$.*
         (B) *If $\tau > \tau s_{\epsilon_j - \epsilon_k}$, then $\beta \in N(w, \tau)$ if and only if $w \ge \tau s_{\epsilon_j - \epsilon_k} s_{2\epsilon_j}$.*

PROOF. As in the proof of Theorem 5.3.1, we shall show that $X_{-\beta} q_{\tau^{(d)}}$ belongs to V_{w, ω_d} for all $1 \le d \le l$, if and only if β satisfies the conditions in the theorem. Again, let X_β be the element of the Chevalley basis given by §3.3.

We have the following (cf. [**95**]).

1. Let $\alpha = \epsilon_j - \epsilon_k, 1 \leq j < k \leq n$. Then

$$X_{-\beta}q_{\tau(d)} = \begin{cases} 0, & \text{if } d < j \text{ or } d \geq k \\ \pm q_{(s_\beta\tau)(d)}, & \text{if } j \leq d < k. \end{cases}$$

2. Let $\alpha = 2\epsilon_j, 1 \leq j \leq n$. Then

$$X_{-\beta}q_{\tau(d)} = \begin{cases} 0, & \text{if } d < j \\ \pm q_{(s_\beta\tau)(d)}, & \text{if } j \leq d \leq n. \end{cases}$$

3. Let $\alpha = \epsilon_j + \epsilon_k, 1 \leq j < k \leq n$. Then

$$X_{-\beta}q_{\tau(d)} = \begin{cases} 0, & \text{if } d < j \\ \pm q_{(s_\beta\tau)(d)}, & \text{if } j \leq d < k \\ \sum_{i=0}^{l_d} c_i q_{\lambda_{di}, \mu_{di}}, & \text{if } k \leq d \leq n \end{cases}$$

where $c_i = \pm 1$, and λ_{di}, μ_{di} are as follows:

Let $\tau = (a_1 \cdots a_n)$, $s = \min\{|a_j|, |a_k|\}$, $r = \max\{|a_j|, |a_k|\}$. Let $\{t_{di}, 0 \leq i \leq l_d\}$ be the set of all integers $s = t_{d0} < t_{d1} < \cdots < t_{dl_d} < r$ with the property that if $l_d > 0$, then $t_{di} \in \{|a_{d+1}|, \ldots, |a_n|\}$, for all $1 \leq i \leq l_d$; in particular, note that $l_d = \#\{t, s < t < r \mid t \notin \{|a_1|, \ldots, |a_d|\}\}$.

For $0 \leq i \leq l_d$, define μ_{di}, λ_{di} as the elements in W^{P_d} given by the d-tuples $\{a_1, \ldots, \hat{a}_j, \ldots, \hat{a}_k, \ldots, a_d, t_{di}, t'_{di+1}\} \uparrow$, $\{a_1, \ldots, \hat{a}_j, \ldots, \hat{a}_k, \ldots, a_d, t_{di+1}, t'_{di}\} \uparrow$, respectively, where $t_{dl_d+1} = r$ and a d-tuple $(b_1 \cdots b_d)$, $\{b_1 \cdots b_d\} \uparrow$ denotes the d-tuple obtained from $\{b_1 \cdots b_d\}$ by arranging the entries in ascending order. Note that (λ_{di}, μ_{di}), $0 \leq i \leq l_d$, are admissible pairs.

The rest of the argument is as in the proof of Theorem 5.3.1. □

5.4.6. The special orthogonal group $SO(2n+1)$. For the statement of our results concerning the membership of $\beta = \tau(\alpha)$, $\alpha = \epsilon_j, \epsilon_j + \epsilon_k$ in $N(w, \tau)$, we need to introduce some specific elements in W which we describe now. Let $\tau = (a_1 \cdots a_n)$.

The elements τ_S^j, $1 \leq j < n$. For $1 \leq i \leq 2n + 1$, let $|a_i| = \min\{a_i, 2n + 2 - a_i\}$. Fix $j, 1 \leq j < n$. Let $|a_j| = r$.

We first define the set I. If $|a_t| < r$, $j < t \leq n$, then the set I is defined to be the empty set. If there exists a $t, j < t \leq n$ such that $|a_t| > r$, then I is defined to be the set $\{i_1 < \cdots < i_l \leq n\}$, where $i_t, 0 \leq t \leq p$ are defined as follows: first, set $i_0 = j$, and then define i_t inductively so that

$$|a_{i_t}| = \max\{|a_m| > r, i_{t-1} < m \leq n\}.$$

Note that $|a_m| < r$ for all $i_l < m \leq n$.

Given any subset S of I, we define τ_S^j as follows. If $S = \emptyset$, then set

$$\tau_S^j = s_\beta\tau,$$

note that this includes the case $I = \emptyset$. Otherwise, if $S \neq \emptyset$, say $S = \{x_1, \ldots, x_m\}$ arranged in ascending order. Denote $x_0 = j (= i_0)$. We define

$$\tau_S^j = \tau \overline{(x_0, x_1)} \, \overline{(x_1, x_2)} \, \cdots \overline{(x_{m-1}, x_m)}$$

where, for $\sigma = (c_1 \cdots c_n) \in W$, and $1 \leq k < l \leq n$, $\sigma \overline{(k, l)}$ is the element in W obtained from σ by replacing c_k, c_l respectively by $|c_l|', |c_k|'$.

The elements $\tau_S^{j,k}$. Let $\alpha = \epsilon_j + \epsilon_k$, $1 \leq j < k \leq n - 1$, and let $\max\{|a_j|, |a_k|\} = r$, $\min\{|a_j|, |a_k|\} = s$.

Define the set I to be the empty set if $|a_t| < r, k < t \leq n$. Otherwise, I is defined to be the set $\{i_1 < \cdots < i_p \leq n\}$: Set $i_0 = k$, and define i_t inductively so that

$$|a_{i_t}| = \max \{|a_m| > r, \, i_{t-1} < m \leq n\},$$

note that $|a_m| < r$, $i_p < m \leq n$, $p \leq p_k$, and $p = 0 \Leftrightarrow p_k = 0$.

Let S be a subset of I. If $S = \emptyset$, then set

$$\tau_S^{j,k} = s_\beta \tau,$$

note that this includes the case $I = \emptyset$. Otherwise, if $S \neq \emptyset$, say $S = \{x_1, \ldots, x_m\}$, arranged in ascending order. Denote $x_0 = k (= i_0)$. We define

$$\tau_S^{j,k} = \tau' \overline{(x_0, x_1)} \, \overline{(x_1, x_2)} \, \cdots \overline{(x_{m-1}, x_m)}$$

where τ' is obtained from τ by replacing a_j, a_k, by s', r', respectively.

5.4.7. Theorem. [100] *Let $w, \tau \in W, w \geq \tau$, and $\beta \in \tau(R^+)$, say $\beta = \tau(\alpha)$, where $\alpha \in R^+$. Note that $s_\beta \tau = \tau s_\alpha$.*

1. *Let $\alpha = \epsilon_k - \epsilon_l$, ϵ_n, or $\epsilon_i + \epsilon_n$. Then $\beta \in N_{w,\tau} \Longleftrightarrow w \geq s_\beta \tau$.*
2. *Let $\alpha = \epsilon_j$, $j < n$.*
   (a) *If $\tau > s_\beta \tau$, then $\beta \in N_{w,\tau}$ (necessarily).*
   (b) *Let $\tau < s_\beta \tau$, and let $|a_j| = r$.*
      (i) *Let $|a_m| < r$, $j < m \leq n$. Then $\beta \in N_{w,\tau} \Longleftrightarrow w \geq s_\beta \tau$.*
      (ii) *Let $|a_m| > r$, for some m, $j < m \leq n$. Then $\beta \in N_{w,\tau} \Longleftrightarrow w \geq \tau_S^j$ (for some S, notation being as above).*
3. *Let $\alpha = \epsilon_j + \epsilon_k$, $j < k \leq n - 1$.*
   (a) *If $\tau > s_\beta \tau$, then $\beta \in N_{w,\tau}$ (necessarily).*
   (b) *Let $\tau < s_\beta \tau$. If τ is $>$ either τs_{ϵ_j} or τs_{ϵ_k}, then $\beta \in N_{w,\tau} \Longleftrightarrow w \geq s_\beta \tau$.*
   (c) *Let $\tau < s_\beta \tau, \tau s_{\epsilon_j}$, and τs_{ϵ_k}. Let $\max \{|a_j|, |a_k|\} = r$.*
      (i) *Let $|a_m| < r$, $k < m \leq n$. Then $\beta \in N_{w,\tau} \Longleftrightarrow w \geq s_\beta \tau$.*
      (ii) *Let $|a_m| > r$, for some m, $k < m \leq n$. Then $\beta \in N_{w,\tau} \Longleftrightarrow w \geq \tau_S^{j,k}$ (for some S, notations being as above).*

PROOF. Let $\beta \in \tau(R^+)$, say $\beta = \tau(\alpha)$, where $\alpha \in R^+$. We have (cf. [96])

1. Let $\alpha = \epsilon_j - \epsilon_k, 1 \leq j < k \leq n$. Then

$$X_{-\beta}q_{\tau(d)} = \begin{cases} 0, & \text{if } d < j \text{ or } d \geq k \\ \pm q_{(s_\beta\tau)^{(d)}}, & \text{if } j \leq d < k. \end{cases}$$

2. Let $\alpha = \epsilon_j, 1 \leq j \leq n$. Then

$$X_{-\beta}q_{\tau(d)} = \begin{cases} 0, & \text{if } d < j \\ \pm q_{\gamma(d,j),\sigma(d,j)}, & \text{if } j \leq d \leq n \end{cases}$$

where for $j \leq d \leq n$, if either $d = n$ or $|a_m| < |a_j|$, for all $d < m \leq n$, then $\sigma(d,j) = \gamma(d,j) = (s_\beta\tau)^{(d)}$; and $\sigma(d,j) = \{a_1, \ldots, \hat{a}_j, \cdots a_d, u_{dj}\} \uparrow$, $\gamma(d,j) = \{a_1, \ldots, \hat{a}_j, \ldots a_d, u'_{dj}\} \uparrow$, u_{dj} being the largest entry u in $\{|a_{d+1}|, \ldots, |a_n|\}$ such that $u_{dj} > |a_j|$, otherwise note that $(\gamma(d,j), \sigma(d,j))$ is an admissible pair.

3. Let $\alpha = \epsilon_j + \epsilon_k, 1 \leq j < k \leq n$. Then

$$X_{-\beta}q_{\tau(d)} = \begin{cases} 0, & \text{if } d < j \\ q_{(s_\beta\tau)^{(d)}}, & \text{if } j \leq d < k \text{ or } d = n \\ 2(\sum_{v=0}^{p_d-1} b_i q_{\theta_{dv},\delta_{dv}}) & \\ \quad + aq_{\xi_d,\nu_d} + \sum_{i=0}^{l_d} c_i q_{\lambda_{di},\mu_{di}}, & \text{if } k \leq d < n \end{cases}$$

where $c_i = \pm 1$, while a, b_i are zero if precisely one of $\{a_j, a_k\}$ is $> n$ and are ± 1 otherwise; $l_d, \mu_{di}, \lambda_{di}$ are defined in the same way as in (3) in the proof of Theorem 5.4.5, and $p_d, \delta_{dv}, \theta_{dv}, \nu_d, \xi_d$ are defined as follows:

Let $\{r_{dv}, 0 \leq v \leq p_d\}$ be the set of all integers $r = r_{d0} < r_{d1} < \cdots < r_{dp_d}$ with the property that if $p_d > 0$, then $r_{dv} \in \{|a_{d+1}|, \ldots, |a_n|\}$, for all $1 \leq v \leq p_d$, in particular, note that $p_d = \#\{tr < t$ and $t \notin \{|a_1|, \ldots, |a_d|\}\}$.

The elements $\{\delta_{dv}, \theta_{dv}\}$ are defined only when $p_d > 0$ in which case they are defined as

$\delta_{dv} = \{a_1, \ldots, \hat{a}_j, \ldots, \hat{a}_k, \ldots, a_d, r_{di}, r'_{di+1}\} \uparrow$,
$\theta_{dv} = \{a_1, \ldots, \hat{a}_j, \ldots, \hat{a}_k, \ldots, a_d, r_{di+1}, r'_{di}\} \uparrow$.

The elements ν_d, ξ_d are defined as

$\nu_d = \{a_1, \ldots, \hat{a}_j, \ldots, \hat{a}_k, \ldots, a_d, r_{dp_d-1}, r_{dp_d}\} \uparrow$,
$\xi_d = \{a_1, \ldots, \hat{a}_j, \ldots, \hat{a}_k, \ldots, a_d, r'_{dp_d}, r'_{dp_d-1}\} \uparrow$

(here, if $p_d = 0$, then $r_{dp_d-1} = t_{dl_d}$, and the sum $\sum_{v=0}^{p_d-1} b_i q_{\theta_{dv},\delta_{dv}}$ is understood to be zero). Note that $(\lambda_{di}, \mu_{di}), (\theta_{dv}, \delta_{dv})$ and (ξ_d, ν_d) are admissible pairs.

The rest of the argument is as in the proof of Theorem 5.3.1 (see [**100**] for details). $\qquad\qquad\qquad\square$

5.4.8. Remark. Recall that τ has been fixed above. Let $\beta = \tau(\alpha)$, $\alpha = \epsilon_j + \epsilon_k$, $j < k \leq n - 1$. Let $\tau < \tau s_\alpha, \tau s_{\epsilon_j}$, and τs_{ϵ_k}. Note that if $p \neq 0$,

p being as above, then the condition that $w \geq \tau_S^{j,k}$ is equivalent to the condition that

$$
w^{(d)} \geq \begin{cases} \{a_1, \dots, \hat{a}_j, \dots, \hat{a}_k, \dots, a_d, a_k'\} \uparrow, & \text{if } j \leq d < k, \\ \{a_1, \dots, \hat{a}_j, \dots, \hat{a}_k, \dots, a_d, |a_{i_{t+1}}|', s'\} \uparrow, & \text{if } i_t \leq d < i_{t+1}, \\ & 0 \leq t < p, \\ (s_\beta \tau)^{(d)}, & \text{if } i_p \leq d \leq n. \end{cases}
$$

5.4.9. The special orthogonal group SO$(2n)$. Here again for the statement of the results, we need to introduce some special elements in W. For this, we first define the integers $\{i_{tm}\}$.

Let $\{i_t, \ 0 \leq t \leq p\}$ be the set of all integers defined as follows:

1. Define $i_0 = k$.
2. If $|a_m| < r$, for all $k < m \leq n$, then define $p = 0$ so there is only one element in the set.
3. If $|a_m| > r$, for some $k < m \leq n$, then i_t is defined inductively so that

$$
|a_{i_t}| = \max \{|a_m| > r, \ i_{t-1} < m \leq n\},
$$

note that $|a_m| < r$, $i_p < m \leq n$; also note that $p \leq p_k$, and $p = 0 \Leftrightarrow p_k = 0$.

If $p > 0$, then for $t, 1 \leq t \leq p$, define $\{i_{tl}, 0 \leq l \leq c_t\}$ inductively as $i_{t0} = i_{t-1}$, and $i_{t1} < \cdots < i_{tc_t}$ are all the indices lying between i_{t-1} and i_t (when this set is nonempty) with the property

$$
|a_{i_{tl}}| > |a_m|, \ \forall m > i_{tl}, \ m \neq i_t, \text{ such that } |a_m| > r.
$$

Note that $i_{10} = i_0 \ (= k)$. Set $i_{p+1\,0} = i_p$.

We represent i_t and $\{i_{tm}\}$ diagrammatically as follows:

$$
j \ \cdots \ \underbrace{i_0 \ \cdots \ i_{11} \ \cdots \ i_{12} \ \cdots \ i_{1c_1} \ \cdots \ i_1} \ \cdots \ i_2 \ \cdots
$$

$$
\underbrace{i_{t-1} \ \cdots \ i_{t1} \ \cdots \ i_{t2} \ \cdots \ i_{tc_t} \ \cdots \ i_t} \ \cdots \ i_p \ \cdots \ n
$$

5.4.10. Theorem. [100] *Let $w, \tau \in W, w \succeq \tau$, and $\beta \in \tau(R^+)$, say $\beta = \tau(\alpha)$, where $\alpha \in R^+$.*

1. *Let $\alpha = \epsilon_l - \epsilon_m$, or $\epsilon_j + \epsilon_k, k = n - 1, n$. Then $\beta \in N(w, \tau) \Longleftrightarrow w \succeq s_\beta \tau$.*
2. *Let $\alpha = \epsilon_j + \epsilon_k, \ j < k < n - 1$.*
   (a) *If $\tau \succ s_\beta \tau$, then $\beta \in N_{w,\tau}$ (necessarily).*
   (b) *Let $s_\beta \tau \succ \tau$.*
      (i) *If precisely one of $\{a_j, a_k\}$ is $> n$, then $\beta \in N(w, \tau) \Longleftrightarrow w \succeq s_\beta \tau$.*

(ii) *Let* $s_\beta\tau \succ \tau$, *and* $a_j, a_k \leq n$. *Let max* $\{|a_i|, |a_j|\} = r$.
Then $\beta \in N(w,\tau) \iff$

$$w \succeq s_\beta\tau, \quad if\ p = 0,$$

and if $p > 0$, *then* $w^{(d)} \succeq \theta^{(d)}$, $\forall j \leq d \leq n$, *where* $\theta^{(d)}$
in W^{P_d} *has the following four descriptions depending on
which segment of* $[j, n]$, *d belongs to; refer to the diagram-
matic representation above of* $[j, n]$:

(5.4.11)

$$\theta^{(d)} = \begin{cases} \{a_1, \dots, \hat{a}_j, \dots, a_d, a_k'\} \uparrow, j \leq d < k, \\ \{a_1, \dots, \hat{a}_j, \dots, \hat{a}_k, \dots, a_d, s', |a_{i_t\,m+1}|'\} \uparrow, \\ \qquad i_{tm} \leq d < i_{t\,m+1}, \ 0 \leq m < c_t, \ 1 \leq t \leq p, \\ \{a_1, \dots, \hat{a}_j, \dots, \hat{a}_k, \dots, a_d, s', |a_{i_{t+1}}|'\} \uparrow, \ i_{tc_t} \leq d < i_t, \ t < p, \\ (s_\beta\tau)^{(d)}, \ i_{pc_p} \leq d \leq n, \end{cases}$$

notation as above.

PROOF. Let $r, s, l_d, p_d, \mu_{di}, \lambda_{di}, \delta_{di}, \theta_{di}, \nu_d, \xi_d$ be as in (3) in the proof of
Theorem 5.4.7. Now (cf. [**106**])

1. Let $\alpha = \epsilon_j - \epsilon_k, 1 \leq j < k \leq n$. Then

$$X_{-\beta}q_{\tau(d)} = \begin{cases} 0, & \text{if } d < j \text{ or } d \geq k \\ \pm q_{(s_\beta\tau)^{(d)}}, & \text{if } j \leq d < k. \end{cases}$$

2. Let $\alpha = \epsilon_j + \epsilon_k, 1 \leq j < k \leq n$.
   (a) If $k = n - 1, n$, then

$$X_{-\beta}q_{\tau(d)} = \begin{cases} 0, & \text{if } d < j \\ \pm q_{(s_\beta\tau)^{(d)}}, & \text{if } j \leq d. \end{cases}$$

   (b) Let $k < n - 1$.
      (i) Let $(l_d, p_d) \neq (0, 0)$. Then

$$X_{-\beta}q_{\tau(d)} = \begin{cases} 0, & \text{if } d < j \\ q_{(s_\beta\tau)^{(d)}}, & \text{if } j \leq d < k, \text{ or } d = n - 1, n \\ aq_{\xi_d, \nu_d} + \phi_d, & \text{if } k \leq d < n - 1 \end{cases}$$

where $\phi_d = 2(\sum_{i=0}^{q_d} b_i q_{\theta_{di}, \delta_{di}}) - \sum_{i=0}^{y_d} c_i q_{\lambda_{di}, \mu_{di}}$, $q_d = p_d - 1$, $y_d = l_d$ or $l_d - 1$ according as $p_d >$ or $= 0$, $c_i = \pm 1$, while $a, b_i, i \leq q_d$ are zero if precisely one of $\{a_j, a_k\}$ is $> n$, and $a, b_i, i < q_d$ are ± 1, $b_{q_d} = \pm 1$ or ± 3, otherwise (and the sum $\sum_{v=0}^{q_d} b_i q_{\theta_{dv}, \delta_{dv}}$ is understood to be zero, if $p_d = 0$).

(ii) Let $(l_d, p_d) = (0, 0)$. Then

$$X_{-\beta} q_{\tau^{(d)}} = \pm q_{\xi_d, \nu_d},$$

where note that $\xi_d = \{a_1, \ldots, \hat{a}_j, \ldots, \hat{a}_k, \ldots, a_d, r', s'\} \uparrow$
$(= (s_\beta \tau)^{(d)})$, $\nu_d = \{a_1, \ldots, \hat{a}_j, \ldots, \hat{a}_k, \ldots, a_d, r, s\} \uparrow$.

The rest of the argument is as in the proof of Theorem 5.3.1 (see [**99, 100**] for details). □

5.4.12. Remark. There is a striking similarity between the condition in 2(b), (ii) of Theorem 5.4.10 and the condition stated in Remark 5.4.8.

5.4.13. Remark. The description in 2(b), (ii) in the above theorem looks quite involved, but this is the best possible description that we could have, given the complicated nature of $\mathbf{D}_n$.

5.5. Tangent space and certain weight multiplicities

In this section we bring out the relationship between $T(w, \tau)$ and the multiplicities of certain weights in the fundamental representations of G. For a weight χ in $V_K(\omega_d)$, let $m_{\omega_d}(\chi)$ (resp. $m_{w, \omega_d}(\chi)$) denote the multiplicity of χ in $V_K(\omega_d)$ (resp. V_{w, ω_d}). Having fixed ω_d in the discussion below, we shall denote $m_{\omega_d}(\chi)$ by just $m(\chi)$, respectively $m_{w, \omega_d}(\chi)$ by $m_w(\chi)$.

The special linear group.

5.5.2. Theorem. *Let G be of Type $\mathbf{A}_l$. Let $w \in W$, $\beta \in R^+$. Then the following are equivalent:*

1. $w \geq s_\beta$.
2. $m_w(\omega_d - \beta) = m(\omega_d - \beta)$, *for all $1 \leq d \leq l$, l being the rank of G.*
3. $\beta \in N(w, \mathrm{id})$.

PROOF. The equivalence of (1) and (3) follows from Theorem 5.3.1(1). We now prove the equivalence of (2) and (3). Given $d, 1 \leq d \leq l$, and $\beta = \epsilon_j - \epsilon_k, 1 \leq j < k \leq n$, from the proof of Theorem 5.3.1 for $SL(n)$, it follows that

$$m(\omega_d - \beta) = \begin{cases} 0, & \text{if } d < j \text{ or } d \geq k \\ 1, & \text{if } j \leq d < k. \end{cases}$$

Hence we obtain that $m_w(\omega_d - \beta) = m(\omega_d - \beta)$, for all $1 \leq d \leq l$ if and only if for all d, $j \leq d < k$, $w \geq s_\beta^{(d)}$. □

More generally, we have

5.5.3. Theorem. *Let G be of Type $\mathbf{A}_l$. Let $w, \tau \in W$, $w \geq \tau$, $\beta \in \tau(R^+)$. Then the following are equivalent:*

1. $w \geq s_\beta \tau$.
2. $m_w(\tau(\omega_d) - \beta) = m(\tau(\omega_d) - \beta)$, *for all $1 \leq d \leq l$, l being the rank of G.*

3. $\beta \in N(w, \tau)$.

The proof is similar to that of Theorem 5.5.2.

For the remaining classical groups, we just give the statements of the results and refer the reader to [99], [100] for proofs.

The symplectic group $Sp(2n)$.

5.5.5. Proposition. *Let $\beta \in R^+$ and let $G = Sp(2n)$.*

1. *Let $\beta = \epsilon_j - \epsilon_k, 1 \leq j < k \leq n$. Then*

$$m(\omega_d - \beta) = \begin{cases} 0, & \text{if } d < j, \text{ or } d \geq k \\ 1, & \text{if } j \leq d < k. \end{cases}$$

2. *Let $\beta = 2\epsilon_j, 1 \leq j \leq n$. Then*

$$m(\omega_d - \beta) = \begin{cases} 0, & \text{if } d < j \\ 1, & \text{if } j \leq d \leq n. \end{cases}$$

3. *Let $\beta = \epsilon_j + \epsilon_k, 1 \leq j < k \leq n$. Then*

$$m(\omega_d - \beta) = \begin{cases} 0, & \text{if } d < j \\ 1, & \text{if } j \leq d < k \\ n + 1 - d, & \text{if } k \leq d \leq n. \end{cases}$$

5.5.6. Corollary. *Let $w \in W$ and $\beta \in R^+$.*

1. *Let $\beta = \epsilon_j - \epsilon_k, \ j < k \leq n, \ 2\epsilon_m, \ 1 \leq m \leq n$. Then $m_w(\omega_d - \beta) = m(\omega_d - \beta)$, for all $1 \leq d \leq n$ if and only if $w \geq s_\beta$.*
2. *Let $\beta = \epsilon_j + \epsilon_k, 1 \leq j < k \leq n$. Then $m_w(\omega_d - \beta) = m(\omega_d - \beta)$, for all $1 \leq d \leq n$ if and only if $w \geq s_\beta$ or $s_{2\epsilon_j} s_{\epsilon_k - \epsilon_n} \ (= (12 \cdots j - 1 j' j + 1 \cdots k - 1 n k + 1 k + 2 \cdots n - 1 k))$.*

5.5.7. Remark. Let $w \in W$, and $\beta \in R^+$. The condition that $m_w(\omega_d - \beta) = m(\omega_d - \beta)$, for all $1 \leq d \leq n$ need not be equivalent to the condition that $\beta \in N(w, \mathrm{id})$. For example, take $w = s_{2\epsilon_j}$ for some $j < n - 1$, $\beta = \epsilon_j + \epsilon_k$ for some $k, \ j < k \leq n - 1$. We have (cf. Theorem 5.3.1(2)), $\beta \in N(w, \mathrm{id})$, but $m_w(\omega_d - \beta) \neq m(\omega_d - \beta), \ k \leq d < n$; note that $m_w(\omega_d - \beta) = 1, \ k \leq d \leq n$, while $m(\omega_d - \beta) = n + 1 - d, \ k \leq d \leq n$. Also, the condition that $w \geq s_\beta$ need not be equivalent to the condition that $m_w(\omega_d - \beta) = m(\omega_d - \beta)$, for all $1 \leq d \leq n$. For example, take $\beta = \epsilon_j + \epsilon_k$, for some $j < k \leq n$, and $w = s_{2\epsilon_j} s_{\epsilon_k - \epsilon_n}$. We have $m_w(\omega_d - \beta) = m(\omega_d - \beta)$, for all $1 \leq d \leq n$ (cf. Corollary 5.5.6), but $w \not\geq s_\beta$. Of course, for $\beta = \epsilon_j - \epsilon_k, \ j < k \leq n, \ 2\epsilon_m, \ 1 \leq m \leq n$, and any $w \in W$, all three conditions of Theorem 5.5.2 are equivalent.

Let l_d, p_d be as in the proof of Theorem 5.4.7, (3). Let $s_{d1} < \cdots < s_{dm_d} < s$ be the set of all integers with the property that if $m_d > 0$, then $s_{du} \in \{|a_{d+1}|, \ldots, |a_n|\}$, for all $1 \leq u \leq m_d$, in particular, note that $m_d = \#\{t, t < s \mid t \notin \{|a_1|, \ldots, |a_d|\}\}$.

5.5.8. Proposition. *Let* $G = Sp(2n)$, $\beta \in \tau(R^+)$, *say* $\beta = \tau(\alpha)$, *where* $\alpha \in R^+$.

1. *Let* $\alpha = \epsilon_j - \epsilon_k, 1 \leq j < k \leq n$. *Then*

$$m(\tau(\omega_d) - \beta) = \begin{cases} 0, & \text{if } d < j, \text{ or } d \geq k \\ 1, & \text{if } j \leq d < k. \end{cases}$$

2. *Let* $\alpha = 2\epsilon_j, 1 \leq j \leq n$. *Then*

$$m(\tau(\omega_d) - \beta) = \begin{cases} 0, & \text{if } d < j \\ 1, & \text{if } j \leq d \leq n. \end{cases}$$

3. *Let* $\alpha = \epsilon_j + \epsilon_k, 1 \leq j < k \leq n$. *Then*

$$m(\tau(\omega_d) - \beta) = \begin{cases} 0, & \text{if } d < j \\ 1, & \text{if } j \leq d < k \\ x_d + l_d + p_d + 1, & \text{if } k \leq d \leq n, \end{cases}$$

where $x_d = m_d$ *or* 0 *according to whether there exists or does not exist a* $t \in \{|a_{d+1}|, \ldots, |a_n|\}$ *such that* $t < s$.

5.5.9. Corollary. *Let* $\beta \in \tau(R^+)$, *say* $\beta = \tau(\alpha)$, $w \in W$.

1. $\alpha = \epsilon_j - \epsilon_k$ *or* $2\epsilon_j$. *Then* $m_w(\tau(\omega_d) - \beta) = m(\tau(\omega_d) - \beta)$, *for all* $1 \leq d \leq n$ *if and only if* $w \geq s_\beta \tau$.
2. $\alpha = \epsilon_j + \epsilon_k$, $1 \leq j < k \leq n$. *Then* $m_w(\tau(\omega_d) - \beta) = m(\tau(\omega_d) - \beta)$, *for all* $1 \leq d \leq n$ *if and only if*

$$w^{(d)} \geq \begin{cases} \{a_1, \ldots, \hat{a}_j, \ldots, a_d, a_k'\} \uparrow, & \text{if } j \leq d < k, \\ \{a_1, \ldots, \hat{a}_j, \ldots, \hat{a}_k, \ldots, a_d, r_{dp_d}, b_d'\} \uparrow, & \text{if } k \leq d \leq n, \end{cases}$$

where $b_d = s_{d1}$ *or* s *according to whether there exists or does not exist a* $t \in \{|a_{d+1}|, \ldots, |a_n|\}$ *such that* $t < s$, *and* r_{dp_d} *is as in Theorem 5.4.7, (3).*

5.5.10. Remark. Let $\beta = \tau(\alpha)$, $\alpha \in R^+$, and $w \in W$, $w \geq \tau$. We have (cf. Proposition 5.0.1) that condition (1) of Theorem 5.5.3 that $w \geq s_\beta \tau$ implies the condition (3) of Theorem 5.5.3 that $\beta \in N(w, \tau)$. Similarly, condition (2) of Theorem 5.5.3 that $m_w(\tau(\omega_d) - \beta) = m(\tau(\omega_d) - \beta)$, for all $1 \leq d \leq n$ implies the condition that $\beta \in N(w, \tau)$, in view of Theorem 5.3.1. Note that if $m_w(\tau(\omega_d) - \beta) = m(\tau(\omega_d) - \beta)$, then $X_{-\beta} q_{\tau(d)} \in V_{w, \omega_d}$.

If $\alpha = \epsilon_j - \epsilon_k$ or $2\epsilon_j$, then the three conditions of Theorem 5.5.3 are equivalent in view of Theorem 5.4.5 and Corollary 5.5.9(1).

Let $\alpha = \epsilon_j + \epsilon_k$. The condition that $w \geq s_\beta \tau$ neither implies nor is implied by the condition that $m_w(\tau(\omega_d) - \beta) = m(\tau(\omega_d) - \beta)$, for all $1 \leq d \leq n$, in general. For example, take $G = Sp(10)$, $\tau = (13452)$, $\alpha = \epsilon_2 + \epsilon_3$, $\beta = \tau(\alpha)$.

1. Let $w = (14'3'52)$ $(= s_\beta \tau)$. Then $w \geq s_\beta \tau$, but $m_w(\omega_3 - \beta) \neq m(\omega_3 - \beta)$; note that $w^{(3)} \not\geq \{1, 5, 2'\}$ (cf. Corollary 5.5.9(2)).
2. Let $w = (12'543)$. Then $m_w(\omega_d - \beta) = m(\omega_d - \beta)$, for all $1 \leq d \leq 5$, but $w \not\geq s_\beta \tau$.

Note that the above two examples also show that condition (3) of Theorem 5.5.3 need not imply (1) or (2) in general.

The special orthogonal groups $SO(2n + 1)$ and $SO(2n)$. We give below the statements of results relating $T(w, \text{id})$ and $m(\omega_d - \beta)$ for $G = SO(2n + 1), SO(2n)$. For the discussion at any other T-fixed point, we refer the reader to [99], [100].

5.5.12. Proposition. *Let $G = SO(2n + 1)$ or $SO(2n)$. Let $\beta \in R^+$. Let $\beta = \epsilon_j + \epsilon_k, 1 \leq j < k < n$, and $k \leq d < n$. Then $m(\omega_d - \beta) = n - d + 2$. In all other cases, we have, $m(\omega_d - \beta) = 0$ or 1.*

5.5.13. Theorem. *Let $G = SO(2n + 1)$ or $SO(2n)$. Let $w \in W$, and $\beta \in R^+$. Then $\beta \in N(w, \text{id})$ if and only if $m_w(\omega_d - \beta) = m(\omega_d - \beta)$, for all $1 \leq d \leq n$.*

5.5.14. Remark. Theorem 5.5.13 need not be true at other T-fixed points. See [100] for details.

5.5.15. Remark. Let $G = SO(2n + 1)$. Let $\beta \in R^+$, $w \in W$. If $\beta = \epsilon_j - \epsilon_k$, $j < k \leq n$, ϵ_n, or $\epsilon_m + \epsilon_n$, $1 \leq m \leq n - 1$, then the three conditions in Theorem 5.5.2 are equivalent. For all $\beta \in R^+$, $w \in W$, the condition that $\beta \in N(w, \text{id})$ is equivalent to the condition $m_w(\omega_d - \beta) = m(\omega_d - \beta)$, for all $1 \leq d \leq n$ (cf. Theorem 5.5.13). The condition that $w \geq s_\beta$ need not be equivalent to the condition that $\beta \in N(w, \text{id})$. For example, take $\beta = \epsilon_j$ for some $j < n$ and $w = s_{\epsilon_j + \epsilon_n}$; we have (cf. Theorem 5.3.1 (3)), $\beta \in N(w, \text{id})$, but $w \not\geq s_\beta$.

5.5.16. Remark. Let $\beta \in R^+$, $w \in W$. If $\beta = \epsilon_j - \epsilon_k$, $j < k \leq n$, or $\epsilon_j + \epsilon_k$, $k = n - 1, n$, then the three conditions in Theorem 5.5.2 are equivalent. For all $\beta \in R^+$, $w \in W$, the condition that $\beta \in N(w, \text{id})$ is equivalent to the condition $m_w(\omega_d - \beta) = m(\omega_d - \beta)$, for all $1 \leq d \leq n$ (cf. Theorem 5.5.13). The condition that $w \geq s_\beta$ need not be equivalent to the condition that $\beta \in N(w, e)$. For example, take $\beta = \epsilon_j + \epsilon_k$, $j < k < n - 1$, and $w = s_{\epsilon_j - \epsilon_n} s_{\epsilon_j + \epsilon_n} s_{\epsilon_k + \epsilon_{n-1}}$; we have (cf. Theorem 5.4.7 (2)), $\beta \in N(w, \text{id})$, but $w \not\geq s_\beta$.

5.6. The B-module $T(w, \mathrm{id})$

In [**33**], it is proved that if G is simply laced, then the K-linear span of the reduced tangent cone of $X(w)$ at e_{id} is generated by the root vectors $X_{-\beta}$, $\beta \in \tau(R^+)$ such that $s_\beta \tau \le w$. It is observed in [**135**] that the same proof (as in [**33**]) gives the following:

> For arbitrary G, the K-linear span of the reduced tangent cone of $X(w)$ at e_{id} is the B-span of the root vectors $X_{-\alpha}$, $\alpha \in R^+$ such that $s_\alpha \le w$.

This raises the question of whether or not the tangent space $T(w, \mathrm{id})$ coincides with the B-span of the root vectors $X_{-\alpha}$, $\alpha \in R^+$ such that $s_\alpha \le w$. The answer is in the affirmative for types **A** and **C**, but not for the remaining types, a result due to Polo (cf. [**135**]).

5.6.1. Theorem. *(cf. [**135**]). Let G be simple. Then $T(w, \mathrm{id})$ is equal to the B-span of $\{X_{-\alpha} \mid w \ge s_\alpha, \alpha \in R^+\}$, for every $w \in W$ if and only if G is of type $\mathbf{A}_n$, or $\mathbf{C}_n$.*

PROOF. Let us set

$$M(w, \mathrm{id}) = B\text{-span of } \{X_{-\alpha} \mid w \ge s_\alpha, \alpha \in R^+\}.$$

First observe that if $w \ge s_\alpha$, then $X_{-\alpha} \in T(w, \mathrm{id})$ (cf. Proposition 5.0.1). Hence we obtain that $M(w, \mathrm{id}) \subset T(w, \mathrm{id})$, since $T(w, \mathrm{id})$ is B-stable.

(1) Let G be of Type **A**. In this case, we have (cf. Theorem 5.4.2), $T(w, \mathrm{id}) = \mathrm{span}$ of $\{X_{-\alpha} \mid w \ge s_\alpha, \alpha \in R^+\}$, and hence $T(w, \mathrm{id}) = M(w, \mathrm{id})$.

(2) Let G be of Type **C**. Let $\alpha \in R^+$ be such that $\alpha \in N(w, \tau)$. We shall now show that $X_{-\alpha} \in M(w, \mathrm{id})$.

If $\alpha = \epsilon_j - \epsilon_k$ or $2\epsilon_i$, then by Theorem 5.4.5(1), it follows that $w \ge s_\alpha$. Hence $X_{-\alpha}$ belongs to $M(w, \mathrm{id})$.

If $\alpha = \epsilon_j + \epsilon_k$, then by Theorem 5.4.5(2), we have either $w \ge s_{\epsilon_j + \epsilon_k}$ or $s_{2\epsilon_j}$. In the former case, we have $X_{-\alpha} \in M(w, \mathrm{id})$. In the latter case, we have that $X_{-2\epsilon_j} \in T(w, \mathrm{id})$ (cf. Proposition 5.0.1). Hence $[X_{\epsilon_j - \epsilon_k}, X_{-2\epsilon_j}]$ belongs to $M(w, \mathrm{id})$, by the B-stability of $T(w, \mathrm{id})$. On the other hand, we have $[X_{\epsilon_j - \epsilon_k}, X_{-2\epsilon_j}] = \pm X_{-\epsilon_j - \epsilon_k}$, since $-2\epsilon_j - (\epsilon_j - \epsilon_k)$ is not a root. Recall (see [**70**] for example) that for $\alpha, \beta \in R, [X_\alpha, X_\beta] = \pm(r + 1)X_{\alpha + \beta}$, where r is the largest integer such that $\beta - r\alpha$ is a root. Hence we obtain that $X_{-\epsilon_j - \epsilon_k}$ $(= X_{-\alpha})$ belongs to $M(w, \mathrm{id})$.

Now let G not be of Type **A** or **C**. This implies that the highest root β is a fundamental weight, say ω. We shall now exhibit a Schubert divisor $X(w)$ in $X(w_0) = G/B$ such that $T(w, \mathrm{id}) \ne M(w, \mathrm{id})$. Let s be the simple reflection and P the maximal parabolic corresponding to ω. Let $w = sw_0$, w_0 being the element of largest length in W. The unique Schubert divisor in G/P is given by $X_P(w)$. We have $V_K(\omega) \cong \mathfrak{g}$, and hence $V_{w,\omega}$ contains $\mathfrak{g}_\alpha$

for $\alpha \in R \setminus \beta$. In particular, $V_{w,\omega}$ contains $\mathfrak{g}_{-\alpha}$ for $\alpha \in S$. This implies, in view of B-stability of $X_P(w)$, that $V_{w,\omega}$ contains H_α $(= [X_\alpha - X_{-\alpha}])$, $\alpha \in S$. Thus $V_{w,\omega}$ contains the Cartan subalgebra $(=$ the zero weight space in $V_{w,\omega})$. In particular, the zero-weight vector $[X_{-\beta}, X_\beta]$ belongs to $V_{w,\omega}$, i.e., $X_{-\beta} q_{\mathrm{id}}$ belongs to $V_{w,\omega}$. Note that the highest-weight vector $q_{\mathrm{id}} = X_\beta$. This fact together with Theorem 5.1.1 implies that $X_{-\beta} \in T_P(w, \mathrm{id})$. Thus we obtain $T_P(w, \mathrm{id}) = T_P(w_0, \mathrm{id})$. On the other hand under the fibration $\pi : G/B \to G/P$, we have $\pi^{-1}(X_P(w)) = X(w)$ and hence $T(w, \mathrm{id}) = T(w_0, \mathrm{id})$. But now, since $s_\beta(\omega)$ $(= s_\beta(\beta)) = -\beta = w_0(\beta)$, we have $s_\beta \in w_0 W_P$; hence $X_P(s_\beta)$ $(= X_P(w_0)) = G/P$ and therefore $X_P(s_\beta) \not\subset X_P(w)$, i.e., $s_\beta \not\leq w$. This implies that $s_\beta \notin M(w, \mathrm{id})$. Hence $M(w, \mathrm{id}) \neq T(w, \mathrm{id})$; note that the set $\{X_{-\alpha},\ \alpha \in R^+ \setminus \beta\}$ spans a proper B-submodule of $T(w, \mathrm{id})$. □

5.7. Two smoothness criteria of Carrell–Kuttler

Let $w, x \in W$, $x \leq w$. Let $E(w, x)$ be the set of T-stable curves in $X(w)$ containing the T-fixed point e_x. The T-stable curves will be discussed in more detail in § 6.2. Let U_α be the one dimensional unipotent subgroup of G, the root subgroup associated to α (cf. §2.3). Then any $C \in E(w, x)$ has the form $\overline{U_\alpha e_x}$ for some α. Moreover, the T-fixed points on C are given by $C^T = \{e_x, e_y\}$, where $y = s_\alpha x$ and s_α denotes the reflection corresponding to α. If $y > x$, then $\alpha < 0$ and we can write $C = \overline{U_\beta e_y}$ with $\beta = -\alpha > 0$; so one can translate the Zariski tangent space $T_y(X(w))$ to $X(w)$ at y along $C \setminus \{e_x\}$ via U_β leaving $X(w)$ invariant. Taking the limit gives a T-stable subspace $\tau_C(X(w), x)$ of $T_x(X(w))$ of dimension $\dim T_y(X(w))$. Set

$$TE(w, x) = \sum_{C \in E(w,x)} T_x(C).$$

If $C = \overline{U_\alpha e_x}$, then $T_x(C)$ is a T-stable line of weight α, so $TE(w, x)$ is a T-submodule of $T_x(X(w))$ such that $\dim TE(w, x) = |E(w, x)|$ (cf. [33]). In particular $\dim T_x(X(w)) \geq |E(w, x)|$. We will call $C \in E(w, x)$ *good* if $X(w)$ is nonsingular along the open T-orbit in C, or, equivalently, if C is not contained in the singular locus of $X(w)$.

5.7.1. Theorem. *([34]) Let the notation be as above. Assume $\dim X(w) \geq 2$. Then a necessary and sufficient condition that $X(w)$ to be nonsingular at e_x is that there exists at least one good $C \in E(w, x)$ such that $\tau_C(X(w), x) = TE(w, x)$.*

Let $\Theta_x(X(w))$ denote the linear span of the reduced tangent cone of $X(w)$ at e_x. If $C \in E(w, x)$ has the form $C = \overline{U_\alpha e_x}$, we will call C *long* or *short* according as α is a long or short root. If G is simply laced, then, by convention, all T-stable curves will be called short. We have

$$TE(w, x) \subset \Theta_x(X(w)) \subset T_x(X(w)).$$

5.7.2. Theorem. [34] *Suppose G contains no G_2 factor. Then the Schubert variety $X(w)$ is smooth at e_x, $x < w$ if and only if $\dim \Theta_y(X(w)) = \dim X(w)$ for all $y \in [x, w]$. In other words, $X(w)$ is smooth at e_x if and only if the reduced tangent cones of $X(w)$ at e_y all $y \in [x, w]$ are linear. In particular, $X(w)$ is smooth if and only if all its reduced tangent cones are linear.*

CHAPTER 6

Rational Smoothness and Kazhdan–Lusztig Theory

In this chapter we describe the notion of rational smoothness à la Kazhdan–Lusztig (cf. [**78**]). Rational smoothness is intuitively an approximation to smoothness defined using cohomological criteria. This is a weaker notion than that of smoothness. In this chapter we gather the known characterizations of rational smoothness of Schubert varieties in terms of Kazhdan–Lusztig polynomials, Bruhat graphs, T-stable curves, Poincaré polynomials, and more. One can also obtain a lower bound for the dimension of the tangent space $T(w, \tau)$ from the combinatorial data in the Bruhat graph. Criteria for rational smoothness and smoothness are also given by Kumar, and for the classical groups criteria are also given in terms of pattern avoidance (see Chapters 7 and 8).

Formally, an irreducible variety X of dimension d is *rationally smooth* if for all $x \in X$, the étale cohomology with values in the constant sheaf $\mathbb{Q}_l$ and support at x is one-dimensional and concentrated in top dimension, i.e.,

$$(6.0.3) \qquad H_x^i(X, \mathbb{Q}_\ell) = \begin{cases} 0 & i \neq 2d \\ \mathbb{Q}_\ell(-d) & i = 2d. \end{cases}$$

A point $x_0 \in X$ is said to be *rationally smooth* if there exists a Zariski-open rationally smooth neighborhood of $x_0 \in X$. In this text we will not work directly with étale cohomology; we will use an equivalent characterization of rational smoothness for Schubert varieties given in Theorem 6.1.19.

If X is a complex projective variety, then a result of McCrory [**126**] implies that X is rationally smooth if and only if the ordinary cohomology $H^*(X)$ over $\mathbb{C}$ admits Poincaré duality. This is the same thing as saying $H^*(X)$ and $IH^*(X)$ coincide, where $IH^*(X)$ denotes the intersection cohomology groups of X (cf. [**59**]); note that this is stronger than saying the Poincaré polynomial of X is symmetric.

Smoothness implies rational smoothness for Schubert varieties, but the converse does not hold in general. For example, in type $\mathbf{C_2}$, $X(s_1 s_2 s_1)$ is rationally smooth but not smooth. In fact, the next theorem states that smoothness and rational smoothness coincide in the case the Dynkin digram of the root system is "simply-laced" (i.e., no multiple bonds).

6.0.4. Theorem. *Let G be of type* **A**, **D**, **E**. *Then $X(w)$ is smooth if and only if it is rationally smooth.*

Theorem 6.0.4 was proven in Type A by Deodhar. Peterson announced a proof for all three types in the mid 1990's; recently, Carrell and Kuttler (cf. [**34**]) have given a proof of Peterson's announced results. Billey has proven the statement for type D using pattern avoidance; this proof appears in Chapter 8.

6.1. Kazhdan–Lusztig polynomials

The Kazhdan–Lusztig polynomials have many interesting properties. Among them, the three we have found most fascinating are: 1) evaluating at $q = 1$ gives a formula for computing the Jordan–Hölder series for Verma modules, 2) they are Poincaré polynomials for local intersection cohomology of Schubert varieties, and this implies 3) these polynomials equal 1 at rationally smooth points of Schubert varieties. In this section we review both the definition of these polynomials in the Hecke algebra and a formula for computing them. Then we give the connections to intersection cohomology, Verma modules, and rational smoothness.

Let A be the ring $\mathbb{Z}[q^{\frac{1}{2}}, q^{-\frac{1}{2}}]$ of Laurent polynomials over $\mathbb{Z}$ in the indeterminate $q^{\frac{1}{2}}$. The *Hecke algebra* $\mathcal{H}$ of W is the algebra over A with basis elements $\{T_w : w \in W\}$ and the following relations:

(6.1.1) $$T_{s_i} T_w = T_{s_i w} \quad \text{if } l(s_i w) > l(w),$$

(6.1.2) $$T_{s_i} T_{s_i} = (q - 1) T_{s_i} + q T_{id}.$$

Note that $T_{id} = 1$ in $\mathcal{H}$. Note also that the Hecke algebra reduces to the group algebra of W if q is set equal to 1.

One feature of $\mathcal{H}$ is that the basis elements T_w are invertible over A. For example, for any simple reflection $s \in S$,

$$T_s^{-1} = q^{-1} T_s - (1 - q^{-1}) T_{id}.$$

Expanding $(T_{w^{-1}})^{-1}$ in the basis $\{T_w\}$, we have

(6.1.3) $$(T_{w^{-1}})^{-1} = (-1)^{l(w)} q^{-l(w)} \sum_{v \leq w} (-1)^{l(v)} R_{v,w}(q) T_v.$$

This defines the *R-polynomials* $R_{v,w}(q) \in \mathbb{Z}[q]$ for $v \leq w$. Set $R_{v,w} = 0$ if $v \nleq w$.

The R-polynomials play an important role in the computation of Kazhdan–Lusztig polynomials and are often easier to work with. The lemma below gives a recursive formula for computing $R_{u,v}$. The proof is a good exercise for the reader.

6.1.4. Lemma. *We have*

1. $R_{w,w} = 1$.

2. *For any $s \in S$ such that $ws < w$,*

$$(6.1.5) \qquad R_{v,w} = \begin{cases} R_{vs,ws} & vs < v \\ (q-1)R_{v,ws} + qR_{vs,ws} & vs > v. \end{cases}$$

3. *$R_{v,w}$ is a polynomial of degree $l(w) - l(v)$.*

Theorem 6.3.4 in § 6.3 states a combinatorial formula for computing a closely related family of polynomials. Below are examples of R-polynomials with $v = $ id and $w \in S_4$:

$$R_{(1234),(1234)} = 1$$
$$R_{(1234),(2134)} = q - 1$$
$$R_{(1234),(1324)} = q - 1$$
$$R_{(1234),(3124)} = q^2 - 2q + 1$$
$$R_{(1234),(2314)} = q^2 - 2q + 1$$
$$R_{(1234),(3214)} = q^3 - 2q^2 + 2q - 1$$
$$R_{(1234),(1243)} = q - 1$$
$$R_{(1234),(2143)} = q^2 - 2q + 1$$
$$R_{(1234),(1423)} = q^2 - 2q + 1$$
$$R_{(1234),(4123)} = q^3 - 3q^2 + 3q - 1$$
$$R_{(1234),(2413)} = q^3 - 3q^2 + 3q - 1$$
$$R_{(1234),(4213)} = q^4 - 3q^3 + 4q^2 - 3q + 1$$
$$R_{(1234),(1342)} = q^2 - 2q + 1$$
$$R_{(1234),(3142)} = q^3 - 3q^2 + 3q - 1$$
$$R_{(1234),(1432)} = q^3 - 2q^2 + 2q - 1$$
$$R_{(1234),(4132)} = q^4 - 3q^3 + 4q^2 - 3q + 1$$
$$R_{(1234),(3412)} = q^4 - 3q^3 + 4q^2 - 3q + 1$$
$$R_{(1234),(4312)} = q^5 - 3q^4 + 4q^3 - 4q^2 + 3q - 1$$
$$R_{(1234),(2341)} = q^3 - 3q^2 + 3q - 1$$
$$R_{(1234),(3241)} = q^4 - 3q^3 + 4q^2 - 3q + 1$$
$$R_{(1234),(2431)} = q^4 - 3q^3 + 4q^2 - 3q + 1$$
$$R_{(1234),(4231)} = q^5 - 3q^4 + 5q^3 - 5q^2 + 3q - 1$$
$$R_{(1234),(3421)} = q^5 - 3q^4 + 4q^3 - 4q^2 + 3q - 1$$
$$R_{(1234),(4321)} = q^6 - 3q^5 + 4q^4 - 4q^3 + 4q^2 - 3q + 1.$$

One can define an involution i on $\mathcal{H}$ by interchanging $q^{\frac{1}{2}}$ and $q^{-\frac{1}{2}}$ and setting $i(T_w) = (T_{w^{-1}})^{-1}$. Using this involution one defines the Kazhdan–Lusztig polynomials $P_{v,w}, v \leq w$, and a new basis for the Hecke algebra.

6.1.6. Theorem. [78] *For each $w \in W$, there exists a unique element $C'_w \in \mathcal{H}$ having the following properties:*

1. $i(C'_w) = C'_w$

2. *Expanding C'_w, one has*

$$C'_w = (q^{-1/2})^{l(w)} \sum_{v \leq w} P_{v,w} T_v$$

where $P_{w,w} = 1$, $P_{v,w}(q) \in \mathbb{Z}[q]$ has degree $\leq \frac{1}{2}(l(w) - l(v) - 1)$ if $v < w$, and $P_{v,w} = 0$ if $v \not\leq w$.

Existence of the basis $\{C'_w\}$ is simply shown by giving a recursive formula for C'_w. For each $w \in W$ and any simple reflection s such that $ws < w$, then

(6.1.7) $$C'_w = C'_{sw} C'_s - \sum \mu(z, sw) C'_z$$

where the sum is over all $z < sw$ for which $sz < z$, $\mu(z, sw)$ is the coefficient of $q^{\frac{1}{2}(l(sw)-l(z)-1)}$ in $P_{z,sw}$.

From Kazhdan–Lusztig's proof of the uniqueness of C'_w, they also obtain several equivalent formulas which can be used to compute $P_{v,w}$ in terms of the $R_{x,y}$ polynomials. One version of this formula is

(6.1.8) $$q^{l(w)-l(v)} \bar{P}_{v,w} - P_{v,w} = \sum_{v < y \leq w} R_{v,y} P_{y,w},$$

where $\bar{P}_{v,w} = P_{v,w}(q^{-1})$. Since the degree of $P_{v,w}$ is at most $\frac{1}{2}(l(w) - l(v) - 1)$, $P_{v,w}$ is obtained from the right-hand side by ignoring all terms of higher degree. The proof of the following lemma is left to the reader as an exercise.

6.1.9. Lemma. *We have the following properties of the Kazhdan–Lusztig polynomials.*

1. *For any $s \in S$ such that $sw < w$,*

(6.1.10) $$P_{v,w} = q^{1-c} P_{sv,sw} + q^c P_{v,sw} - \sum \mu(z, sw) q^{\frac{1}{2}(l(w)-l(z))} P_{v,z}$$

where the sum is over all $z < sw$ for which $sz < z$, $\mu(z, sw)$ is the coefficient of $q^{\frac{1}{2}(l(sw)-l(z)-1)}$ in $P_{z,sw}$ (highest possible degree term which can be zero), and

$$c = \begin{cases} 0 & v < sv \\ 1 & v > sv. \end{cases}$$

2. *$P_{v,w}(0) = 1$ for all $v \leq w$.*
3. *If $l(w) - l(v) \leq 2$, then $P_{v,w} = 1$.*

Below are all Kazhdan–Lusztig polynomials with $v = \text{id}$ and $w \in S_5$ which are different from 1:

w	$P_{\mathrm{id},w}$
(14523) (15342) (24513) (25341) (34125) (34152) (35124) (35142) (35241) (35412) (41523) (42315) (42351) (42513) (42531) (43512) (45132) (45213) (51342) (52314) (52413) (52431) (53142) (53241) (53421) (54231)	$q+1$
(34512) (45123) (45231) (53412)	$2q+1$
(52341)	$q^2 + 2q + 1$
(45312)	$q^2 + 1$

The Kazhdan–Lusztig conjecture states that the coefficients of $P_{v,w}(q)$ are all non-negative integers even in the case when W is an arbitrary Coxeter group. Kazhdan and Lusztig used the theorem below to prove their conjecture in the case when W is a Weyl group. The conjecture has been proven for many cases including affine Weyl groups [**79**] and the universal Coxeter groups [**42**]; however it remains open for general Coxeter groups at this time.

6.1.11. Theorem. [**79**] *Let G be a semisimple, simply connected algebraic group. Let W be the Weyl group of G. For $w \in W$, let $\mathcal{H}^*(w)$ be the intersection cohomology sheaf (with respect to middle perversity) of the Schubert variety $X(w)$. Then for $v \in W$, $v \leq w$, we have*

$$(6.1.12) \qquad P_{v,w}(q) = \sum_{j \geq 0} \dim(\mathcal{H}^{2j}(w)_v)\, q^j$$

where $\mathcal{H}^{2j}(w)_v$ denotes the stalk of $\mathcal{H}^{2j}(w)$ at the point e_v.

The proof of Theorem 6.1.11 is quite deep. It requires Deligne's proof of the Weil conjectures and the Hard Lefschetz theorem. We refer the reader to the original manuscript [**79**]. For background on intersection cohomology, we refer the reader to [**21, 59, 82, 149**]. This theorem has two important corollaries which we include here.

6.1.13. Corollary. *The coefficients of the Kazhdan–Lusztig polynomial $P_{v,w}(q)$ are all nonnegative if W is a Weyl group of a semisimple, simply connected algebraic group.*

6.1.14. Corollary. [**79**, Cor. 4.9] *Let* $w \in W$. *The i-th component (of the intersection cohomology ring)* $\mathcal{H}^i(X(w)) = 0$ *for i odd. Furthermore,*

$$(6.1.15) \qquad \sum_i \dim \mathcal{H}^{2i}(X(w))q^i = \sum_{v \leq w} q^{l(v)} P_{v,w}(q).$$

A long standing open problem is to find a combinatorial description of the coefficients of $P_{v,w}$. See Section 6.3 below for some of the known results. Such a combinatorial formula would also be of interest in the representation theory of Verma modules (see [**70, 74**] for background). Let $M(\lambda)$ be the Verma module with highest weight $\lambda \in \mathfrak{h}^*$ and let $L = L(\lambda)$ be the unique irreducible quotient of $M(\lambda)$. For any $\mathfrak{g}$-module M, let $[M : L(\lambda)]$ be the multiplicity of $L(\lambda)$ in M. The formal character $\mathrm{ch}(M)$ is given by $\mathrm{ch}(M) = \sum [M : L(\lambda)] \,\mathrm{ch} L(\lambda)$. Define a "shifted" action of W on $\mathfrak{h}^*$ by $w \cdot \lambda = w(\lambda + \rho) - \rho$ for $w \in W$, $\lambda \in \mathfrak{h}^*$, and $\rho = \frac{1}{2}\sum_{\alpha \in R^+} \alpha$. Combining results of Verma [**156**], Bernstein–Gelfand–Gelfand [**10**], and van den Hombergh [**155**], one has $[M(v \cdot \lambda) : L(\mu)] \neq 0$ if and only if $\mu = w \cdot \lambda$ for some w with $v \leq w$ and some dominant regular weight λ. Furthermore, $[M(v \cdot \lambda_0) : L(w \cdot \lambda_0)]$ is independent of λ_0 [**74, 75**]. Hence, we can define $m(v, w)$ to be $[M(v \cdot \lambda_0) : L(w \cdot \lambda_0)]$ for any $v \leq w$ in W and any dominant weight λ_0.

6.1.16. Theorem. *The multiplicity* $m(v, w)$ *is equal to* $P_{v,w}(1)$.

Theorem 6.1.16 was originally conjectured in [**78**] and proved independently by Beilinson–Bernstein [**9**] and Brylinski–Kashiwara [**31**]. This theorem has since been extended in several ways. We refer the reader to [**39**] for a more complete history and generalizations.

From the table of $P_{v,w}(q)$ for S_5 above, one might suppose that there is a suitable characterization of the set of all polynomials with constant term 1 which can be Kazhdan–Lusztig polynomials. However, this is not possible due to the following theorem.

6.1.17. Theorem. [**136**] *Given any polynomial of the form* $1 + a_1 q + \cdots + a_d q^d$ *with nonnegative integer coefficients and* $a_d \neq 0$, *there exist permutations* $v, w \in S_n$ *such that the Kazhdan–Lusztig polynomial*

$$(6.1.18) \qquad P_{v,w}(q) = 1 + a_1 q + \cdots + a_d q^d.$$

In particular, let $v, w \in S_{2+d+a_1+\cdots+a_d}$ *be the unique permutations with codes given by*

$$\mathrm{code}(v) = (0, 1^{a_1}, 2^{a_2}, \ldots, d^{a_d}, d-1, d-2, \ldots, 1, 0, 0)$$
$$\mathrm{code}(w) = (2^{a_1}, 3^{a_2}, \ldots, (d+1)^{a_d}, d+1, d, \ldots, 2, 0, 0),$$

where $\mathrm{code}(x) = (c_1, \ldots, c_n)$ *if* $c_i = \#\{j > i : x_j < x_i\}$. *Then* (6.1.18) *holds.*

Finally, we relate the Kazhdan–Lusztig polynomials with rational smoothness. Almost all of the known results on rational smoothness of Schubert varieties rely on this theorem as the definition of rational smoothness.

6.1.19. Theorem. [78] *The following are equivalent for any $v \leq w$ in W.*

1. $X(w)$ *is rationally smooth at e_v.*
2. $P_{x,w} = 1$ *for all $v \leq x \leq w$.*
3. *For all $v \leq x \leq w$,*

$$\sum_{x \leq y \leq w} R_{x,y} = q^{l(w)-l(x)}.$$

In fact whenever the Kazhdan–Lusztig polynomials are known to have nonnegative coefficients, then

$$P_{x,w} = 1 \ \forall \ v \leq x \leq w \iff P_{v,w} = 1.$$

See Corollary 3 in the next section for the proof.

It is interesting to note how the coefficients of the Kazhdan–Lusztig polynomials change when the second indexing Weyl group element is fixed. We know that the locus of T-fixed points that are not rationally smooth in a Schubert variety is a lower order ideal in the Bruhat–Chevalley order. Therefore, $P_{v,w}(q) \neq 1$ implies $P_{u,w}(q) \neq 1$. The following theorem says that the singularities are in fact getting worse as one descends in Bruhat–Chevalley order.

6.1.20. Theorem. [25, 73] *For $u \leq v \leq w \in W$, the polynomial $P_{u,w}(q) - P_{v,w}(q)$ always has nonnegative integer coefficients.*

6.2. Carrell–Peterson's criteria

Carrell and Peterson have given several criteria for determining rational smoothness in terms of T-stable curves, the Bruhat graph, Kazhdan–Lusztig polynomials, Deodhar's inequality, and symmetry of certain Poincaré polynomials. Many of these criteria are useful for computations. We first introduce the T-stable curves and then its combinatorial abstraction called the Bruhat graph. The first theorem gives conditions for the entire Schubert variety to be rationally smooth. The second theorem gives conditions for particular points to be rationally smooth.

For a root $\alpha > 0$, let Z_α denote the $SL(2)$-copy in G corresponding to α; note that Z_α is simply the subgroup of G generated by U_α and $U_{-\alpha}$. Given $x \in W$, precisely one of $\{U_\alpha, U_{-\alpha}\}$ fixes the point e_x. Thus $Z_\alpha \cdot e_x$ is a T-stable curve in G/B (note that $Z_\alpha \cdot e_x \cong \mathbf{P}^1$), and conversely any T-stable curve in G/B is of this form [33]. Now $s_\alpha \in Z_\alpha$, so $Z_\alpha \cdot e_x$ also contains the point $e_{s_\alpha x}$. Therefore, corresponding to each $\alpha > 0$, the number of T-stable curves in G/B is $\frac{1}{2}(\#W)$, and thus the total number of T-stable curves in G/B is $\frac{1}{2}(\#W)N$, N being the number of positive roots.

The T-stable curve $Z_\alpha \cdot e_x$ is contained in a Schubert variety $X(w)$ if and only if $e_x, e_{s_\alpha x}$ are both in $X(w)$. Given $y \leq w$, let

$$r(y, w) = \#\{r \in \mathcal{R} \mid ry \leq w\},$$

where $\mathcal{R}$ denotes the set of all reflections in W. Then there are precisely $r(y, w)$ T-stable curves in $X(w)$ passing through e_y. The inequality of Theorem 6.2.1 below shows $r(y, w) \geq l(w)$ (taking $x = e$). Thus there are at least $l(w)$ T-stable curves in $X(w)$ passing through a T-fixed point e_y. More generally, it can be shown [**32**, Lemma, Sec. 2] that for any projective variety Y with a torus action, if the number of fixed points of Y is finite, then every one of the fixed points lies on at least $\dim Y$ distinct curves that are closures of one-dimensional T-orbits in Y.

6.2.1. Theorem. [**43**] **Deodhar's inequality.** *Let (W, S) be an arbitrary Coxeter system. For $x \leq y \leq w$,*

(6.2.2) $$\#\{r \in \mathcal{R} \mid x \leq ry \leq w\} \geq l(w) - l(x).$$

Theorem 6.2.1 was first conjectured by Deodhar and proved by him in the Type A case. Then Dyer [**43**] proved the general case using the nil Hecke ring. It was also proved by Polo [**135**] for finite Weyl groups using properties of Schubert varieties and by Carrell–Peterson for crystallographic groups [**32**]. Dyer's proof appears in Chapter 7 as a corollary to Theorem 7.1.11.

The question of rational smoothness boils down to determining when $r(y, w) = l(w)$ for all y. This combinatorial data given by the T-stable curves can be abstracted away from the geometry to obtain the Bruhat graph as follows:

6.2.3. Definition. The *Bruhat graph* $\Gamma(u, w)$ is the graph with vertex set $\{v \in W : u \leq v \leq w\}$ and edges between x and y if $x = y\sigma$ for some reflection σ (not necessarily a simple reflection). Note, this graph contains more edges than the corresponding interval in the Hasse diagram of the Bruhat poset. We refer to $\Gamma(\mathrm{id}, w)$ as the *Bruhat graph* of w.

Using this terminology, Carrell and Peterson give the following list of equivalent conditions for rational smoothness at every point in a Schubert variety in G/B. Note that one can also test rational smoothness at every point by considering just the point corresponding to the identity.

6.2.4. Theorem. [**33**] *Assume that the coefficients of all Kazhdan–Lusztig polynomials $P_{u,v}$ are known to have nonnegative coefficients for all $u, v \in W$. For a Schubert variety $X(w)$ the following are equivalent:*

1. *$X(w)$ is rationally smooth at every point.*
2. *The Poincaré polynomial $p_w(q)$ of $X(w)$ is symmetric.*

3. *The Bruhat graph* $\Gamma(\mathrm{id}, w)$ *is regular, i.e., every vertex has the same number of edges, namely* $l(w)$.

PROOF OF THEOREM 6.2.4. Recall that the Poincaré polynomial of $X(w)$ is $p_w(q) = \sum_{v \leq w} q^{l(v)}$. Kazhdan and Lusztig showed that the Poincaré polynomial for the full intersection cohomology is $\sum_{v \leq w} P_{v,w}(q) q^{l(v)}$ (see Theorem 6.1.14). This polynomial is always symmetric since intersection cohomology has Poincaré duality. Therefore, when $P_{v,w} = 1$ for all $v \leq w$, the two Poincaré polynomials agree so $p_w(q) = \sum_{v \leq w} q^{l(v)}$ is symmetric, proving that Condition 1 implies Condition 2.

For each $y \in W$, $l(y) = \#\{r \in \mathcal{R} \mid ry < y\}$. Each edge of the Bruhat graph meets two points y, ry for some $r \in \mathcal{R}$, so that

(6.2.5)
$$\sum_{y \leq w} l(y) = \sum_{y \leq w} \#\{r \in \mathcal{R} \mid ry < y \leq w\} = \sum_{y \leq w} \#\{r \in \mathcal{R} \mid y < ry \leq w\}.$$

Therefore, by symmetry of the Poincaré polynomial, we have

(6.2.6)
$$\sum_{y \leq w} l(w) - l(y) = \sum_{y \leq w} l(y) = \sum_{y \leq w} \#\{r \in \mathcal{R} \mid y < ry \leq w\}.$$

Now apply Deodhar's inequality (6.2.2) with $x = y$ to show that, for each $y \leq w$, $\#\{r \in \mathcal{R} \mid y < ry \leq w\} \geq l(w) - l(y)$. Hence, (6.2.6) implies $l(w) - l(y) = \#\{r \in \mathcal{R} \mid y < ry \leq w\}$ which, proves Condition 2 implies Condition 3.

Finally, we will show Condition 3 implies Condition 1. Apply induction on $k = l(w) - l(y)$. Clearly, if $k = 0$, then $P_{w,w} = 1$ so assume $y < w$ and $P_{z,w} = 1$ for all $l(w) - l(z) < k$. Define

(6.2.7)
$$f(q) = q^{l(w)-l(y)} \left[P_{y,w}(q^{-2}) - 1 \right].$$

Since $\deg P_{y,w}(q) \leq \frac{1}{2}(l(w) - l(y) - 1)$ and $P_{y,w}(0) = 1$ for all $y \leq w$, $f(q)$ must be a polynomial with no constant term.

Using clever manipulations, Deodhar has shown in [41] that for each $y \leq w$,

(6.2.8)
$$\frac{d}{dt} \left[q^{l(w)-l(y)} P_{y,w}(q^{-2}) \right] \Big|_{q=1} = \sum_{\substack{r \in \mathcal{R} \\ y < ry \leq w}} P_{ry,w}(1).$$

Therefore,

(6.2.9)
$$f'(1) = \sum_{\substack{r \in \mathcal{R} \\ y < ry \leq w}} P_{ry,w}(1) - [l(w) - l(y)].$$

By the induction hypothesis, $P_{ry,w}(1) = 1$ for $y < ry$ and by assumption $l(w) - l(y) = \#\{r \in \mathcal{R} \mid y < ry \leq w\}$ so $f'(1) = 0$. Provided the coefficients of the Kazhdan–Lusztig polynomials are nonnegative, then $f'(1) = 0$ implies $f(q)$ is a constant, so $P_{y,w} = 1$. $\qquad\square$

The following theorem gives several criteria for Schubert varieties to be rationally smooth at a particular T-fixed points.

6.2.10. Theorem. *Assume that the coefficients of all Kazhdan–Lusztig polynomials $P_{u,v}$ are known to have nonnegative coefficients for all $u, v \in W$. For a Schubert variety $X(w)$ and a T-fixed point $e_x \in X(w)$ the following are equivalent:*

1. *$X(w)$ is rationally smooth at e_x.*
2. *$P_{y,w} = 1$ for all $y \in [x, w]$.*
3. *$P_{x,w} = 1$.*
4. *$\#\{r \in \mathcal{R} \mid y < ry \leq w\} = l(w) - l(y)$, $\forall y \in [x, w]$.*
5. *The Bruhat graph $\Gamma(x, w)$ is regular, i.e., every vertex has $l(w) - l(x)$ edges.*
6. *For $x \leq y \leq w$, $r(y, w) = l(w)$, i.e., e_y lies on exactly $l(w)$ T-invariant curves on $X(w)$.*

The first three conditions are equivalent by Theorem 6.1.19. Condition 3 was first proved by Deodhar in [**38**] for $W = S_n$. It was extended by Carrell and Peterson to all Weyl groups in [**33**] along with Conditions 4, 5 and 6. Condition 4 is originally due to Jantzen [**74**] and was proven using facts in the representation theory of Verma modules.

PROOF. Following [**33**], we will show Condition 3 is equivalent to the first three. The remaining conditions can be proven using the same techniques as in Theorem 6.2.4. We leave the details as an exercise.

By Theorem 6.1.19, we need to show $P_{v,w} = 1$ implies $P_{x,w} = 1$ for all $v \leq x \leq w$. Using the notation in the proof of Theorem 6.2.4, if $P_{v,w} = 1$, then $f(q) = 1$, so by (6.2.9) we have

$$0 = f'(1) = \sum_{\substack{r \in \mathcal{R} \\ v < rv \leq w}} P_{rv,w}(1) - [l(w) - l(v)].$$

Assuming the Kazhdan–Lusztig polynomials have nonnegative coefficients we see that all $P_{rv,w}$ are constant. Furthermore, by Deodhar's Inequality (6.2.2), we must have $l(w) - l(v) = \#\{r \in \mathcal{R} \mid v < rv \leq w\}$. Therefore, $P_{rv,w} = 1$ for all $v < rv \leq w$ which in turn proves the corollary. $\qquad\square$

Combining Theorem 6.2.10 and Lemma 6.1.9, we obtain the following corollary.

6.2.11. Corollary. *The set of e_τ that are not rationally smooth on $X(w)$ is a B-stable closed subvariety of $X(w)$ codimension at least 3.*

Finally, in this section we discuss an application of the Bruhat graph on computing the dimension of the tangent space to a Schubert variety. Let $w, \tau \in W$. Let $R(w, \tau) = \{\beta \in \tau(R^+) \mid w \geq s_\beta \tau\}$, $r(\tau, w) = \#R(w, \tau)$. The following proposition gives a lower bound for $\dim T(w, \tau)$.

6.2.12. Proposition. [33, 135] **A lower bound for** $\dim T(w, \tau)$ Let w, τ be as above. Let $\beta \in R(w, \tau)$. Then $X_{-\beta} \in T(w, \tau)$. In particular, $\dim T(w, \tau) \geq r(\tau, w)$.

PROOF. The hypothesis that $w \geq s_\beta \tau$ implies that the (T-stable) curve $Z_\beta e_\tau$ ($\subset G/B$) is in fact contained in $X(w)$. Let $Z_{\beta,\tau}$ be the tangent space to $Z_\beta e_\tau$ at e_τ; we have that $Z_{\beta,\tau}$ is simply the one-dimensional span of $X_{-\beta}$. On the other hand, $Z_{\beta,\tau} \subset T(w, \tau)$, since $Z_\beta e_\tau \subset X(w)$. This implies that $X_{-\beta} \in T(w, \tau)$. □

6.3. Combinatorial formulas for Kazhdan–Lusztig polynomials

As was shown in §6.1, the Kazhdan–Lusztig polynomials can be used to determine whether or not a Schubert variety is rationally smooth at a point. Due to the large number of formulas related to Kazhdan–Lusztig theory in the literature, we will not attempt to give this subject complete coverage, but instead we will summarize a selection of combinatorial algorithms and give references to the appropriate literature.

6.3.1. Brenti's results. In this section we describe some of the algorithms for computing $R_{u,v}(q)$ and $P_{u,v}(q)$ formulated by Francesco Brenti. These algorithms have been developed in several papers. We refer the reader to [27] for a more complete overview of the history, consequences, and conjectures related to the formulas below. These results are stated here only for S_n. See Remark 6.3.16 for extensions to other Weyl groups.

Brenti's algorithm for computing $R_{u,v}$ is stated in terms of a related family of polynomials $\tilde{R}_{u,v}$. Define $\tilde{R}_{u,v}$ to be the unique polynomial satisfying

$$(6.3.2) \qquad R_{u,v} = q^{\frac{l(v)-l(u)}{2}} \tilde{R}_{u,v}(q^{1/2} - q^{-1/2}).$$

6.3.3. Definition. Given $u \in S_n$, let $C_{i,j}(u)$ be the set of all sequences of the form $(u(i_1), \ldots, u(i_k))$ where $1 \leq k \leq n$, $i = i_1 < \cdots < i_k = j$, and, $u(i_1) < \cdots < u(i_k)$. If $u < v \in S_n$, let $C(u, v)$ be $C_{v^{-1}(d), u^{-1}(d)}(u)$ where $d = \max\{1 \leq i \leq n : u^{-1}(i) \neq v^{-1}(i)\}$, i.e., $v^{-1}(d)$ is the position in v of the largest number that does not appear in the same position in both u and v.

For example, $C_{1,5}(1427635) = \{(1,6), (1,4,6), (1,2,6)\}$. We think of the elements in $C_{i,j}(u)$ as permutations in cycle notation consisting of single cycles. Multiplying a permutation on the left by a cycle rotates the numbers to the left, e.g., $(1,4,6)1427635 = 4627135$.

6.3.4. Theorem. *Let $u, v \in S_n$, then*

$$(6.3.5) \qquad \tilde{R}_{u,v} = \begin{cases} \sum_{w \in C(u,v)} q^{k(w)-1} \tilde{R}_{w^{-1}u,v}(q) & u < v \\ 1 & u = v \\ 0 & u \not\leq v \end{cases}$$

where $k(w)$ is the number of elements in the cycle w.

To illustrate the algorithm we compute $\tilde{R}_{u,v}$ for $u = 1427635$, $v = 6457213$. First, $d = 6$ since 7 appears in the same position in both u and v. Then, $v^{-1}(6) = 1$ and $u^{-1}(6) = 5$ so $C(u,v) = C_{1,5}(1427635) = \{(1,6), (1,4,6), (1,2,6)\}$. Hence,

$$\tilde{R}_{1427635,6457213} = q\tilde{R}_{6427135,6457213} + q^2 \tilde{R}_{6127435,6457213} + q^2 \tilde{R}_{6417235,6457213}$$
$$(6.3.6) \qquad = q(q^3) + q^2(2q^4 + q^6) + q^2(q^2 + q^4)$$
$$= 2q^4 + 3q^6 + q^8.$$

6.3.7. Remark. $\tilde{R}_{u,v}(q)$ is clearly a polynomial in q with nonnegative integer coefficients. The recursion in (6.3.5) will terminate after at most n steps.

Next, we begin describing Brenti's algorithm for the Kazhdan–Lusztig polynomials. This requires the introduction of polynomials $\gamma_{\beta_1,\dots,\beta_s}(q)$ and operators U_j.

6.3.8. Definition. Let U_j be the operator acting on the space of all Laurent polynomials in q with real coefficients given by

$$(6.3.9) \qquad U_j\left(\sum_{i \in \mathbb{Z}} a_i q^i\right) = \sum_{i \geq j} a_i q^i$$

for any $j \in \mathbb{Q}$. For any composition $\beta = \beta_1, \dots, \beta_s$, $\beta_i \in \{1, 2, 3, \dots\}$, define $\gamma_{\beta_1,\dots,\beta_s}$ recursively by

$$(6.3.10)$$

$$\gamma_{\beta_1,\dots,\beta_s} = \begin{cases} -1 & \beta_1 = 1, s = 1 \\ (q-1)U_{\frac{|\beta|}{2}}\left(q^{|\beta|-1}\gamma_{\beta_2,\dots,\beta_s}\left(\frac{1}{q}\right)\right) & \beta_1 = 1, s \geq 2 \\ (q-1)\gamma_{\beta'}(q) + U_{\frac{|\beta|+1}{2}}\left(q^{|\beta|-1}(1-q)\gamma_{\beta'}\left(\frac{1}{q}\right)\right) & \beta_1 > 1, \end{cases}$$

where $\beta' = \beta_1 - 1, \beta_2, \dots, \beta_s$ and $|\beta| = \beta_1 + \beta_2 + \cdots + \beta_s$.

One may verify the following examples:

$$(6.3.11) \qquad \begin{array}{rcl} \gamma_1 & = & -1 \\ \gamma_{1,1} & = & -q^2 + q \\ \gamma_2 & = & q^2 + q + 1 \\ \gamma_{2,1} & = & -q^3 + q^2 - q. \end{array}$$

6.3.12. Definition. Define $B(u,v)$ to be a labeled directed graph with vertex set $\{x \in S_n : u \le x \le v\}$ and directed edges from x to y labeled by $\lambda(x,y) = j - i + \sum_{k=1}^{i-1}(n-k)$ if and only if $y = x(i,j)$ for some $1 \le i < j \le n$ and $l(y) > l(x)$. For any directed path $\Gamma = (x_0, x_1, \ldots, x_k)$ with $x_i < x_{i+1}$, we say i is a *descent* in Γ if $\lambda(x_{i-1}, x_i) > \lambda(x_i, x_{i+1})$. Assign Γ a composition

$$(6.3.13) \qquad\qquad D(\Gamma) = (\beta_1, \ldots, \beta_s)$$

if Γ has $s - 1$ descents, $|\beta| = k$ and for each i, β_i is the length of the path between the $(i-1)$-st and i-th descents.

For example, take $u = 15243$ and $v = 45231$. There are 22 paths in $B(u,v)$ below from u to v.

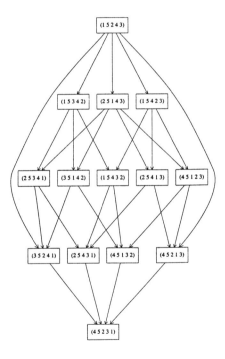

Among these paths is $\Gamma = 15243 \xrightarrow{9} 15342 \xrightarrow{2} 35142 \xrightarrow{9} 35241 \xrightarrow{3} 45231$; hence $D(\Gamma) = (1,2,1)$. The 22 paths give rise to a multiset of 22 compositions, namely $\{(2), (1,1), (4), (1,3)^2, (2,2)^4, (3,1)^3, (1,1,2)^3, (1,2,1)^4, (2,1,1)^2, (1,1,1,1))\}$.

6.3.14. Theorem. [27] *Let $u < v \in S_n$, then*

$$(6.3.15) \qquad P_{u,v}(q) = (-1)^{l(v)-l(u)} \sum_{\Gamma} q^{\frac{1}{2}(l(v)-l(u)-|D(\Gamma)|)} \gamma_{D(\Gamma)}(q),$$

where the sum is over all directed paths in $B(u,v)$ from u to v.

Note that $\frac{1}{2}(l(v) - l(u) - |D(\Gamma)|)$ is always a nonnegative integer since $l(v) - l(u)$ and $|D(\Gamma)|$ must have the same parity.

Continuing with the example above and applying Theorem 6.3.14 we obtain

$$\begin{aligned}
P_{u,v} = (-1)^{8-4}(&q\,\gamma_{(2)} + q\,\gamma_{(1,1)} + \gamma_{(4)} + 2\gamma_{(1,3)} + 4\gamma_{(2,2)} + 3\gamma_{(3,1)} \\
&+ 3\gamma_{(1,1,2)} + 4\gamma_{(1,2,1)} + 2\gamma_{(2,1,1)} + \gamma_{(1,1,1,1)})
\end{aligned}$$

where

$$\begin{aligned}
\gamma_{(11)} &= -q^2 + q \\
\gamma_{(2)} &= q^2 - q + 1 \\
\gamma_{(4)} &= q^4 - 3q^3 + 2q^2 - 3q + 1 \\
\gamma_{(13)} &= -q^4 + 3q^3 - 2q^2 \\
\gamma_{(22)} &= q^4 - 2q^3 + q^2 \\
\gamma_{(31)} &= -q^4 + 3q^3 - 2q^2 + q \\
\gamma_{(112)} &= 0 \\
\gamma_{(121)} &= -q^3 + q^2 \\
\gamma_{(211)} &= 0 \\
\gamma_{(1111)} &= 0
\end{aligned}$$

Therefore,

$$P_{u,v} = q + 1.$$

6.3.16. Remark. Brenti's algorithm for Kazhdan–Lusztig polynomials extends to other finite Coxeter groups as well [27]. Simply replace the edge labels in $B(u,v)$ with the reflection s_β if $x = ys_\beta$. Fix an order on the set of all reflections by choosing a reduced word $s_{i_1} s_{i_2} \dots s_{i_p}$ for the unique longest element of the Weyl group. Then every reflection can be written as $s_{i_1} \dots s_{i_{j-1}} s_{i_j} s_{i_{j-1}} \dots s_{i_1}$ for some $1 \le j \le p$. This naturally gives rise to an ordering on reflections where s_{i_1} is the smallest and $s_{i_1} \dots s_{i_{p-1}} s_{i_p} s_{i_{p-1}} \dots s_{i_1}$ is the largest. Now any chain in $B(u,v)$ can be partitioned into increasing segments by breaking the chain whenever two adjacent edges are labeled by reflections where the first is bigger than the second.

6.3.17. Deodhar's algorithm. Deodhar has defined a statistic on subexpressions of reduced expressions call the *defect*. This statistic naturally arises in the Hecke algebra and in a formula for Kazhdan–Lusztig polynomials.

6.3.18. Definition. Fix a reduced expression $\mathbf{r} = r_1 r_2 \ldots r_p = s_{j_1} s_{j_2} \ldots s_{j_p}$ for $w \in W$. Let $\sigma = (\sigma_0, \sigma_1, \ldots, \sigma_p)$ be a sequence of Weyl group elements such that $\sigma_{j-1}^{-1} \sigma_j \in \{id, r_j\}$ for all $1 \le j \le p$, i.e., σ corresponds with some subexpression $r_{i_1} r_{i_2} \ldots r_{i_k}$. Let $\mathcal{S}$ be the set of all such sequences for the fixed reduced expression $\mathbf{r}$. Define the *defect* of $\sigma \in \mathcal{S}$ by

$$(6.3.19) \qquad d(\sigma) = \#\{1 \le j \le p : \sigma_{j-1} r_j < \sigma_{j-1}\}.$$

Let $\pi(\sigma) = \sigma_p$.

For example, in Type A with $\mathbf{r} = s_1 s_2 s_3 s_2 s_1 s_2$, three of the 2^6 possible σ's are

σ	$d(\sigma)$	$\pi(\sigma)$
$(id, s_1, s_1 s_2, s_1 s_2 s_3, s_1 s_2 s_3, s_1 s_2 s_3, s_1 s_2 s_3)$	0	$s_1 s_2 s_3$
$(id, s_1, s_1 s_2, s_1 s_2, s_1 s_2, s_1 s_2 s_1, s_1 s_2 s_1, s_1 s_2 s_1)$	1	$s_1 s_2 s_1$
$(id, id, s_2, s_2, s_2, s_2, s_2)$	2	s_2

6.3.20. Theorem. [40] *Assume the coefficients of all Kazhdan–Lusztig polynomials for W are nonnegative. Then for each pair $x, y \in W$, there exists a set $E_{\min} \subset \mathcal{S}$ such that*

$$(6.3.21) \qquad P_{x,y} = \sum_{\substack{\sigma \in E_{min} \\ s.t. \ \pi(\sigma) = x}} q^{d(\sigma)}.$$

In [40], Deodhar gives a recursive algorithm for computing the set $E_{\min}$ starting from $\mathcal{S}$. However, this algorithm depends on the expansion of the basis elements C_w' into T_w's, which is equivalent to computing the Kazhdan–Lusztig polynomials via the known formulas.

6.3.22. Theorem. [40] *Choose a reduced expression $y = r_1 \ldots r_p$ for each $y \in W$ and let $D_y' = C_{r_1}' C_{r_2}' \ldots C_{r_p}'$. Then $\{D_y' : y \in W\}$ form a basis for the Hecke algebra and*

$$(6.3.23) \qquad D_y' = q^{-l(y)/2} \sum_{\sigma \in \mathcal{S}} q^{d(\sigma)} T_{\pi(\sigma)}.$$

6.3.24. Remark. The element D_y' depends on the chosen reduced word for y. Different reduced expressions can give different elements in the Hecke algebra. However, the set $\{D_y' : y \in W\}$ is always a basis.

Deodhar's algorithm is especially easy to apply if the set $E_{\min}$ includes every subword of a fixed reduced word for w. In [15], Billey and Warrington characterize all permutations w with this property as being *321-hexagon-avoiding*. (See Chapter 8 for the definition of pattern avoidance.) These permutations avoid the five permutations

$$(6.3.25) \qquad \begin{matrix} 321 & 46718235 & 46781235 \\ & 56718234 & 56781234. \end{matrix}$$

These permutations actually have at least six equivalent characterizations.

6.3.26. Theorem. *Let* $\mathbf{a} = s_{i_1} \cdots s_{i_r}$ *be a reduced expression for* $w \in A_n$. *The following are equivalent:*

1. w *is 321-hexagon-avoiding.*
2. *For all* $x \leq w$,
$$P_{x,w} \sum_{\substack{\sigma \in \mathcal{S} \\ s.t.\ \pi(\sigma)=x}} q^{d(\sigma)}$$
where $P_{x,w}$ *is the Kazhdan–Lusztig polynomial of* x, w.
3. *The Poincaré polynomial for the full intersection cohomology group of* $X(w)$ *is*
$$\sum_i \dim(\mathrm{IH}^{2i}(X(w)))q^i = (1+q)^{l(w)}.$$
4. *The Kazhdan–Lusztig basis element* C'_w *satisfies* $C'_w = C'_{s_{i_1}} \cdots C'_{s_{i_r}}$.
5. *The Bott–Samelson resolution of* $X(w)$ *is small (defined in §9.1).*
6. $\mathrm{IH}_*(X(w)) \cong H_*(Y)$, *where* Y *is the Bott–Samelson resolution of* $X(w)$.

6.3.27. Remark. For concreteness, this theorem refers only to A_n. However, Theorem 6.3.26 holds for the other "nonbranching" Weyl groups B_n, F_4, G_2. One need simply replace "321-avoiding" by "short-braid-avoiding" in any statements made (e.g., "321-hexagon-avoiding" $\mapsto$ "short-braid-hexagon-avoiding"). Short-braid-avoiding implies that no reduced word contains the substring $s_i s_{i+1} s_i$ for any i.

6.3.28. Lascoux's Algorithm. Lascoux and Schützenberger [**117**] have given an algorithm for computing the Kazhdan–Lusztig polynomials in terms of paths along trees for the "Grassmannian" permutations in the symmetric group, i.e., permutations with at most one descent in one-line notation. Lascoux [**114**] has extended this algorithm to the set of vexillary permutations. Boe [**17**] has generalized the above results to all Hermitian symmetric spaces while keeping the same flavor of the algorithm. In an effort to find a geometric proof of Lascoux and Schützenberger's algorithm, Zelevinsky [**157**] later found small resolutions for Schubert varieties in the Grassmannian and as a consequence, Kazhdan–Lusztig polynomials are realized as the Poincaré polynomials of certain fibers of these resolutions. This result is described in Chapter 9.

We describe Lascoux's algorithm for computing the Kazhdan–Lusztig polynomials $P_{w_0w,w_0x}(q)$ in the case x is a *vexillary permutation*. The vexillary permutations, by definition, avoid the pattern 2143. (See Chapter 8 for the definition of pattern avoidance). Therefore, w_0x avoids the pattern 3412 that one might refer to as "covexillary". Note, w_0x will correspond with a smooth Schubert variety unless it contains a 4231-pattern according to Theorem 8.1.1.

6.3.29. Theorem. [114] *Given any vexillary permutation x and an arbitrary permutation $w \geq x$, then the Kazhdan–Lusztig polynomial*

$$P_{w_0 w, w_0 x} = \sum q^{|T|},$$

where the sum is over all edge labelings T in $EL(x, w)$, described below, and $|T|$ denotes the sum of the edge labels.

The set $EL(x, w)$ is computed in the following three steps. Note, only the third step depends on w.

Step 1: *Construct the tree for $x = (x_1 \ldots x_n)$.*

Let $c_i = \#\{j > i : x_j < x_i\}$. Then $(c_1, \ldots, c_n)$ is called the *code of* x. Sorting the code into increasing order gives a partition $\lambda(x) = \lambda_1 \leq \lambda_2 \leq \cdots \leq \lambda_k$ called the *shape* of x. This shape can be viewed as a Ferrers shape, i.e., a left justified stack of squares with λ_1 squares in the top row, λ_2 squares in the second row, etc. From the Ferrers diagram of shape $\lambda(x)$ we obtain a parenthesis–word in the letters ")" and "(" by walking along the northeast border of the diagram and recording a "(" for each horizontal step and a ")" for each vertical step. This parenthesis–word, in turn, gives a rooted tree: pair up the parentheses from the closest pairs and work out, these pairs are the nodes of the tree, the leaves of the tree are inner most parentheses, and a pair encloses its children. Note, unpaired parenthesis do not contribute to the tree.

For example, if $x = (14752368) \in S_8$, has code$(x) = (0, 2, 4, 2, 0, 0, 0, 0)$, $\lambda(x) = (2, 2, 4)$. The corresponding Ferrers shape is

and the northeast boarder walk is h, h, v, v, h, h, v which gives (())((). This parenthesis–word gives rise to a tree with two leaves:

Step 2: *Label the leaves of tree(x) with bigrassmannian permutations.*
This labeling requires three bijections.

1. *Leaves of tree(x)* $\longrightarrow$ *distinct numbers in code(x).* Each leaf of tree(x) corresponds to a corner in the Ferrers diagram: the corner square ending a row of length r is mapped to $r \in$ code(x).

2. *Distinct numbers in code(x)* $\longrightarrow$ *Crossings(x).* Here Crossings$(x) =$ $\{(i, j, j + 1, k) : w_{j+1} < w_k < w_i < w_j, w_i = w_k + 1, i \le j < k\}$. This set can be viewed geometrically as the set of all pairs of line segments $((A_1, A_2), (B_1, B_2))$ in the plane with endpoints $A_1 = (i, w_i)$, $A_2 = (k, w_k)$, (assuming $w_k = w_i - 1$), $B_1 = (j, w_j)$, $B_2 = (j + 1, w_{j+1})$ (assuming $w_j > w_{j+1}$, $i \le j < k$), such that the line segments $\overline{A_1 A_2}$ and $\overline{B_1 B_2}$ intersect. Note that two line segments are allowed to intersect at an endpoint.

    Label the crossing $(i, j, j + 1, k)$ by $\min(c_i, c_j)$. Note, if $c_a = c_b$ and $a < b$, then b never appears as the first index of a crossing, since x was assumed to be vexillary. Therefore, no value in the code is ever the label for two different crossings and this labeling is a bijection.

3. Crossings$(x) \longrightarrow$ *Maximal bigrassmannians below X.*

    A *bigrassmannian permutation* is a permutation such that there is at most one index i and one index j, so that $ws_i < w$ and $s_j w < w$, i.e., at most one left and one right descent. The code of a bigrassmannian permutation is of the form $(0, 0, \ldots 0, c, c, ..c, 0, 0, \ldots)$. A bigrassmannian is denoted by $[a, b, c, d]$ if the code has a initial zeros, b copies of c, and $c+d$ final zeros. Note a *Grassmannian permutation* is a permutation with at most one right descent, so bigrassmannian permutations are Grassmannian.[1]

    A crossing $(i, j, j+1, k)$ in Crossings(x) is mapped to the bigrassmannian permutation denoted by $[h, j - h, x_k - h, n - x_k - j + h]$ where $h = \#\{p < j : x_p < x_k\}$.

In the example, the left leaf corresponds with 2 in code(14752368) and the right leaf corresponds with 4. The numbers 2 and 4 map to crossings $(2, 4, 5, 6)$ and $(3, 3, 4, 7)$, respectively. The crossing $(2, 4, 5, 6)$ then maps to the bigrassmannian $[1, 3, 2, 2]$ and $(3, 3, 4, 7)$ maps to the bigrassmannian $[2, 1, 4, 1]$.

Step 3: *Compute maximal bounds on the edge labels depending on w.*

Choose $w \in S_n$ such that $w \ge x$.

Say a leaf of tree(x) is labeled by the bigrassmannian $[a, b, c, d]$, then the maximum edge label along any path leading to this leaf is

$$\text{distance}([a, b, c, d], w) = \max\{n \ge 0 \mid [a - n, b + n, c + n, d - n] \le w\}.$$

---

[1]Lascoux and Schützenberger first used the bigrassmannian permutations to embed the Bruhat–Chevalley order in a distributive lattice [**120**].

6.3.30. Definition. Let $EL(x, w)$ be the set of all labelings of tree(x) such that the labels increase along every path from the root to a leaf and the leaf labels cannot exceed the leaf bounds above.

Finishing the example for $x = (14752368)$, take $w = (87654321)$, then the leaf bounds for both $[1, 3, 2, 2]$ and $[3, 3, 4, 7]$ are 1. Therefore, $EL(x, w)$ contains the following 6 trees:

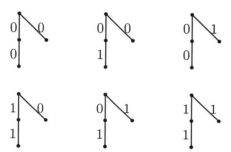

Therefore, with $w_0 x = (85247631)$ and $w_0 w = (12345678)$ we have that
$P_{(12345678),(85247631)} = 1 + 2q + 2q^2 + q^3$

CHAPTER 7

Nil-Hecke Ring and the Singular Locus of $X(w)$

The most general test for smoothness of Schubert varieties (and rational smoothness as well) has been given by Shrawan Kumar [91]. In this section, we state and prove Kumar's criteria for smoothness and rational smoothness at T-fixed points. The proof hinges on the relation between the nil-Hecke ring and the formal T character of the ring of functions on the scheme theoretic tangent cone $T_{e_v}(X_w)$. We also include Dyer's proof of Deodhar's inequality here since it uses the nil-Hecke algebra. We begin by introducing the nil-Hecke ring and fixing notation. The results in this chapter extend to the case where G is an arbitrary Kac–Moody group.

Throughout this chapter we assume the field K is equal to $\mathbb{C}$.

7.1. The nil-Hecke ring

Let $S(\mathfrak{h}^*)$ be the symmetric algebra of $\mathfrak{h}^*$ (the dual of the Cartan subalgebra $\mathfrak{h}$), and let $Q(\mathfrak{h}^*)$ be the quotient field of $S(\mathfrak{h}^*)$ viewed as a W-field. Let Q_W denote the $Q(\mathfrak{h}^*)$-module with basis $\{\delta_w,\ w \in W\}$ endowed with a ring structure defined as follows:

$$f\delta_v \cdot g\delta_w = fv(g)\delta_{vw}$$

where $v, w \in W$, $f, g \in Q(\mathfrak{h})$. Q_w is called the *nil-Hecke ring*.

7.1.1. Definition. For each simple reflection $s_i \in W$, let x_i be the element of Q_W defined as

$$(7.1.2) \qquad x_i = -(\delta_{id} + \delta_{s_i})\frac{1}{\alpha_i},$$

where id is the identity element in W. For any $w \in W$ and any reduced expression $s_{i_1} \cdots s_{i_r} = w$, define $x_w = x_{i_1} \cdots x_{i_r}$.

Kostant and Kumar originally introduced the elements x_w in [86] and show they are well-defined, form a basis for Q_W, and are closely related to the Schubert cycles in the cohomology ring of G/B. The x_w's play a role analogous to the divided difference operators in the theory of Bernstein–Gelfand–Gelfand [11] and Demazure [37]. In fact, the critical aspect of the $\{x_w\}$ for testing smoothness and computing the cup product in $H^*(G/B)$ is the coefficients relating this basis of Q_W with the original basis elements δ_w, which we describe next.

Expanding x_w in the basis of $\{\delta_v\}$ uniquely defines the coefficients $c_{w,v} \in Q(\mathfrak{h}^*)$, namely

$$(7.1.3) \qquad x_w = \sum_v c_{w,v} \delta_v.$$

Note that what we call $c_{w,v}$, Kumar calls $c_{w^{-1},v^{-1}}$. Similarly, the inverse matrix of coefficients will be denoted by $d_{w,v}$;

$$(7.1.4) \qquad \delta_w = \sum_v d_{w,v} x_v.$$

An explicit expression for $c_{w,v}$ can be derived directly from the definition of x_w.

7.1.5. Theorem. [6, 91] *Fix a reduced expression for w, say $s_{a_1} s_{a_2} \cdots s_{a_p}$, then*

$$(7.1.6) \qquad c_{w,v} = (-1)^{l(w)} \sum s_{a_1}^{\epsilon_1} \alpha_{a_1}^{-1} s_{a_2}^{\epsilon_2} \alpha_{a_2}^{-1} \cdots s_{a_p}^{\epsilon_p} \alpha_{a_p}^{-1},$$

where the sum is over all sequences $(\epsilon_1 \epsilon_2 \cdots \epsilon_p)$ such that $\epsilon_i = 0, 1$ for all i, $v = s_{a_1}^{\epsilon_1} s_{a_2}^{\epsilon_2} \cdots s_{a_p}^{\epsilon_p}$ (not necessarily reduced), and each s_{a_i} acts on all $\alpha_{a_j}^{-1}$ for $j \geq i$.

7.1.7. Corollary. *In particular,* $c_{w,w} = \displaystyle\prod_{\beta \in R^+ \cap wR^-} \frac{1}{\beta} = \prod_{\substack{\beta \in R^+ \\ s_\beta w < w}} \frac{1}{\beta}$.

For example, take $w = s_1 s_2 s_1 \in S_n$ and $v = \mathrm{id}$. Say we want to compute the coefficient of δ_{id} in $x_w = x_1 x_2 x_1$. Then (7.1.6) gives

$$(7.1.8) \qquad c_{w,v} = (-1)^3 \left[\frac{1}{\alpha_1} \cdot \frac{1}{\alpha_2} \cdot \frac{1}{\alpha_1} + \frac{1}{s_1 \alpha_1} \cdot \frac{1}{s_1 \alpha_2} \cdot \frac{1}{s_1^2 \alpha_1} \right]$$

$$(7.1.9) \qquad = -\left(\frac{1}{\alpha_1^2 \alpha_2} - \frac{1}{\alpha_1^2 (\alpha_1 + \alpha_2)} \right) = \frac{-1}{\alpha_1 \alpha_2 (\alpha_1 + \alpha_2)}$$

$$(7.1.10) \qquad = \text{coefficient of } \delta_{\mathrm{id}} \text{ in } x_w.$$

In fact this computation holds for any two simple reflections in any W that do not commute.

Dyer has given a nice description of the $c_{w,v}$'s in terms of a homogeneous polynomial function, $g_{w,v}$, times the inverse of a product of positive roots. Using this description Dyer was able to prove Deodhar's inequality (Theorem 7.1.24) in any Coxeter group by relating it to the degree of $g_{w,v}$, which must be nonnegative.

7.1.11. Theorem. [43] *For any $x \leq w$ in W there exist unique nonzero homogeneous polynomials $g_{w,x}$ and $g'_{w,x}$ of $S(\mathfrak{h}^*)$ such that*

$$(7.1.12) \qquad c_{w,x} = g_{w,x} c_{x,x} \prod_{\substack{\alpha \in R^+ \\ x < s_\alpha x \leq w}} \alpha^{-1} = g_{w,x} \prod_{\substack{\alpha \in R^+ \\ s_\alpha x \leq w}} \alpha^{-1}$$

$$(7.1.13) \qquad d_{w,x} = g'_{w,x} d_{w,w} \prod_{\substack{\alpha \in R^+ \\ x \leq s_\alpha w < w}} \alpha^{-1} = g'_{w,x} \prod_{\substack{\alpha \in R^+ \\ x \not\leq s_\alpha w}} \alpha$$

where

$$(7.1.14) \qquad deg\ g_{w,x} = \#\{\beta \in R^+ | x < s_\beta x \leq w\} - l(w) + l(x)$$

$$(7.1.15) \qquad deg\ g'_{w,x} = \#\{\beta \in R^+ | x \leq s_\beta x < w\} - l(w) + l(x).$$

Dyer's proof requires two lemmas, which the reader may prove as an exercise.

7.1.16. Lemma. [86] *For any linear polynomial $a \in \mathfrak{h}^*$,*

$$(7.1.17) \qquad a x_w = x_w(w^{-1}a) - \sum a(\gamma) x_{s_\gamma w}$$

where the sum is over all $\gamma \in R+$ such that $l(s_\gamma w) = l(w) - 1$, and note $a(\gamma) \in K$.

7.1.18. Lemma. *Let $v < w$ and let $a \in \mathfrak{h}^*$. Then we have*

$$(7.1.19) \qquad (a - vw^{-1}(a))c_{w,v} = -\sum a(\gamma)c_{s_\gamma w,v},$$

where the sum is over all $\gamma \in R+$ such that $l(s_\gamma w) = l(w) - 1$.

PROOF OF THEOREM 7.1.11. We prove (7.1.12) by induction on $l(w) - l(x)$ and leave (7.1.13) as an exercise. Note if $x = w$, then (7.1.12) holds by Theorem 7.1.5. So, assume $x < w$. Then for any $\alpha \in R^+$, one has $x < s_\alpha x < w$ if and only if there exists an element $u = s_\gamma w$ such that $l(u) = l(w) - 1$ and $x < s_\alpha x \leq u$. Let $a \in \mathfrak{h}^*$ as in Lemma 7.1.18. Applying the induction hypothesis to (7.1.19) we have

$$(7.1.20) \quad c_{w,x} = (a - xw^{-1}(a))^{-1} \sum_{\substack{\gamma \in R+ \\ l(s_\gamma w) = l(w) - 1}} -a(\gamma)\ c_{s_\gamma w,x}$$

$$(7.1.21) \qquad = (a - xw^{-1}(a))^{-1} \sum_{\substack{\gamma \in R+ \\ l(s_\gamma w) = l(w) - 1}} -a(\gamma)\ g_{s_\gamma w,x} \prod_{\substack{\alpha \in R^+ \\ s_\alpha x \leq s_\gamma w}} \alpha^{-1}$$

$$(7.1.22) \qquad = (a - xw^{-1}(a))^{-1}\ g_a \prod_{\substack{\alpha \in R^+ \\ s_\alpha x < w}} \alpha^{-1}$$

for some nonzero homogeneous polynomial $g_a \in S(\mathfrak{h}^*)$.

It remains to be shown that $(a - xw^{-1}(a))^{-1} g_a$ is a polynomial independent of a. Let n be the dimension of the image of the linear transformation $\text{id} - xw^{-1}$ of $\mathfrak{h}^*$. Since $x \neq w$, $n \geq 1$. If $n = 1$, then $xw^{-1} = s_\gamma$ for some positive root γ, in which case $a - xw^{-1}(a) = a(\gamma)\gamma$. Choosing a so that $a(\gamma) > 0$, we can define $g_{w,x} = a(\gamma)^{-1} g_a$ (the γ gets absorbed into the product).

If $n > 1$, choosing dominant weights $a_1, a_2 \in \mathfrak{h}^*$, we must have

$$(7.1.23) \qquad \frac{g_{a_1}}{a_1 - xw^{-1}(a_1)} = \frac{g_{a_2}}{a_2 - xw^{-1}(a_2)},$$

i.e., $g_{a_1}(a_2 - xw^{-1}(a_2)) = g_{a_2}(a_1 - xw^{-1}(a_1))$. Since $S(\mathfrak{h}^*)$ is a unique factorization domain, both sides of (7.1.23) must be polynomials; so let $g_{w,x}$ be this polynomial. $\qquad\square$

7.1.24. Corollary. [38, 43] Deodhar's Inequality. *Let (W, S) be an arbitrary Coxeter system. For $x \leq y \leq w$,*

$$\#\{r \in \mathcal{R} \mid x \leq ry \leq w\} \geq l(w) - l(x),$$

where $\mathcal{R}$ denotes the set of all reflections in W.

PROOF. It suffices to prove the inequality in the case the root system is finite since any Bruhat interval is contained in a finitely generated parabolic subgroup. From Theorem 7.1.11, we have for any $x \leq y \leq w$

$$0 \leq \deg(g_{w,y}) + \deg(g'_{y,x}) = \#\{\beta \in R^+ : x \leq s_\beta y \leq w\} - (l(w) - l(x)).$$

$\qquad\square$

Corollary 7.1.24 was first conjectured by Deodhar and proved by him in the Type A case and other several special cases [38]. In addition to Dyer's proof of the general case, it was also proved by Polo [135] for finite Weyl groups using properties of Schubert varieties and by Carrell–Peterson for crystallographic groups [32].

7.2. Criteria for smoothness and rational smoothness

The following criteria can be used to determine the entire singular locus (and rational singular locus) of any Schubert variety in G/B. Recall rational smoothness is covered in Chapter 6.

7.2.1. Theorem. [91, Th. 5.5] *Let $v, w \in W$, $v \leq w$ and let $S(w, v) = \{\alpha \in R^+ \mid s_\alpha v \leq w\}$.*

1. *$X(w) \subset G/B$ is smooth at e_v if and only if*

$$(7.2.2) \qquad c_{w,v} = (-1)^{l(w) - l(v)} \prod_{\beta \in S(w,v)} \beta^{-1}.$$

2. $X(w)$ *is rationally smooth at* e_v *if and only if for each* $v \leq x \leq w$ *there exists a positive integer* d_x *such that*

$$(7.2.3) \qquad c_{w,x} = d_x(-1)^{l(w)-l(x)} \prod_{\beta \in S(w,x)} \beta^{-1}.$$

7.2.4. Remark. Theorem 7.2.1 holds in arbitrary characteristic (see [**28**], [**91**, Remark 8.10]), and for all Kac–Moody groups G. Brion's proof (cf. [**28**]) uses equivariant Chow group theory for torus actions. In addition, Kumar's theorem has been extended beyond the realm of Schubert varieties by Brion [**29**] and Arabia [**8**].

We present Kumar's proof of Theorem 7.2.1 in §7.4 after introducing a key theorem, also due to Kumar, on the character of the graded local ring at $e_v \in X(w)$. The rest of this section will develop consequences of Theorem 7.2.1.

It is interesting to compare the formulas (7.2.2), (7.2.3) and (7.1.12). In words, $X(w)$ is smooth at e_v if and only if $g_{w,v} = \pm 1$ and is rationally smooth if and only if $g_{w,x}$ is constant for each $u \leq x \leq w$.

Explicit computations using Kumar's theorem are not efficient due to the fact that $c_{w,v}$ is a rational function. Below we state a more efficient algorithm for applying Kumar's theorem in the case where G is semisimple using the inverse matrix, given by

$$(7.2.5) \qquad \delta_w = \sum_v d_{w,v} x_v.$$

This algorithm is still however exponential in the length of the Weyl group element. It is known that $d_{v,w}$ is a polynomial in $S(\mathfrak{h}^*)$ and $d_{w,v} = 0$ unless $v \leq w$ [**12**, **86**] (see Theorem 7.2.11 below).

7.2.6. Theorem. [**87**] *Let* G *be semisimple, and let* w_0 *be the unique element of* W *of longest length. Given any* $v, w \in W$ *such that* $v \leq w$, *we have*

$$(7.2.7) \qquad c_{w,v} \prod_{\alpha > 0} \alpha = (-1)^{l(w)-l(v)} d_{vw_0, ww_0}.$$

7.2.8. Corollary. *Given any* $v, w \in W$ *such that* $v \leq w$, *the Schubert variety* X_{vw_0} *is smooth at* e_{ww_0} *if and only if*

$$(7.2.9) \qquad d_{w,v} = \prod_{\gamma \in Z(w,v)} \gamma,$$

where $Z(w, v) = \{\gamma \in R_+ : v \not\leq s_\gamma w\}$.

PROOF. Combining Theorems 7.2.1 and 7.2.6 we have X_{vw_0} is smooth if and only if

$$(7.2.10) \qquad d_{w,v} = \frac{\prod_{\alpha>0} \alpha}{\prod_{\beta \in S(vw_0, ww_0)} \beta} = \prod_{\beta \in S(vw_0, ww_0)^c} \beta.$$

Here $S(x,y)^c$ is the complement of $S(x,y)$ in R^+. Note that $x \leq y$ in Bruhat order if and only if $xw_0 \geq yw_0$. Therefore,

$$\begin{aligned} S(vw_0, ww_0) &= \{\alpha \in R_+ : ws_\beta w_0 \leq vw_0\} \\ &= \{\beta \in R_+ : v \leq s_\beta w\} \\ &= Z(w,v)^c. \end{aligned}$$

$\square$

Furthermore, there is an explicit formula for $d_{w,v}$. Let $R(w)$ be the set of all sequences $\mathbf{b} = b_1 \ldots b_p$ such that $s_{b_1} s_{b_2} \ldots s_{b_p}$ is a reduced expression for w.

7.2.11. Theorem. [12] *Let $v, w \in W$ and fix a reduced word $\mathbf{b} = b_1 b_2 \cdots b_p \in R(w)$. Then*

$$(7.2.12) \qquad d_{w,v} = \sum_{b_{i_1} b_{i_2} \ldots b_{i_k} \in R(v)} r_\mathbf{b}(i_1) r_\mathbf{b}(i_2) \cdots r_\mathbf{b}(i_k),$$

where $r_\mathbf{b}(j)$ is defined by

$$(7.2.13) \qquad r_\mathbf{b}(j) = r_{b_1 b_2 \ldots b_p}(j) = s_{b_1} s_{b_2} \ldots s_{b_{j-1}}(\alpha_{b_j}),$$

and the sum is over all sequences $1 \leq i_1 < i_2 < \ldots < i_k \leq p$ such that $b_{i_1} b_{i_2} \ldots b_{i_k} \in R(v)$. Furthermore, the sum in (7.2.12) is independent of the choice of $\mathbf{b} \in R(w)$.

The following special case for smoothness was first proved by C. K. Fan [**46**] and follows easily from Kumar's criterion in Theorem 7.2.1 and Theorem 7.1.5. Let $w \in W$ have the property that none of its reduced expressions contain the substring sts where s, t are simple reflections that do not commute; then w is called *short-braid avoiding*. Note that short-braid avoiding is equivalent to saying that every two occurrences of s in a reduced word for w are separated by at least two elements t, t' that do not commute with s.

7.2.14. Theorem. Fan's result. *Assume $w \in W$ is short-braid avoiding. Then $X(w)$ is nonsingular if and only if w is a product of distinct simple reflections.*

The following theorems were originally proved by Carrell (cf. [**32**]); now they also follow easily from Kumar's criteria. Recall the simple reflections are labeled as in the Dynkin diagrams on page 207.

7.2.15. Theorem. [**32**, Proposition 4.12] *For any simple reflection $s \in S$, the Schubert variety $X(w_0 s)$ in G/B is smooth if and only if one of the following holds.*

1. *Type A: $G = SL(n)$ and s is either s_1 or s_n.*
2. *Type C: $G = Sp(2n)$ and s is s_1, i.e., the simple reflection corresponding to the minuscule fundamental dominant weight.*

7.2.16. Theorem. [**32**, Prop. 4.13] [**91**, Prop. 7.5] *For any simple root $\alpha_i \in S$, the Schubert variety $X(w_0 s_i)$ in G/B is smooth at e_v if and only if $\omega_i - v^{-1} w \omega_i$ is a positive root, ω_i being the fundamental weight associated to α_i.*

7.3. Representation-theoretic results on the tangent cone

Assume G is simply connected, in particular, the character group $X(T)$ coincides with the weight lattice Λ. Let $R(T)$ be the group algebra of $X(T)$ with basis $\{e^\lambda\}_{\lambda \in \Lambda}$. Let $Q(T)$ be the quotient field of $R(T)$. Any T-module M is determined by its weight spaces M_λ, $\lambda \in \mathfrak{h}^*$. This information is summarized in the *formal character* defined by

$$\mathrm{ch} M = \sum_{e^\lambda \in X(T)} (\dim M_\lambda) e^\lambda.$$

In analogy with the discussion in §7.1, we introduce a family of coefficients $b_{w,v}$ in $Q(T)$ and relate them both to the $c_{w,v}$'s and to the T-action on the tangent cones to Schubert varieties.

For each simple reflection s_i, let y_i be the element of $Q_W(T)$ (the smash product of $\mathbb{Z}[W]$ with $Q(T)$) defined as

$$y_i = (\delta_{\mathrm{id}} + \delta_{s_i}) \frac{1}{1 - e^{-\alpha_i}}.$$

For any $w \in W$ and any reduced expression $s_{i_1} \cdots s_{i_r} = w$, define $y_w = y_{i_1} \cdots y_{i_r}$. Expansion of y_w in the basis $\{\delta_v\}$ uniquely defines the coefficients $b_{w,v} \in Q(T)$, namely

$$y_w = \sum_v b_{w,v} \delta_v.$$

Again, note that what we call $b_{w,v}$, Kumar calls $b_{w^{-1}, v^{-1}}$. The elements y_w are shown to be well-defined in [**87**, Prop. 2.4].

The relationship between the coefficients $b_{w,v}$ and $c_{w,v}$ is dependent on the following bracket operation. If $a = \sum n_\lambda e^\lambda \in R(T)$ and k is a nonnegative integer, then define $(a)_k$ to be the homogeneous component of a of degree k assuming each λ has degree 1, i.e.,

(7.3.1) $$(a)_k = \sum n_\lambda \frac{\lambda^k}{k!} \in S^k(\mathfrak{h}^*).$$

The square bracket operation is defined by

$$(7.3.2) \qquad\qquad [a] = (a)_{k_0}$$

where k_0 is the smallest nonnegative integer such that $(a)_{k_0} \neq 0$. If a is just a scalar, then $[a] = a$. If $q = \frac{a}{b} \in Q(T)$ for some $a, b \in R(T)$, then

$$(7.3.3) \qquad\qquad [q] = \frac{[a]}{[b]} \in Q(\mathfrak{h}^*).$$

For example,

$$(7.3.4) \qquad\qquad \left[\frac{1}{1 - e^\alpha}\right] = \frac{[1]}{-[\alpha + \alpha^2/2! + ...]} = \frac{1}{-\alpha}.$$

Define an involution $*$ on $Q(T)$ taking $e^\lambda \to e^{-\lambda}$. The following lemma follows from the example.

7.3.5. Lemma. *As elements of $Q(\mathfrak{h}^*)$, $[*b_{w,v}] = c_{w,v}$.*

Let $\mathrm{gr}(\mathcal{O}(v, X(w))$ denote the associated graded ring to the local ring $\mathcal{O}(v, X(w))$ at $e_v \in X(w)$. Then define $T_v(X(w))$ to be the (scheme-theoretic) tangent cone $\mathrm{Spec}(\mathrm{gr}(\mathcal{O}(v, X(w)))$. The main result of this section is the following:

7.3.6. Theorem. [**91**, Th. 2.2] *For any $v \leq w$ in W, $\mathrm{gr}(\mathcal{O}(v, X(w)))$ has a canonical T-action and decomposes into direct sum of finite-dimensional T-weight spaces. Furthermore,*

$$(7.3.7) \qquad\qquad ch\, \mathrm{gr}(\mathcal{O}(v, X(w))) = *b_{w,v} \in Q(T).$$

Kumar's proof of Theorem 7.3.6 requires several detailed computations which ultimately show that $\mathrm{gr}\mathcal{O}(\mathrm{id}, v^{-1}X(w))$ is isomorphic to $\mathrm{gr}\mathbb{C}[v^{-1}\overline{BwB} \cap U^-]$, where recall that U^- is the unipotent part of B^-, the Borel subgroup opposite to B. Furthermore, using the Demazure character formula, one shows that the character of $\mathrm{gr}\mathbb{C}[v^{-1}\overline{BwB} \cap U^-]$ has the suggested form involving $b_{w,v}$. The interested reader should consult [**91**].

7.4. Proof of smoothness criterion

In this section we present Kumar's proof of Theorem 7.2.1 from [**91**]. We begin by constructing a nice affine neighborhood of a T-fixed point e_v and proving two related lemmas. The main proof then follows easily.

Let $\exp : \mathfrak{u}^- \tilde{\to} U^-$ be the exponential map, where $\mathfrak{u}^-$ is the Lie algebra of U^-. Let

$$(7.4.1) \qquad\qquad Y := \exp^{-1}(U^- e_{\mathrm{id}} \cap v^{-1}X(w))$$

be the closed irreducible subvariety of $\mathfrak{u}^-$, where we identify U^- with $U^- e_{\mathrm{id}}$. Fix nonzero root vectors $X_{-\beta}$ for $\beta \in R^+$ (as in Chapter 2, §2.3). For any $\alpha \in R^+$, let $f_\alpha : \mathfrak{u}^- \to \mathbb{C}$ be the linear map defined by $\sum_{\beta \in R^+} t_\beta X_{-\beta} \mapsto t_\alpha$, and let f_α^Y be the restriction of f_α to Y. Define the subvariety Z of Y, with the reduced scheme structure, by

$$Z = \{x \in Y : f_\alpha^Y(x) = 0, \text{ for all } \alpha \in S(w^{-1}, v^{-1})\},$$

where $S(w^{-1}, v^{-1}) = \{\alpha \in R^+ \mid s_\alpha v^{-1} \leq w^{-1}\}$

7.4.2. Lemma. *With notation as above, the variety $Z = \{0\}$.*

PROOF. Clearly $0 \in Z$. We will show that any irreducible component Z^o of Z through 0 does not contain any one-dimensional T-stable closed irreducible subvariety V. Therefore, Z^o must be zero-dimensional.

Any irreducible component $Z^o \subset Z$ is clearly T-stable under the adjoint action of the maximal torus T on $\mathfrak{u}^-$. Furthermore, any one-dimensional T-stable closed irreducible subvariety of $\mathfrak{u}^-$ is of the form $\mathbb{C}X_{-\beta} \subset \mathfrak{u}^-$, for some $\beta \in R^+$. In particular, if $V = \mathbb{C}X_{-\beta_0}$ for some $\beta_0 \in R^+$, then

$$\exp(\mathbb{C}X_{-v\beta_0})e_v \subset X(w).$$

Now if $-v\beta_0 \in R^+$, then $vs_{\beta_0} < v \leq w$, so $\beta_0 \in S(w^{-1}, v^{-1})$. If $v\beta_0 \in R^+$, then

$$\exp(\mathbb{C}X_{v\beta_0})\exp(\mathbb{C}X_{-v\beta_0})e_v \subset X(w).$$

For a root $\alpha > 0$, recall that Z_α denotes the $SL(2)$-copy in G corresponding to α; note that Z_α is simply the subgroup of G generated by U_α and $U_{-\alpha}$. In particular, for the subgroup $Z_{vs_{\beta_0}v^{-1}} \subset G$ generated by $\exp(\mathbb{C}X_{-v\beta_0})$ and $\exp(\mathbb{C}E_{v\beta_0})$, we have the inclusion $Z_{vs_{\beta_0}v^{-1}}e_v \subset X(w)$. Again this gives that $\beta_0 \in S(w^{-1}, v^{-1})$; see §6.2. In either case, $V = \mathbb{C}X_{-\beta_0}$, for some $\beta_0 \in S(w^{-1}, v^{-1})$. But, by the definition of the variety Z, such a V is not contained in Z. This contradiction establishes the claim that Z^o does not contain any one-dimensional T-stable closed irreducible subvariety.

Embed $i : \mathfrak{u}^- \hookrightarrow G/B$ via the map $X \mapsto (\exp X)e_{\mathrm{id}}$. The map i is clearly a T-equivariant open immersion. Take the Zariski closure $\overline{Z^o}$ of $i(Z^o)$ in G/B. We have shown that Z^o does not contain any one-dimensional T-stable closed irreducible subvarieties, but we also know there are at least $\dim Z^o$ such curves passing through the origin (see the discussion in §6.2). Hence, $\dim Z^o = 0$.

Since any irreducible component of Z is T-stable (and closed) in $\mathfrak{u}^-$, and any closed T-stable subset of $\mathfrak{u}^-$ contains 0, we get that any irreducible component of Z passes through 0. In particular, $Z = \{0\}$. $\square$

7.4.3. Lemma. *Assume $\#S(w, v) = l(w) = \dim Y$ where Y is the variety defined in (7.4.1). View f_γ^Y as elements of the local ring $\mathcal{O}_{0,Y}$ and let I be the ideal generated by $\{f_\gamma^Y\}_{\gamma \in S}$ inside the local ring $\mathcal{O}_{0,Y}$. Then*

1. $\sum_{m \geq 0} I^m / I^{m+1} \xrightarrow{\sim} \frac{\mathcal{O}_{0,Y}}{I}[X_1, X_2, \ldots, X_\ell].$

2. $[\mathrm{ch}(\mathrm{gr}(\mathcal{O}(v, X(w)))] = d(-1)^{l(w)-l(v)} \displaystyle\prod_{\beta \in S(w,v)} \beta^{-1}$, where $d \in \mathbb{N}$.

3. As a vector space $\dim \mathcal{O}_{0,Y}/I = d$.

PROOF. By Lemma 7.4.2, $Z = \{0\}$. Since I is the image of the ideal defining Z, there exists an integer $r > 0$ such that $\mathfrak{m}^r \subset I \subset \mathfrak{m}$, where $\mathfrak{m} \subset \mathcal{O}_{0,Y}$ is the maximal ideal. By assumption, $\#S(w^{-1}, v^{-1}) = \#S(w, v) = \ell(w) = \dim Y$, so we can enumerate the elements of $S(w^{-1}, v^{-1})$ as $\{\gamma_1, \dots, \gamma_\ell\}$, where $\ell = \ell(w)$.

Recall that Schubert varieties are Cohen–Macaulay (cf. §2.9), hence the variety Y is Cohen–Macaulay. By [**48**, Lemma (a), Section 2.4] and the fact that $\dim Z = 0$, the elements $\{f^Y_{\gamma_j}\}_{1 \leq j \leq \ell}$ form a regular sequence in $\mathcal{O}_{0,Y}$. Moreover, by [**48**, Lemma (b), Section 2.4], the canonical ring homomorphism

$$(7.4.4) \qquad \frac{\mathcal{O}_{0,Y}}{I}[X_1, X_2, \dots, X_\ell] \longrightarrow \sum_{m \geq 0} I^m/I^{m+1},$$

which takes x_j to the image of $f^Y_{\gamma_j}$ in I/I^2, is an isomorphism. In particular,

$$(7.4.5)$$

$$\mathrm{ch}\left(\mathrm{gr}\left(\mathcal{O}_{\mathrm{id},v^{-1}X(w)}\right)\right) = \mathrm{ch}\left(\mathrm{gr}\left(\mathcal{O}_{0,Y}\right)\right) = \mathrm{ch}(\mathbb{C}[Y]) \quad (Y \text{ being affine})$$

$$(7.4.6) \qquad = \mathrm{ch}\left(\mathcal{O}_{0,Y}/I\right) \prod_{j=1}^{\ell}(1 - e^{\gamma_j})^{-1}, \quad \text{by } (7.4.4).$$

Since $\mathcal{O}_{0,Y}/I$ corresponds to a zero-dimensional variety, it is finite-dimensional vector space over $\mathbb{C}$, and hence

$$(7.4.7) \qquad [\mathrm{ch}\left(\mathcal{O}_{0,Y}/I\right)] = \dim\left(\mathcal{O}_{0,Y}/I\right).$$

By (7.4.5) and (7.4.7) we get

$$(7.4.8) \qquad [\mathrm{ch}\left(\mathrm{gr}\left(\mathcal{O}_{\mathrm{id},v^{-1}X(w)}\right)\right)] = (-1)^{\ell}d \prod_{j=1}^{\ell} \gamma_j^{-1},$$

where $d = \dim\left(\mathcal{O}_{0,Y}/I\right)$. Using the fact that the reflection over $v\alpha$ is $s_{v\alpha} = vs_\alpha v^{-1}$, a straightforward computation shows that

$$\prod_{\alpha \in S(w^{-1},v^{-1})} v\alpha = (-1)^{l(v)} \prod_{\beta \in S(w,v)} \beta.$$

Therefore, applying v to both sides we get

$$(7.4.9) \qquad [\mathrm{ch}\,(\mathrm{gr}\,(\mathcal{O}(v, X(w))))] = (-1)^{\ell}\, d \prod_{j=1}^{\ell} (v\gamma_j)^{-1}$$

$$(7.4.10) \qquad = (-1)^{\ell(w)-\ell(v)} d \prod_{\beta \in S(w,v)} \beta^{-1}.$$

$\square$

Finally, we complete the proof of Theorem 7.2.1.

PROOF. First we will prove Part 2 of Theorem 7.2.1. By Theorem 6.2.10, Condition 5 we know $e_v \in X(w)$ is a rationally smooth point if and only if for each $v \le x \le w$, $\#S(w,x) = l(w)$. Furthermore, the latter condition holds if and only if $\deg(g_{w,x}) = 0$ for each $v \le x \le w$ by Theorem 7.1.11. In other words

$$(7.4.11) \qquad c_{w,x} = \text{constant} \cdot \prod_{\beta \in S(w,x)} \beta^{-1}.$$

Lemma 7.4.3 determines the sign of the constant in (7.4.11) and completes the proof for Part 2.

We will use the following characterization of smooth points in proving both directions of Part 1. Recall that $\mathrm{gr}\mathcal{O}(v, X(w))$ is the associated graded ring to the local ring $\mathcal{O}(v, X(w))$ at $e_v \in X(w)$. In general,

$$(7.4.12) \qquad \mathrm{gr}(\mathcal{O}(v, X(w)) \cong \mathrm{Sym}(\mathrm{gr}_1\mathcal{O}(v, X(w)))/I$$

for some ideal I. In particular, $X(w)$ is smooth at e_v if and only if $\mathrm{gr}(\mathcal{O}(v, X(w))$ is a polynomial algebra generated by degree one elements, i.e., I is trivial [62, Lecture 14].

In order to prove the first implication in Part 1, assume $e_v \in X(w)$ is smooth. Write out the character

$$(7.4.13) \qquad \mathrm{ch}(\mathrm{gr}_1\mathcal{O}(v, X(w))) = \sum_{\gamma \in S_v} e^{\gamma},$$

for some subset $S_v \subset vR^+$. (Note that all weight spaces are one-dimensional). Then by the smoothness assumption and (7.4.12) we have that

$$(7.4.14) \quad \mathrm{ch}(\mathrm{gr}(\mathcal{O}(v, X(w))) = \mathrm{ch}(\mathrm{Sym}(\mathrm{gr}_1\mathcal{O}(v, X(w)))) = \prod_{\gamma \in S_v} \frac{1}{1 - e^{\gamma}}.$$

Therefore, by Theorem 7.3.6

$$(7.4.15) \qquad c_{w,v} = [\mathrm{ch}(\mathrm{gr}(\mathcal{O}(v, X(w)))]$$

$$(7.4.16) \qquad = \left[\prod_{\gamma \in S_v} \frac{1}{1 - (1 + \gamma + \gamma^2/2! + \dots)} \right] = \prod_{\gamma \in S_v} (-\gamma)^{-1},$$

which shows that $c_{w,v}$ is a product of certain roots inverted. Also, since smoothness implies rational smoothness, Part 2 of Theorem 7.2.1 gives

$$(7.4.17) \qquad c_{w,v} = (-1)^{l(w)-l(v)} d_v \prod_{\beta \in S(w,v)} \beta^{-1}$$

for some constant $d_v \in \mathbb{N}$. Equating products gives

$$(7.4.18) \qquad d_v \prod_{\beta \in S(w,v)} \beta^{-1} = \pm \prod_{\gamma \in S_v} \gamma^{-1}.$$

Let p be a prime divisor of d_v. Reading equation 7.4.18 mod p, we conclude that $d_v = 1$ (observe that no root mod p is 0 in $Q \otimes_{\mathbb{Z}} \mathbb{F}_p$, Q being the root lattice and $\mathbb{F}_p$ the prime field of order p), and $S_v = S(w,v)$.

Conversely, assume $c_{w,v} = (-1)^{l(w)-l(v)} \prod_{\beta \in S(w,v)} \beta^{-1}$. Then, $\#S(w,v) = l(w)$ since we know that $\deg(c_{w,v}) = -l(w)$. Applying Lemma 7.4.3, the coefficient $d = \dim \mathcal{O}_{0,Y}/I = 1$ so $\mathcal{O}_{0,Y}/I$ is a field, which implies $\mathrm{gr}(\mathcal{O}(v, X(w)))$ is a polynomial ring. Therefore, one concludes that e_v is a smooth point of $X(w)$. $\qquad \square$

7.4.19. Remark. (cf. [91]) The integer d_x in Theorem 7.2.1, (2) is the multiplicity of the point $e_x \in X(w)$ as defined in §4.7.

CHAPTER 8

Patterns, Smoothness and Rational Smoothness

In this chapter we describe several combinatorial tests for smoothness and rational smoothness in terms of avoiding certain patterns in permutations. These algorithms run in polynomial time in the size of the permutations, and are therefore the most efficient tests known at this time. Kumar's criteria and computations with the tangent space bases are all exponential. We also state the conjecture due to Lakshmibai–Sandhya for determining the irreducible components of singular loci in the case $G = SL(n)$.

Recall that if e_v is singular in $X(w)$, then so is e_u for each $u \leq v$. Therefore, to compute the singular locus one only needs to find the maximal elements $v \leq w$ in the Bruhat–Chevalley order that are singular.

Currently the best known way to compute these maximal elements is to search over all elements below w in the Bruhat–Chevalley order and check if each is singular or lies below a known singular point. At each point e_v one can use any of the methods discussed so far in this book. For example, one can use Kumar's criteria from Chapter 7, or one can compute the dimension of the tangent space using techniques from Chapter 5. Needless to say, these methods are not efficient.

A big open problem in the field of computing singularities of Schubert varieties is to find a more efficient way to directly compute the maximal elements in the singular locus. This problem is open even in type A.

We have included tables in this text of the singular locus for all elements of S_7, $\mathfrak{B}_4$, $\mathfrak{C}_4$, $\mathfrak{D}_4$. The tables can be read as follows:

Perm:(2 4 5 6 1 3) **Sing:**(2 4 1 5 3 6) (2 1 5 4 3 6) (2 1 4 6 3 5)

means $\mathrm{Sing}(X(245613))$ consists of the union of three Schubert varieties, namely $X(241536), X(215436)$, and $X(214635)$.

8.1. Type A: criterion in terms of patterns

Recall that for $G = SL(n)$, $W = S_n$. First consider $SL(4)$. In this case $X(3412), X(4231)$ are the only singular Schubert varieties. The situation for a general n turns out to be nothing more than this, as given by the following theorem.

8.1.1. Theorem. ([107] *Let $w \in S_n$, say $w = (a_1, \ldots, a_n)$. Then $X(w)$ is singular if and only if the following property holds:*

(8.1.2) $\left\{ \begin{array}{l} \text{there exist } i, j, k, l, \ 1 \leq i < j < k < l \leq n \text{ such that} \\ \text{either (1) } a_k < a_l < a_i < a_j \text{ or (2) } a_l < a_j < a_k < a_i \end{array} \right\}.$

The above theorem was originally proved using Theorem 5.4.2. An alternate proof will also follow, as a special case, from the proofs of Theorems 8.3.16, 8.3.17 below. Permutations that do not satisfy the properties in (8.1.2) are said to *avoid the patterns* 3412 and 4231 since they do not contain any length 4 subsequence with the same relative order as either of these two sequences.

8.2. Conjecture in type A

Continuing with $SL(4)$, the two singular Schubert varieties have the following singular loci: $\mathrm{Sing}X(3412) = X(1324)$ and $\mathrm{Sing}X(4231) = X(2143)$. Conjecture 8.2.13 below states that for a general n, the situation is just a variation on these two cases.

Assume $X(w)$ is singular, G being $SL(n)$. Let $w = (a_1 \cdots a_n)$. Since $X(w)$ is singular, by Theorem **8.1.1, there exist $i, j, k, m, \ 1 \leq i < j < k < m \leq n$ such that

(8.2.1)   either $a_k < a_m < a_i < a_j$ or $a_m < a_j < a_k < a_i$.

It is shown in [107] that in the former case, if w' is obtained from w by replacing a_i, a_j, a_k, a_m, respectively by a_k, a_i, a_m, a_j, then $e_{w'} \in \mathrm{Sing}X(w)$; and in the latter case, if w' is obtained from w by replacing a_i, a_j, a_k, a_m, respectively by a_j, a_m, a_i, a_k, then $e_{w'} \in \mathrm{Sing}X(w)$.

For $w \in W$, let P_w, resp. Q_w, be the maximal element of the set of parabolic subgroups that leave $\overline{BwB}$ (in G) stable under multiplication on the left, resp. right.

8.2.2. Definition. Given parabolic subgroups P, Q, we say that $\overline{BwB}$ is P-Q stable if $P \subset P_w$ and $Q \subset Q_w$.

Since every parabolic subgroup P corresponds with a subset of the simple reflections S_P, this is equivalent to saying that the left descents of w are contained in S_P and the right descents are contained in S_Q.

The following lemma holds for any semisimple G.

8.2.3. Lemma. *Let G be semisimple. Let $w \in W$. Then*

(8.2.4) $S_{P_w} = \{ \alpha \in S \,|\, w^{-1}(\alpha) < 0 \}$

(8.2.5) $S_{Q_w} = \{ \alpha \in S \,|\, w(\alpha) < 0 \}.$

One can find a proof of this theorem in [104], for example. In the case $G = SL_n(K)$, the above statement takes the following form.

8.2.6. Corollary. *Let $w = (a_1 \ldots a_n) \in S_n$. Then*

$$(8.2.7) \qquad S_{Q_w} = \{\epsilon_i - \epsilon_{i+1} | a_i > a_{i+1}\}$$

$$(8.2.8) \qquad S_{P_w} = \{\epsilon_i - \epsilon_{i+1} | i+1 \text{ appears before } i \text{ in } (a_1, a_2, \ldots, a_n)\}$$

8.2.9. Remark. In combinatorial terms, S_{Q_w} is indexed by the descents of w, and S_{P_w} is indexed by the descents of w^{-1}.

8.2.10. Remark. If $X(\tau)$ is an irreducible component of $\text{Sing} X(w)$, then $\overline{B\tau B}$ is P_w-Q_w stable.

8.2.11. Definition. The set F_w. Let $w = (a_1 \ldots a_n) \in \mathcal{S}_n$. We first define E_w to be the set of all $\tau' \leq w$ such that either 1) or 2) below holds.

(1) There exist i, j, k, m, $1 \leq i < j < k < m \leq n$, such that
  (a) $a_k < a_m < a_i < a_j$, i.e., w contains 3412.
  (b) If $\tau' = (b_1 \ldots b_n)$, then there exist i', j', k', m', $1 \leq i' < j' < k' < m' \leq n$ such that $b_{i'} = a_k$, $b_{j'} = a_i$, $b_{k'} = a_m$, $b_{m'} = a_j$, i.e., τ' contains 1324 using the numbers in the 3412 pattern of w.
  (c) Let τ be the element obtained from w by replacing a_i, a_j, a_k, a_m by a_k, a_i, a_m, a_j, and similarly let w' be the element obtained from τ' by replacing $b_{i'}, b_{j'}, b_{k'}, b_{m'}$ by $b_{j'}, b_{m'}, b_{i'}, b_{k'}$. Then, $\tau' \geq \tau$ and $w' \leq w$.

(2) There exist i, j, k, m, $1 \leq i < j < k < m \leq n$, such that
  (a) $a_m < a_j < a_k < a_i$ (i.e., w contains 4231).
  (b) If $\tau' = (b_1 \ldots b_n)$, then there exist i', j', k', m', $1 \leq i' < j' < k' < m' \leq n$ such that $b_{i'} = a_j$, $b_{j'} = a_m$, $b_{k'} = a_i$, $b_{m'} = a_k$ (i.e., τ' contains 2143 using the numbers in the 4231 pattern of w.)
  (c) Let τ be the element obtained from w by replacing a_i, a_j, a_k, a_m by a_j, a_m, a_i, a_k, and let w' be obtained from τ' by replacing $b_{i'}, b_{j'}, b_{k'}, b_{m'}$ by $b_{k'}, b_{i'}, b_{m'}, b_{j'}$. Then, $\tau' \geq \tau$ and $w' \leq w$.

Define the set

$$(8.2.12) \qquad F_w = \{\tau \in E_w \mid \overline{B\tau B} \text{ is } P_w\text{-}Q_w \text{ stable}\}.$$

8.2.13. Conjecture. [107] *For $w \in \mathcal{S}_n$, the singular locus of $X(w)$ is equal to $\cup_\lambda X(\lambda)$, where λ runs over the maximal (under the Bruhat–Chevalley order) elements of F_w.*

For example, the conjecture says that $\text{Sing} X(35142)$ is $X(13254)$ (which is confirmed by the tables in Chapter 13 also). Note that the bad pattern in (35142) is 3512. These numbers are rearranged to get 1325 even though they are not in exactly the same positions in (13254).

Examples (1) and (2) below show the necessity of the two conditions $\tau \leq \tau'$ and $w' \leq w$.

Example 1. Let $w = (52431) \in S_5$. We have from the tables that $\mathrm{Sing}X(w) = X(21543)$. A 4231 pattern occurs in the subsequence (5231), and the corresponding τ is given by $\tau = (21453)$. Set $\tau' = (42153)$ and consider the corresponding $w' = (45231)$. We have $\tau \leq \tau'$, but $w' \not\leq w$. Further, $\tau' \notin \mathrm{Sing}X(w)$, thus supporting the conjecture.

Example 2. Let $w = (456312) \in S_6$. We have from the tables that $\mathrm{Sing}X(w) = X(415326) \cup X(154326) \cup X(146325)$. A 3412 pattern occurs in the subsequence (4512), and the corresponding τ is given by $\tau = (146325)$. Set $\tau' = (146253)$ and consider the corresponding $w' = (456123)$. We have $w' \leq w$, but $\tau \not\leq \tau'$. Further, $\tau' \notin \mathrm{Sing}X(w)$, thus supporting the conjecture.

Example 3. Let $w = (35142) \in S_5$. We have from the tables that $\mathrm{Sing}X(w) = X(13254)$. A 3412 pattern occurs in the subsequence (3512), and the corresponding τ is given by $\tau = (13245)$. Set $\tau' = (13254)$ and consider the corresponding $w' = (35124)$. We have $\tau \leq \tau'$, $w' \leq w$. Further, $\tau' \in \mathrm{Sing}X(w)$, thus supporting the conjecture.

8.2.17. Remark. This conjecture has been verified for all elements of S_7. See the tables of singular loci in Chapter 13.

8.2.18. Remark. This conjecture has been verified to be true for a certain class of Schubert varieties, namely the class of Schubert varieties related to ladder determinantal varieties (cf. [**56**] and Chapter 12) and 321-hexagon-avoiding permutations [**15**] (cf. §6.3.17).

8.2.19. Remark. Recently, Gasharov (cf. [**53**]) has proved that each element in F_w is singular.

8.3. Types B, C, D: criterion in terms of patterns

This section contains the analogs of Theorem 8.1.1 for Schubert varieties of types B, C and D. In these cases, we give characterizations for smoothness and rational smoothness in terms of patterns.

Recall from Chapter 3 that the Weyl group of type C is

$$(8.3.1) \qquad \mathfrak{C}_n = \{(a_1...a_{2n}) \in S_{2n} \mid a_i = 2n + 1 - a_{2n+1-i}, \ 1 \leq i \leq 2n\}.$$

The Weyl group of types B and C are isomorphic and the Weyl group of type D is contained in $\mathfrak{C}_n$ as a subgroup.

8.3.2. Definition. Signed permutations. An element $(a_1, \ldots, a_{2n}) \in \mathfrak{C}_n$ can be represented by $(a_1, \ldots, a_n)$ or equivalently by $(a_{n+1}, \ldots, a_{2n})$. We will use the latter representation to get the *signed permutation* notation.

Namely, $(a_1, \ldots, a_{2n}) = (b_1 \ldots b_n)$ where

$$b_i = \begin{cases} a_{n+i} - n & \text{if } a_{n+i} > n \\ a_{n+i} - n - 1 & \text{if } a_{n+i} \leq n. \end{cases}$$

For example, $(241635) \in \mathfrak{C}_3$ becomes $3\bar{1}2$ in the signed notation. Note, $\bar{1}$ denotes -1 (negative signs lead to messy notation like $3 - 12$).

We will use the signed permutations throughout this section for both the Weyl groups of type B and C. In addition, the Weyl group of type D is the subgroup of the signed permutations with an even number of barred elements. The cost of using this more compact notation is that we will need to relabel the simple reflections and the roots.

Reflections. The reflections in any Weyl group are the set of all elements of the form $u s_i u^{-1}$ for any $u \in W$ and any simple reflection s_i. In particular, the reflections in $\mathfrak{C}_n$ are transpositions t_{ij} and signed transpositions s_{ij}: for $i < j$ and $w = (w_1 \ldots w_i \ldots w_j \ldots w_n)$,

$$(8.3.4) \qquad w t_{ij} = \ldots w_j \ldots w_i \ldots$$
$$(8.3.5) \qquad w s_{ij} = \ldots \overline{w_j} \ldots \overline{w_i} \ldots$$
$$(8.3.6) \qquad w s_{ii} = \ldots \overline{w_i} \ldots.$$

Note that the transposition $t_{i(i+1)}$ corresponds with the simple reflection we usually denoted by s_{n-i}. In order to use a more natural notation for the signed permutations, we will denote s_{n-i} by t_i throughout this section in addition to using the notation for reflections as above. Also, let t_0 be the simple reflection s_n. Then $(w_1, \ldots, w_n)t_0 = (\overline{w_1}, \ldots, w_n)$.

Roots. If $\epsilon_1, \epsilon_2, \ldots, \epsilon_n$ is the standard basis for K^n, then let $e_1, \ldots, e_n$ be the new basis obtained just by reversing the order, i.e., $e_i = \epsilon_{n+1-i}$. For example, the simple roots in type C_n are now $2e_1, e_2 - e_1, \ldots, e_n - e_{n-1}$.

Proctor's test for Bruhat–Chevalley order. The test for Bruhat–Chevalley order given in §3.3 can easily be translated into the signed notation as well. Namely,

$$v = (v_1, \ldots, v_n) \leq w = (w_1, \ldots, w_n)$$

if and only if for all $1 \leq i \leq n$

$$\{v_i, \ldots, v_n\} \uparrow \, \leq \, \{w_i, \ldots, w_n\} \uparrow.$$

Recall that for a k-tuple $(a_1 \ldots a_k)$ of distinct integers, $\{a_1 \ldots a_k\} \uparrow$ denotes the ordered k-tuple obtained from $\{a_1, \ldots, a_k\}$ by arranging its elements in ascending order. Two sequences of increasing integers are then compared entry by entry as in the partial order on $I_{k,n}$ given by (3.1.5).

For example, take $w = \bar{6}\bar{5}4\bar{1}23$ and $v = \bar{4}63\bar{2}51$ (in one-line notation). Then we compute the following table of sorted lists

(8.3.9)	$\bar{3}$	$<$	1
(8.3.10)	$\bar{3}2$	$<$	15
(8.3.11)	$\bar{3}\bar{1}2$	$<$	$\bar{2}15$
(8.3.12)	$\bar{3}\bar{1}24$	$<$	$\bar{2}135$
(8.3.13)	$\bar{5}\bar{3}\bar{1}24$	$<$	$\bar{2}1356$
(8.3.14)	$\bar{6}\bar{5}\bar{3}\bar{1}24$	$<$	$\bar{4}\bar{2}1356$

Hence, $v \leq w$ as elements of B_6.

Proctor's test for $v \leq w$ in the Bruhat–Chevalley order on D_n [**137**, Thm. 5D] is more complicated. Namely, if $v, w \in D_n$, then $v \leq w$ as elements of D_n if and only if

- For each $1 \leq i \leq n$ we have $\{v_i, \ldots, v_n\} > \{w_i, \ldots, w_n\}$.
- For each $1 \leq k \leq n$, let $\mathbf{a} = a_1, \ldots, a_k$ and $\mathbf{b} = b_1, \ldots, b_k$ be the initial segments of $\{v_i, \ldots, v_n\}$ and $\{w_i, \ldots, w_n\}$, respectively as sorted lists. If the sets $\{|a_1|, \ldots, |a_k|\}$ and $\{|b_1|, \ldots, |b_k|\}$ are both equal to $\{1, 2, \ldots, k\}$, then the number of negative elements in $\mathbf{a}$ and $\mathbf{b}$ have the same parity (both even or both odd).

In order to extend Theorem 8.1.1, we define a more general version of pattern avoidance in terms of the following function which *flattens* any subsequence into a signed permutation.

8.3.15. Definition. Flattening function and pattern avoidance. For any sequence $a_1 a_2 \cdots a_k$ of distinct nonzero real numbers, define the sequence $\mathrm{fl}(a_1 a_2 \cdots a_k)$ to be the unique element $b = b_1 b_2 \cdots b_k$ in $\mathfrak{C}_k$ such that

- for all j, both a_j and b_j have the same sign;
- for all i, j, we have $|b_i| < |b_j|$ if and only if $|a_i| < |a_j|$.

For example, $\mathrm{fl}(\bar{6}, 3, \bar{7}, 1) = \bar{3}2\bar{4}1$. Any word containing the subsequence $\bar{6}, 3, \bar{7}, 1$ does not avoid the pattern $\bar{3}2\bar{4}1$. In particular, $w = 8\bar{6}23\bar{7}451 \in \mathfrak{C}_8$ (in signed notation) does not avoid $\bar{3}2\bar{4}1$.

Another way to describe pattern avoidance is with the signed permutation matrices. Namely, a signed permutation matrix w avoids the pattern v if no submatrix of w is the matrix v.

The following theorem extends Theorem 8.1.1 to all four classical groups allowing one to test both smoothness and rational smoothness by testing pattern avoidance.

8.3.16. Theorem. [**13**] *Fix the root system to be one of the types A_{n-1}, B_n, C_n or D_n. Let W_4 be the subgroup of $\mathfrak{C}_4$ corresponding to this type. If*

$X(w)$ *is a Schubert variety of the fixed type, then the following conditions are equivalent:*

1. *$X(w)$ is rationally smooth.*
2. *For each subsequence $i_1 < i_2 < i_3 < i_4$ such that $v = \mathrm{fl}(w_{i_1}w_{i_2}w_{i_3}w_{i_4})$ $\in W_4$, $X(v)$ is a rationally smooth Schubert variety.*
3. *w avoids certain patterns; see tables in Chapter 13 for the minimal list of bad patterns.*

Equivalently, in types B, C, D the second condition can be stated in the S_{2n} notation: For each subsequence $i_1 < i_2 < \cdots < i_8$ such that $v = \mathrm{fl}(w_{i_1}w_{i_2}\ldots w_{i_8}) \in S_8$ is also in W_4, then $X(v)$ is a rationally smooth Schubert variety.

The proof of Theorem 8.3.16 for types A, B, C follows directly from Lemmas 8.3.21 and 8.3.26 below. Type D is proved similarly but the extra patterns and conditions make the proof quite long, so we have chosen not to present it here.

8.3.17. Theorem. *Let w be any element of the Weyl group of type A, B, C or D.*

1. *In types A and D, $X(w)$ is smooth if and only if $X(w)$ is rationally smooth.*
2. *In type B, $X(w)$ is smooth if and only if $X(w)$ is rationally smooth and w avoids $\bar{2}\bar{1}$ (or 3412 in the notation of S_{2n}).*
3. *In type C, $X(w)$ is smooth if and only if $X(w)$ is rationally smooth and w avoids $1\bar{2}$ (or 4231 in the notation of S_{2n}).*

8.3.18. Remark. Theorem 8.3.17 does not substantially reduce the minimal number of patterns one needs to check in each case. For example, there are 26 patterns in the minimal list of bad patterns for type B rational smoothness. After eliminating all of the patterns containing $\bar{2}\bar{1}$ there are still 19 remaining patterns.

8.3.19. Remark. As mentioned in Chapter 6, Deodhar ([**34**]) was the first to prove that smoothness and rational smoothness are equivalent in type A. Peterson has announced a proof in types A, D, E, which has been recently proved by Carrell–Kuttler ([**34**]). We have included our proof in the cases A and D in the proof of Theorem 8.3.17, which appears at the end of this section.

We begin with an easy lemma on the Bruhat–Chevalley order.

8.3.20. Lemma. *Suppose two signed permutations v and w in $\mathfrak{C}_n$ agree everywhere except in positions $i_1 < \cdots < i_k$. Then $v \le w$ if and only if $\mathrm{fl}(v_{i_1}\ldots v_{i_k}) \le \mathrm{fl}(w_{i_1}\ldots w_{i_k})$.*

PROOF. Using the Deodhar–Proctor criteria for Bruhat–Chevalley order, we know that $v \le w$ if and only if the set $\{v_j, \ldots, v_n\}$ is greater than

$\{w_j, \ldots, w_n\}$ for each j under the partial order on sets. Since v, w agree everywhere except positions $i_1, i_2, \ldots, i_n$, we only need to show that given any two sets, we have $\{x, a_1, a_2, \ldots, a_k\} < \{x, b_1, \ldots, b_k\}$ if and only if $\{a_1, a_2, \ldots, a_k\} < \{b_1, \ldots, b_k\}$. Then the lemma follows from the fact that the flattening function maintains the relative order of its arguments.

Without loss of generality we can assume $a_1 \leq \cdots \leq a_k$ and $b_1 \leq \cdots \leq b_k$. Say $a_i < x \leq a_{i+1}$ and $b_j < x \leq b_{j+1}$, then the following are equivalent:

1. $\{a_1, a_2, \ldots, a_k\} < \{b_1, \ldots, b_k\}$.
2. $i \geq j$, $a_m < b_m$ for all $1 \leq m \leq j$ and all $i+1 \leq m \leq k$, and $a_{j+1} \leq \cdots \leq a_i < x \leq b_{j+1} \leq \cdots \leq b_i$.
3. $\{x, a_1, a_2, \ldots, a_k\} < \{x, b_1, \ldots, b_k\}$.

Now the equivalence of 1 $\iff$ 2 and 2 $\iff$ 3 is easy to check. $\square$

8.3.21. Lemma. *If $w \in \mathfrak{C}_n$ contains a pattern corresponding with a Schubert variety in $\mathfrak{C}_4$ which is not rationally smooth, then $X(w)$ is not rationally smooth. Hence, $X(w)$ is not smooth.*

PROOF. Let $d_w(v)$ be the *degree* of the vertex v in the Bruhat graph for w, i.e., the number of edges incident to v. We will show that the Bruhat graph is not regular by showing that there exists an explicit $v \leq w$ such that $d_w(v)$ is strictly greater than $d_w(w) = l(w)$.

For any $u \in \mathfrak{C}_n$ such that $u \leq w$, define $E_w(u)$ to be the set of all transpositions or signed transpositions $\widetilde{t}_{ij}$ such that $u\widetilde{t}_{ij} \leq w$. $E_w(u)$ is isomorphic to the set of edges emanating from u in the Bruhat graph. Furthermore, define

$$(8.3.22) \qquad d_w^6(u) = |\{\widetilde{t}_{ij} \in E_w(u) : |\{i, j, i_1, i_2, i_3, i_4\}| = 6\}|$$

$$(8.3.23) \qquad d_w^5(u) = |\{\widetilde{t}_{ij} \in E_w(u) : |\{i, j, i_1, i_2, i_3, i_4\}| = 5\}|$$

$$(8.3.24) \qquad d_w^4(u) = |\{\widetilde{t}_{ij} \in E_w(u) : |\{i, j, i_1, i_2, i_3, i_4\}| = 4\}|.$$

Then, the degree of u in the Bruhat graph breaks up into three summands as follows:

$$(8.3.25) \qquad\qquad d_w(u) = d_w^4(u) + d_w^5(u) + d_w^6(u).$$

Say w contains a bad pattern in positions $i_1 < i_2 < i_3 < i_4$. Let w' be the signed permutation $\mathrm{fl}(w_{i_1} w_{i_2} w_{i_3} w_{i_4})$ in $\mathfrak{C}_4$. By computer verification on $\mathfrak{C}_4$, there exists an element $v' \leq w'$ such that the number of edges incident to v' in the Bruhat graph is greater than the number of edges incident to w'. Now, define $v \in \mathfrak{C}_n$ to be the signed permutation which agrees with w everywhere, except in positions $i_1 i_2 i_3 i_4$ and for which $\mathrm{fl}(v_{i_1} v_{i_2} v_{i_3} v_{i_4})$ equals v'.

By Lemma 8.3.20 and the definition of v above, one sees that in order to determine if $v\widetilde{t}_{ij} \leq w$ we only need to compare the flattened signed permutations in positions i, j, i_1, i_2, i_3, i_4. For each such pair of flattened elements, $v'', w'' \in \mathfrak{C}_6$, a computer verification has shown that $d^4_{w''}(v'') > d^4_{w''}(w'')$, $d^5_{w''}(v'') \geq d^5_{w''}(w'')$, and $d^6_{w''}(v'') = d^6_{w''}(w'')$. From this one can see that $d^4_w(v) > d^4_w(w)$, $d^5_w(v) \geq d^5_w(w)$, and $d^6_w(v) = d^6_w(w)$ by examining the disjoint summands: for pairs $i < j$ such that $|\{i, j, i_1, i_2, i_3, i_4\}| = 6$ and distinct i such that $|\{i, j, i_1, i_2, i_3, i_4\}| = 5$. Hence, $d_w(v)$ is strictly greater than $d_w(w)$.

Therefore, by Theorem 6.2.4, $X(w)$ is not rationally smooth if w contains a bad pattern. $\square$

8.3.26. Lemma. *If $w \in \mathfrak{C}_n$ avoids all patterns in §13.3, then $P_w(t) = \sum_{v \leq w} t^{l(v)}$ factors completely into symmetric factors of the form $(1 + t + \ldots + t^k)$; hence $X(w)$ is rationally smooth.*

PROOF. In Theorem 11.2.5, it will be shown that under certain conditions

$$P_w(t) = (1 + t + \cdots + t^k)P_{w'}(t).$$

These conditions can be embodied in the following rules for types B and C. Let $w \in \mathfrak{C}_n$, and assume $w_d = \pm n$ and $w_n = \pm e$.

1. If $w_d = +n$ and $w_d > w_{d+1} > \cdots > w_n$, then p_w factors with $w' = wt_d \cdots t_{(n-1)}$ and $\mu = n - d$.
2. If w contains a consecutive sequence ending in $w_n = e$ for $e > 0$, then p_w factors with $w' = t_{n-1} \cdots t_{e+1} t_e w$ and $\mu = n - e$.
3. If each w_i is negative and $w_1 > w_2 > \ldots \widehat{w_d} \cdots > w_n$ (decreasing after removing w_d), then p_w factors with $w' = wt_{d-1} \cdots t_1 t_0 t_1 \cdots t_{n-1}$ and $\mu = d + n - 1$.
4. If each w_i is negative and $w_1 > w_2 > \cdots > w_{n-1}$, then p_w factors with $w' = t_{n-1} \cdots t_1 t_0 t_1 \cdots t_{e-1} w$ and $\mu = e + n - 1$.
5. If each w_i is positive except for $w_d = \bar{n}$ and $w_1 > w_2 > \cdots > w_d$, then p_w factors with $w' = wt_{d-1} \cdots t_1 t_0$ and $\mu = d$.

First, if w avoids all bad patterns, then $P_w(t)$ factors according to at least one of the rules above. This follows from a careful analysis of cases:

1. If $w_d = \bar{n}$, then $w_1 > w_2 > \cdots > w_d$. Everything else is forbidden by the patterns

$$\begin{array}{ccc} \bar{1}2\bar{3} & 1\bar{2}\bar{3} & 12\bar{3} \\ \bar{2}\bar{1}3 & 2\bar{1}\bar{3} & 2\bar{1}\bar{3} \end{array}.$$

2. If $w_d = \bar{n}$, either all w_i are negative or all positive if $i < d$. From the above list of forbidden sequences one can see the only allowable patterns of length 3 ending in $\bar{3}$ are $21\bar{3}$ and $\bar{1}2\bar{3}$.
3. If $w_d = \bar{n}$ and all w_i are positive for $i < d$, then $w_i > 0$ for all $i > d$ and w avoids $1\bar{3}2$ and $2\bar{3}\bar{1}$. By Rule 2 above, we also have $w_1 > w_2 > \cdots > w_d$. Hence, $P_w(t)$ factors using Rule 5.

4. If $w_d = \bar{n}$ and all w_i are negative, then $w_{d+1} > w_{d+2} > \cdots > w_n$ if w avoids $\overline{3}\overline{2}\overline{1}$. By Rule 1, we also have $w_1 > w_2 > \cdots > w_d$, hence, $P_w(t)$ factors using Rule 3 or Rule 4.

5. If $w_d = n$ and $w_{d+1} > w_{d+2} > \cdots > w_n$, then $P_w(t)$ factors using Rule 1.

6. If $w_d = n$, w is not decreasing after position d and w avoids all bad patterns, then the patterns containing $n, n-1, \ldots, w_n$ must all be one of the following forms:

(8.3.27)
$$\tilde{1}4\tilde{2}3 \quad \tilde{2}4\tilde{1}3 \quad 4\tilde{1}\tilde{2}3$$
$$4\tilde{1}32 \quad 42\tilde{1}3 \quad 43\tilde{1}2$$

where $\tilde{i}$ is either i or $\bar{i}$. Therefore, w must contain a consecutive sequence ending in w_n and w_n must be positive. Hence, $P_w(t)$ factors using Rule 2.

7. If $w_d = \bar{n}$, $w_1, \ldots, \widehat{w_d}, \ldots w_n$ are not all positive or all negative, and w avoids all bad patterns, then the patterns containing $n, n-1, \ldots w_n$ must all be one of the forms given in (8.3.27) or one of the forms below:

(8.3.28)
$$\bar{1}\bar{2}43 \quad \bar{1}\bar{3}42 \quad \bar{1}4\tilde{2}3 \quad \bar{1}432$$
$$\bar{2}\bar{3}41 \quad \bar{2}4\tilde{1}3 \quad \bar{3}4\tilde{1}2 \quad \bar{3}421$$
$$4\bar{1}23 \quad 4\bar{1}\bar{2}3 \quad 4\bar{1}32 \quad 4\bar{2}13$$
$$\bar{4}312 \quad \bar{4}321$$

In particular, w must contain a consecutive sequence ending in w_n and w_n must be positive. Hence, $P_w(t)$ factors using Rule 2.

Second, we claim that if w avoids all bad patterns and $P_w(t) = (1 + \cdots + t^k)P_{w'}$, then so does w'. This fact follows from the construction of w' in each case. In Rules 1, 2, 3 and 4, all flattened patterns in w' appear as patterns in w. In Rule 5, new patterns are created in w' all starting with n. Examining the list of patterns in §13.3 one sees that 4231 is the only bad pattern that could have been created in w' since $w' \in S_n$. If the pattern 4231 appears in w', then one of the following patterns must have been in w: $\bar{4}231$, $2\bar{4}31$, $23\bar{4}1$, or $231\bar{4}$. However, each of these four patterns are bad patterns themselves or they contain the bad pattern $12\bar{3}$, contradicting the assumption that w avoids all bad patterns.   □

PROOF OF THEOREM 8.3.16. By Theorem 6.1.19, rational smoothness only depends the Weyl group. Recall that the Weyl groups of types B and C are the same. Furthermore, the Weyl group of type C contains the Weyl group of type A as the subgroup with no signed entries. Therefore, to prove the theorem in types A, B and C, we only need to prove it for type C. This follows easily from Lemmas 8.3.21 and 8.3.26. In type D, the analog of Lemma 8.3.21 just requires an additional computer verification.

The analog of Lemma 8.3.26, on the other hand, requires even more cases and therefore has been omitted (or equivalently, left to the reader).   □

PROOF OF THEOREM 8.3.17. Throughout this proof we will use the fact that smoothness implies rational smoothness. In each case, we use the explicit description of the tangent space from Theorem 5.3.1 and the fact that if the dimension of the tangent space at e_{id} is equal to the dimension of the Schubert variety, then we have a smooth Schubert variety.

Case 1: Type A.   Recall from Theorem 5.3.1 that a positive root $\beta \in N(w, \mathrm{id})$ if and only if $s_\beta \leq w$. If $X(w)$ is rationally smooth then the Bruhat graph on $\{v \leq w\}$ is regular. Hence $\#\{\beta \in N(w, \mathrm{id}) : s_\beta \leq w\} = l(w) = \dim X(w)$, so $X(w)$ is smooth.

Case 2: Type B.   In the following cases, we use the results of Theorem 8.3.16 and 6.2.4 that rational smoothness in types B and C is equivalent to avoiding the patterns in the tables of §13.3 which is also equivalent to $\#S(v, w) := \#\{\beta \in R^+ : v s_\beta \leq w\} = l(w)$ for all $v \leq w$. In particular, whenever $X(w)$ is rationally smooth, $l(w) = \#\{\beta \in R^+ : s_\beta \leq w\}$ and w must avoid the specific patterns

$$\bar{2}\bar{1}3 \quad 3\bar{2}1 \quad 3\bar{2}\bar{1}$$
$$\bar{3}\bar{2}1 \quad \bar{3}\bar{2}\bar{1}.$$

Recall from Theorem 5.3.1 that a positive root β is in $N(w, \mathrm{id})$ if and only if at least one of the following is true:

1. $s_\beta \leq w$.
2. $\beta = e_i$ for $i > 1$ and $s_{e_i + e_1} \leq w$.
3. $\beta = e_i + e_j$ for $j > i > 1$ and $s_{e_j} s_{e_i + e_1} \leq w$.

Assume that $X(w)$ is rationally smooth but not smooth; then $\#N(w, \mathrm{id}) > l(w) = \#\{\beta \in R^+ : s_\beta \leq w\}$. Therefore, there exists a $\beta \in R^+$ such that 1 does not hold and either (2) or (3) does hold. Using Proctor's test for the Bruhat–Chevalley order, one can show that any time $s_{e_i + e_j} \not\leq w$ and $s_{e_j} s_{e_i + e_1} \leq w$ then w contains the pattern $\bar{2}\bar{1}3$, hence $X(w)$ is not rationally smooth. Since we have assumed $X(w)$ is rationally smooth, it cannot be the case that β satisfies the conditions of (3) but not (1). Hence, there exists $\beta = e_i$ for some $i > 1$ such that $s_{e_i} \not\leq w$ but $s_{e_i + e_1} \leq w$. Again by Proctor's rules, $s_{e_i + e_1} \leq w$ implies that there exists a $1 \leq k \leq n$ such that $w_k \leq -i$ and a $i \leq j \leq n$ such that $k \neq j$ and $w_j < 0$. Since $w \not\geq s_{e_j}$, one must have $-1 \geq w_j > -i \geq w_k$ and $k < j$, hence w contains $\mathrm{fl}(w_k w_j) = \bar{2}\bar{1}$.

Conversely, assume w contains the pattern $\bar{2}\bar{1}$ and $X(w)$ is rationally smooth. Then $l(w) = \#\{\beta \in R^+ : s_\beta \leq w\}$. We want to show there exists an $i > 1$ such that $s_{e_i} \not\leq w$ and $s_{e_i + e_1} \leq w$. Assume $\bar{2}\bar{1}$ occurs in positions k, l with $1 \leq k < l \leq n$ and $w_k < w_l < 0$. Then no entry to the left of k contains a number with larger absolute value than w_k since w must avoid $\pm 3\bar{2}1$; hence $|w_k| \geq k$ and in fact $|w_k| \geq k + 1$ since w_l lies to the right of position k. Hence, by Proctor's rules $w \geq s_{e_{w_k}+e_1}$. If k, l are chosen so that there are no negative entries between w_k and w_l, then since w avoids the pattern $\bar{2}1\bar{3}$ we can assume $w_j > w_k$ for all $j > k$. Therefore, $s_{e_{w_k}} \not\leq w$.

Case 3: Type C. Recall from Theorem 5.3.1 that a positive root β is in $N(w, \mathrm{id})$ if and only if at least one of the following is true:

1. $s_\beta \leq w$.
2. $\beta = e_i + e_j$ (for $i < j$) and $s_{2e_j} \leq w$.

Assume that $X(w)$ is rationally smooth but not smooth; then $\#N(w, \mathrm{id}) > l(w) = \#\{\beta \in R^+ : s_\beta \leq w\}$. Therefore, there exists a $\beta = e_i + e_j \in R^+$, $s_{e_i+e_j} \not\leq w$, and $s_{2e_j} \leq w$. By Proctor's rules, this implies there exists a $k \geq j > 1$ such that $w_k \leq -j$ and no other entry $l \geq i$ exists such that $w_l \leq -i$.

In order to show w contains the pattern $1\bar{2}$, we need to show that there exists an $m < k$ such that w_m is positive and $w_m < |w_k|$. Since w must avoid all patterns of the form $\pm 3\bar{2} \pm 1$, there either exists some $l > k$ such that $|w_l| < |w_k|$ and no $m < k$ such that $|w_m| > |w_k|$, or vice-versa. Either way, there must be at least $j - 1 \geq 1$ entries to the left of k, which are all smaller in absolute value than $|w_k|$. By examining the table of patterns, one can verify that the only allowable length 3 patterns ending in $\bar{3}$ are $21\bar{3}$ and $\bar{1}2\bar{3}$. By the pigeon-hole principle, if all entries less than $|w_k|$ and to the left of k, are negative, then there must exist at least one $i \leq p < k$ such that $w_p < -i$, contradicting the assumption that $s_{e_i+e_j} \not\leq w$. Therefore, all smaller entries to the left of k must be positive, and since this set is necessarily nonempty, w contains the pattern $1\bar{2}$.

Conversely, assume w contains $1\bar{2}$ and $X(w)$ is rationally smooth. Computer verification shows that the only allowable length 4 patterns in w contain at most one negative, hence by the Proctor criteria $s_{e_i+e_j} \not\leq w$ for any $i < j$. However, $s_{e_2} \leq w$ since w contains $1\bar{2}$ and therefore $e_1 + e_2 \in N(w, \mathrm{id})$ by the second condition.

Case 4: Type D. Note that some of the patterns in Table 13.3.9 are not in the group D_n because they have an odd number of negative signs.

Theorem 5.3.1 shows that $T(w, e_{\mathrm{id}})$ is spanned by the set of all $\beta \in R^+$ such that either

1. $s_\beta \leq w$.
2. $\beta = e_i + e_j$ and $s_{e_i-e_1} s_{e_i+e_1} s_{e_j+e_2} \leq w$.

If w is rationally smooth, then by Lemma 8.3.29 (below) the second condition never contributes anything to the dimension of $T(w)$. Therefore, we have $\dim(T(w)) = \#\{\beta \in R^+ : s_\beta \leq w\} = l(w)$ by the Carrell–Peterson theorem on the regularity of the Bruhat graph (cf. Theorem 6.2.4) which implies $X(w)$ is smooth. $\qquad\square$

8.3.29. Lemma. *Let $W = D_n$ and assume $w \in W$ is rationally smooth. Let $v_{ij} = s_{e_i - e_1} s_{e_i + e_1} s_{e_j + e_2}$. Then $v_{ij} \leq w$ implies $s_{e_i + e_j} \leq w$.*

PROOF. In one-line notation $v_{ij} = \bar{1}\bar{j}\dots\bar{i}\dots\bar{2}$, e.g., in D_7 we have $v_{57} = \bar{1}73\bar{4}\bar{5}6\bar{2}$. If $v_{ij} \leq w$, then by Proctor's test for the Bruhat–Chevalley order in Type D, w has at least four negative entries, say $w_{i_1} w_{i_2} w_{i_3} w_{i_4} = \bar{r}\bar{s}\bar{t}\bar{u}$, such that $i_1 < i_2 < i_3 < i_4$, $i_3 \geq i$, $i_4 \geq j$, and

$$(8.3.30) \qquad \begin{aligned} \{u\} &\geq \{2\} \\ \{ut\} &\geq \{2i\} \\ \{stu\} &\geq \{2ij\} \\ \{rstu\} &\geq \{12ij\}. \end{aligned}$$

In order to show $s_{e_i + e_j} \leq w$, we need to show

1. $u \geq i$.
2. $\{tu\} \geq \{ij\}$.

Since w is rationally smooth, the only three allowable patterns of four negatives are $\bar{1}\bar{2}\bar{3}\bar{4}$, $\bar{4}\bar{1}\bar{2}\bar{3}$, $\bar{2}\bar{3}\bar{4}\bar{1}$. Furthermore, if $\mathrm{fl}(\bar{r}\bar{s}\bar{t}\bar{u}) = \bar{2}\bar{3}\bar{4}\bar{1}$, then $u = 1$ since the following patterns are not allowed:

$$\begin{array}{ccc} 1\bar{3}\bar{2} & \bar{3}\bar{4}1\bar{2} & \bar{3}\bar{4}\bar{2}1 \\ \bar{1}\bar{3}\bar{2} & \bar{3}\bar{4}\bar{1}\bar{2} & \bar{3}\bar{4}\bar{2}\bar{1}. \end{array}$$

If $u = 1$, that contradicts the first line of (8.3.30). Therefore, one can assume $\mathrm{fl}(\bar{r}\bar{s}\bar{t}\bar{u})$ is one of the other two patterns in which case $s < t < u$. Now, from (8.3.30), one notes that $\{stu\} \geq \{2ij\}$ which implies $\{tu\} \geq ij$, i.e., $t \geq i$ and $u \geq j$. $\qquad\square$

8.4. Type C results of Lakshmibai–Song using permutations

As in Chapter 3, §3.3, let $V = K^{2n}$, together with a nondegenerate skew-symmetric bilinear form. Let $H = SL(V)$, $G = Sp(V)$. We shall follow the notation of Chapter 3, §3.3. For $w \in W_H$ resp. W_G, let e_w, resp. f_w, denote the point wB, resp. wB^σ, in H/B, resp. G/B^σ; let $X(w) = \overline{BwB}(\mathrm{mod}\ B)$, resp. $Y(w) = \overline{B^\sigma w B^\sigma}(\mathrm{mod}\ B^\sigma)$, be the associated Schubert variety in H/B, resp. G/B^σ. For $w \in W_H$, resp. W_G, let $l_H(w)$, resp. $l_G(w)$, denote the length of w in W_H (resp. W_G), and let $T(w)$, resp. $S(w)$, denote the Zariski tangent space to $X(w)$, resp. $Y(w)$, at e_{id}, resp. f_{id}. For a root $\alpha \in R_H$, resp. R_G, we shall denote the associated reflection by r_α, resp. s_α. Let

$$N_H(w) = \{\alpha \in R_H^+ \mid w \geq r_\alpha\}.$$

Then we have [**109**] that $T(w)$ is spanned by $\{X_{-\alpha}, \alpha \in N_H(w)\}$ where $\alpha \in R_H$ and X_α denotes the element of the Chevalley basis of $\mathrm{Lie}(H)$, associated to α. In particular we have that $X(w)$ is smooth if and only if $\#N_H(w) = l_H(w)$.

Let $w = (a_1 \cdots a_{2n}) \in W_G$. Let

(8.4.1) $$m_w = \#\{i, \ 1 \leq i \leq n \mid a_i > n\},$$

(8.4.2) $$c_w = \#\{i, \ 1 \leq i \leq n \mid w \geq s_{2\varepsilon_i}\},$$

(8.4.3) $$\delta_w = \dim T(w) - \ell_H(w).$$

8.4.4. Theorem. [94] *With notation as above, σ induces an involution (which is also denoted by just σ) on $T(w)$, and $S(w)$ is simply the orbit space of $T(w)$ modulo the action of σ.*

Let
$$N_G(w) = \{\alpha \in R_G^+ \mid X_{-\alpha} \in S(w)\}.$$
Here X_α denotes the element in the Chevalley basis of $\mathrm{Lie}(G)$, associated to α.

Then using the explicit description of $S(w)$ as given in Theorem 5.4.5, we have the following theorem.

8.4.5. Theorem. [111] *Let $n_G(w) = \#N_G(w)$, $n_H(w) = \#N_H(w)$. Then*

(8.4.6) $$n_G(w) = \frac{1}{2}(n_H(w) + c_w).$$

Note the resemblance of the formula (8.4.6) to the length formula $l_G(w) = \frac{1}{2}(l_H(w) + m_w)$ (refer to [94] for the length formula).

8.4.7. Theorem. [111] *Let $w \in W_G$. Then $Y(w)$ is singular if and only if $\delta_w > m_w - c_w$.*

8.4.8. Remark. It is shown in [111] that the condition $\delta_w > m_w - c_w$ is equivalent to the following:

$(**)$ $\begin{cases} \text{either} & (1) \ m(w) < c(w), \\ \text{or} & (2) \ m(w) = c(w), \ \text{and property } (*) \text{ of} \\ & \qquad\qquad \text{Theorem 8.1.1 holds for } w, \\ \text{or} & (3) \ m(w) > c(w), \ \text{and } \delta(w) > m(w) - c(w). \end{cases}$

Let $y \leq w$ and let $n(w, y) = \dim T(w, y)$. Let $r(w, y) = \#\{\beta \in y(R^+) \mid w \geq s_\beta y\}$. Then we have

$$n(w, y) \geq r(w, y) \geq l(w).$$

The first inequality follows from Proposition 6.2.12, and the second inequality follows from Deodhar's inequality (cf. Corollary 7.1.24, with $x = \mathrm{id}$). Denote $n(w, \mathrm{id}), r(w, \mathrm{id}), l(w)$ by A_w, B_w, C_w, respectively. In [148], Song has given a criterion for the equality $A_w = B_w$, as well as for the equality $B_w = C_w$.

8.4.9. Theorem. [148] *Let $W = W(Sp(2n))$, and $w \in W$. Let $m = \min\{i, 1 \leq i \leq n \mid w \geq s_{2e_i}\}$. Then $A_w = B_w$ if and only if $w \geq s_{e_m + e_{m+1}}$.*

Let $W = W(Sp(2n))$, and $w \in W$, say $w = (a_1 \cdots a_n)$. Let $a_i = 1$, this implies that $a_{2n+1-i} = 2n$. Let u be the element of $W(Sp(2n-2))$ obtained from w by dropping out $1, 2n$, and replacing a_t by $a_t - 1$, $t \neq i, 2n+1-i$.

8.4.10. Remark. [148] One can give an iterated condition for the equality $B_w = C_w$ using the notation above: $B_w = C_w$ if and only if $B_w - C_w = B_u - C_u$, and $B_u = C_u$.

CHAPTER 9

Minuscule and cominuscule G/P

In this chapter we present the results for the minuscule and cominuscule G/P. We first present the results of Zelevinsky ([**157**]) and Sankaran–Vanchinathan ([**143, 144**]) on small resolutions for Schubert varieties in the minuscule and cominuscule cases. We then present the results of Brion and Polo (cf. [**30**]) on the tangent space to Schubert varieties in the minuscule and cominuscule cases. The results of Lakshmibai–Weyman (cf. [**112**]) on the irreducible components of the singular loci of Schubert varieties as well as recursive formulae for the multiplicity and the Hilbert polynomial at a singular point are then presented. We have also included two closed formulas due to Kreiman–Lakshmibai ([**89**]), Rosenthal–Zelevinsky [**142**] for the multiplicity at a singular point for Schubert varieties in the Grassmannian. We have also included a closed formula for the Hilbert polynomial at a singular point due to Kreiman–Lakshmibai ([**89**]).

Let G, B, T, W, R, S, R^+, etc, be as in Chapter 2.

9.0.11. Definition. A maximal parabolic subgroup $P \supset B$ with ω as the associated fundamental weight is of *minuscule type*, if ω satisfies $(\omega, \beta^*) \leq 1$, for all $\beta \in R^+$. The fundamental weight ω is said to be *minuscule*.

Recall (cf. Chapter 2) the following list of minuscule fundamental weights:

- Type $\mathbf{A}_n$: All fundamental weights are minuscule.
- Type $\mathbf{B}_n$: ω_n.
- Type $\mathbf{C}_n$ $(n \geq 3)$: ω_1.
- Type $\mathbf{D}_n$ $(n \geq 4)$: $\omega_1, \omega_{n-1}, \omega_n$.
- Type $\mathbf{E}_6$: ω_1, ω_6.
- Type $\mathbf{E}_7$: ω_7.
- There are no minuscule fundamental weights in types E_6, F_4 and G_2.

Geometric Interpretation. Let ω be a fundamental weight with P as the associated maximal parabolic subgroup. It is easily checked that ω is minu-scule if and only if for any pair (w, τ) in W^P, where $w \geq \tau$, $l(w) = l(\tau) + 1$, i.e., $X_P(\tau)$ is a Schubert divisor in $X_P(w)$, the Chevalley multiplicity (cf. Chapter 4, §4.8.1), $m_\omega(w, \tau) = 1$. Let p_w be the extremal weight vector in $H^0(G/Q, L_\omega)$ of weight $-w(\omega)$, and let H_w be the zero set in G/P of p_w. As seen in Chapter 4, §4.8 we have that ω is minuscule if and only if for any pair (w, τ) in W^P such that $X_P(\tau)$ is a Schubert divisor in $X_P(w)$, $X_P(\tau)$ occurs with multiplicity 1 in the intersection $X(w) \cap H_w$, i.e., $i(X_P(w), H_w; X_P(\tau)) = 1$.

Representation-theoretic Interpretation. Let ω, P be as above. For each $w \in W^P$, fix a generator u_w for the one-dimensional weight space of weight $w(\omega)$ in the Weyl module $V_K(\omega)$; note that u_w is simply an extremal weight vector in $V_K(\omega)$ of weight $w(\omega)$. It is easily seen that ω is minuscule if and only if the set of extremal weight vectors $\{u_w, w \in W^P\}$ is a basis for $V_K(\omega)$, equivalently, W permutes the set of weights in (the T-module) $V_K(\omega)$.

Recall the well-known fact (see [23], for example) that there exists a unique root $\beta \in R$ such that for any other root $\alpha \in R, \beta > \alpha$, i.e., $\beta - \alpha$ is a positive integral combination of positive roots; note in particular that $\beta > 0$, i.e, β belongs to R^+. The root β is called the *highest root*.

9.0.14. Definition. A maximal parabolic subgroup P with α as the associated simple root is of *cominuscule type* if α occurs with coefficient 1 in the highest root β. The associated fundamental weight ω is said to be *cominuscule*.

Following is the list of cominuscule fundamental weights:

- In Types $\mathbf{A}_n$, $\mathbf{D}_n$, $\mathbf{E}_6$, and $\mathbf{E}_7$ the list of minuscule fundamental weights (as described above) is also the list of cominuscule fundamental weights.
- Type $\mathbf{B}_n$: ω_1.
- Type $\mathbf{C}_n$ $(n \geq 3)$: ω_n.
- There are no cominuscule fundamental weights in types E_6, F_4 and G_2.

Thus ω_1 in Type $\mathbf{B}_n$ and ω_n in Type $\mathbf{C}_n$ $(n \geq 3)$ are the only cominuscule fundamental weights are not minuscule.

For $G = Sp(2n)$, resp. $SO(2n), P = P_n$, G/P is called the *symplectic Grassmannian*, resp. the *orthogonal Grassmannian*; it consists of the maximal totally isotropic subspaces of K^{2n} for the skew symmetric form, resp. the symmetric form, defining $Sp(2n)$, resp.$SO(2n)$.

The importance of the study of the class of the minuscule and cominuscule G/P's stems from the following considerations.

The affine space $\mathbb{A}^n$ and the projective space $\mathbb{P}^n$ are the best known (and the simplest) of all algebraic varieties. As observed in Chapter 3, $\mathbb{P}^n$ may be identified with the Grassmannian $G_{1,n+1}$. Thus, next to $\mathbb{P}^n$, the Grassmannians are the best understood examples of projective varieties, and as seen in Chapter 3, the Grassmannian $G_{d,n}$ gets identified with $SL(n)/P_d$ for a suitable maximal parabolic subgroup of $SL(n)$. Thus the Grassmannian varieties are precisely the minuscule G/P's, G being the special linear group. As we saw in Chapter 3, the Schubert varieties in $G_{d,n}$ are well-understood. The following gives a list of some of the nice properties of Schubert varieties in the Grassmannian $G_{d,n}$.

1. Any Schubert divisor is a moving divisor (cf. Definition 5.2.4).
2. For a pair $w, w' \in W^{P_d}$ such that $X(w')$ is a Schubert divisor in $X(w)$, the Chevalley multiplicity $m_{\omega_d}(w, w')$ equals 1. Further, we have the *Pieri formula*

$$X(w) \cap H_w = \cup X(w') \text{ (scheme-theoretically)},$$

   where H_w as above is the zero set in $G_{d,n}$ of p_w (the Plücker coordinate associated to w), and the union on the right hand side is taken over **all** Schubert divisors in $X(w)$.
3. The Bruhat graph of $W^{P_d}(= I_{d,n})$ is a *distributive lattice*.

Recall that a *lattice* is a partially ordered set $(\mathcal{L}, \leq)$ such that, for every pair of elements $x, y \in \mathcal{L}$, there exist elements $x \vee y$ and $x \wedge y$, called the *join*, respectively the *meet* of x and y, defined by:

$$x \vee y \geq x, \ x \vee y \geq y, \text{ and if } z \geq x \text{ and } z \geq y, \text{ then } z \geq x \vee y,$$
$$x \wedge y \leq x, \ x \wedge y \leq y, \text{ and if } z \leq x \text{ and } z \leq y, \text{ then } z \leq x \wedge y.$$

A lattice is called *distributive* if the following identities hold:

$$x \wedge (y \vee z) = (x \wedge y) \vee (x \wedge z)$$
$$x \vee (y \wedge z) = (x \vee y) \wedge (x \vee z).$$

Sketch of proofs of (1), (2), (3). The result (1) follows from the fact that if $X(w')$ is a Schubert divisor in $X(w)$, say $w = \{i_1, \cdots, i_d\}, w' = \{j_1, \cdots, j_d\}$, then there exists a unique t, $1 \leq t \leq d$ such that $i_r = j_r, r \neq t$, $i_t = j_t + 1$; hence $w = s_\alpha w'$, where $\alpha = \epsilon_{j_t} - \epsilon_{j_t+1}$.

The result (2) is nothing but the geometric interpretation above of minuscule fundamental weights.

PROOF OF (3). Let $\tau = (i_1, \ldots, i_d)$, $\phi = (j_1, \ldots, j_d)$ be two noncomparable elements in $I_{d,n}$. Set

$$\lambda = (k_1, \ldots, k_d), \ \mu = (l_1, \ldots, l_d)$$

where $k_t = \max\{i_t, j_t\}$, $l_t = \min\{i_t, j_t\}$, $1 \leq t \leq d$. Note that $(k_{t-1}, k_t) = (i_{t-1}, i_t)$ or (i_{t-1}, j_t) or (j_{t-1}, i_t) or (j_{t-1}, j_t), from which it follows that $k_{t-1} < k_t$. Similarly, $l_{t-1} < l_t$. Thus λ, μ are well-defined elements of $I_{d,n}$. Further from the definition of λ, μ, it is easily seen that λ, resp. μ, is the unique minimal, resp. maximal element of $I_{d,n}$ which is greater, resp. less, than both τ and ϕ.

It is quite a striking fact that all of the above three properties hold more generally for Schubert varieties in a minuscule G/P; (2) is nothing but the geometric interpretation above of minuscule fundamental weights. For a proof of (1), (3), see [**112**, **65**], respectively; it is shown in [**65**] that

G/P is minuscule if and only if the Bruhat graph of W^P is a distributive lattice. In fact, many more geometric results such as the description of the tangent space (at a T-fixed point), the description of singular points, the expression for the multiplicity at a singular point, etc., are exactly the same for the case of a Grassmannian and that of more general minuscule G/P's. Thus, next to $\mathbb{P}^n$, the minuscule G/P's (generalizing the Grassmannians) may be considered as the simplest examples of projective varieties. Among the nonminuscule G/P's, the best understood are the cominuscule G/P's; even though the geometry and the combinatorics in the cominuscule case are more complicated than the minuscule case, nevertheless, the results on tangent space, singular locus, etc., for the cominuscule case resemble closely the corresponding results for the minuscule case. Thus we present results for the minuscule and cominuscule G/P's on small resolutions for Schubert varieties, description of tangent spaces to Schubert varieties, description of the irreducible components of the singular loci of Schubert varieties, description of the multiplicity at singular points on Schubert varieties, expressions for the Kazhdan–Lusztig polynomials, the Hilbert polynomials at singular points on Schubert varieties due to the several authors mentioned in the beginning of this chapter.

9.1. Results on small resolutions

The results of Lascoux–Schützenberger (cf. §6.3.28) on the Kazhdan–Lusztig polynomials associated with Grassmannians opened the way for an understanding of the Kazhdan–Lusztig polynomial in the minuscule case, which took two directions. On the one hand, the results of Lascoux–Schützenberger were extended combinatorially by Boe in [17], who gave a description of all Kazhdan–Lusztig polynomials in all minuscule, including the exceptional groups. His results are the most complete on the subject. The other direction, developed by Zelevinsky in [157] for Grassmannians and extended to the symplectic and orthogonal Grassmannians by Sankaran and Vanchinathan [143, 144], consisted in understanding geometrically the results of Lascoux-Schützenberger in terms of small resolutions.

In this section we introduce small resolutions and explain how they relate to Kazhdan–Lusztig theory. In short, given a Schubert variety $X(w) \subset G/B$, if we can find a small resolution $p : \tilde{X} \to X(w)$, then for $\tau \le w$, the Kazhdan–Lusztig polynomial $P_{\tau,w}(q)$ is equal to the Poincaré polynomial $P_t(p^{-1}(e_\tau))$. Zelevinsky (cf. [157]) constructed small resolutions for Schubert varieties in the Grassmannian. Toward generalizing Zelevinsky's results, Sankaran and Vanchinathan (cf. [143, 144]) describe a class of Schubert varieties in the symplectic and orthogonal Grassmannians for which they construct small resolutions. They also describe a class of Schubert varieties in the symplectic Grassmannian not admitting any small resolutions! We recall these results below.

9.1.1. Definition. Small resolutions. Let X be an irreducible complex variety. A proper, birational morphism $p : \tilde{X} \to X$, with $\tilde{X}$ smooth is called a *resolution of X*. Such a resolution is said to be *small* if

$$\text{codim}_X\{x \in X \mid \dim p^{-1}(x) \geq i\} > 2i, \ \forall i > 0.$$

9.1.2. Theorem. [59] *Let* $p : \tilde{X} \to X$ *be a small resolution. Then for any* $i \geq 0$*, for the intersection cohomology sheaf* $\mathcal{H}(X)$ *with respect to the middle perversity, the stalk* $\mathcal{H}^i(X)_x$ *is isomorphic to the singular cohomology group* $H^i(p^{-1}(x), \mathbb{C})$.

Combining Theorem 9.1.2 with Theorem 6.1.11, we obtain

9.1.3. Theorem. *Let* $p : \tilde{X} \to X(w)$ *be a small resolution of a Schubert variety* $X(w)$ *in* G/B*, G being semisimple. Let* $X(\tau) \subset X(w)$*. Then the Poincaré polynomial* $P_t(p^{-1}(e_\tau))$ *of the fiber* $p^{-1}(e_\tau)$ *is equal to the Kazhdan–Lusztig polynomial* $P_{\tau,w}(q)$*, with* $q = t^2$.

9.1.4. Resolutions of singularities ([22], [37], [61]). We describe a particular resolution of singularities for Schubert varieties. Note this is somewhat different than the more standard Bott–Samelson resolution.

Let Q be a parabolic subgroup. For a simple root α, we shall denote the associated maximal parabolic by P_α. If $\alpha = \alpha_i$, we shall denote P_α by just P_i. Let $X(w) \subset G/Q$ be any Schubert variety, and P_w the largest subgroup of G which leaves $X(w)$ stable for the action of G on G/Q by left multiplication. Clearly, P_w is a parabolic subgroup containing B. We shall refer to P_w as the *stabilizer* of $X(w)$. Note that it is possible to find a parabolic subgroup $P \subset P_w$ and a Schubert variety $X(w')$, $w' < w$, such that $P_w X(w') = PX(w') = X(w)$; for example, suppose P_w is a rank one parabolic with α as the associated simple root, then we may take w' to be $s_\alpha w$. Let $R_0 = P_w \cap P_{w'}$ and $P_w \times^{R_0} X(w') = P_w \times X(w')/ \sim$, where $\sim$ is the equivalence relation $(gr, rx) \sim (g, x)$, for $g \in P_w$, $r \in R_0, x \in X(w')$. Then the map $\pi_0 : P_w \times^{R_0} X(w') \to X(w)$, $(g, x) \mapsto gx$ is surjective and P_w-equivariant (for the P_w-action $p \cdot (g, x) = (pg, x)$, $p \in P_w$), but not birational in general. However it is possible to choose P and $w' < w$ such that if $R_1 = P \cap P_{w'}$, then $\dim P/R_1$ equals the codimension of $X(w')$ in $X(w)$, and so $\pi_1 : P \times^{R_1} X(w') \to X(w)$ is P_w-equivariant and birational; for example, if α is a simple root such that the rank one parabolic $P_\alpha \subseteq P_w$, then we can choose $w' = s_\alpha w$ and $P = P_\alpha$. Since any one-dimensional Schubert variety is smooth, iterating this construction we obtain a P_w-equivariant resolution

$$p : P_{(1)} \times^{R_1} P_{(2)} \times \cdots \times^{R_{r-2}} P_{(r-1)} \times^{R_{r-1}} X(w^r) = \tilde{X}(w) \to X(w),$$

where $w^1 = w$, $w^2 = w'$, $\ldots$, $P_{(r)} = P_{w^r}$, $P_{(i)} \subset P_{w^i}$, $1 \leq i \leq r$, $R_i = P_{(i)} \cap P_{(i+1)}$, $1 \leq i \leq r-1$, and $X(w^r)$ is smooth. In fact, at each step, $P_{(i)}$ may be chosen to be the stabilizer of $X(w^i)$.

The above resolution is usually referred to as a *Bott–Samelson resolution* even though it is not quite the same as the usual Bott–Samelson resolution. When $X(w) \subset G/B$, such resolutions were obtained by H. Hansen (cf. [**61**]) and M. Demazure (cf. [**37**]).

9.1.5. An example of a Bott–Samelson resolution.
Take $G = SL(4)$, $w = (2341)$. We have $w = s_1 s_2 s_3$, $P_w = P_1$. We may take $w^2 = s_2 s_3$, $P_{(2)} = P_2$, $w^3 = s_3$, $P_{(3)} = P_3$, $w^4 = \mathrm{id}$.

9.1.6. Small resolution on Grassmannians in Type A.
Zelevinsky has given determined an explicit small resolution on Grassmannians in Type A. Section 9.1.17 explains how this resolution can be used to compute Kazhdan–Lusztig polynomials.

9.1.7. Theorem.
[**157**] *Let $G = SL(n)$, and let P be a maximal parabolic subgroup so that G/P is a Grassmannian variety. Let X be a Schubert variety in G/P. There exist Schubert varieties*

$$X = X(w^1) \supset X(w^2) \supset \cdots \supset X(w^r),$$

such that

$$P_{(1)} \times^{R_1} P_{(2)} \times \cdots \times^{R_{r-2}} P_{(r-1)} \times^{R_{r-1}} X(w^r) = \tilde{X}(w^1) \to X(w^1)$$

is a $P_{(1)}$-equivariant small resolution of $X(w^1)$, where $P_{(i)}$ is the stabilizer of $X(w^i)$, $1 \leq i \leq r$ and $R_i = P_{(i)} \cap P_{(i+1)}$, $1 \leq i \leq r-1$.

In order to explicitly determine the elements $w^1, w^2, \ldots, w^r$ in Zelevinsky's resolution, we need to introduce another notation for W_P.

9.1.8. Definition. The $2 \times m$ matrix notation for w (cf. [**143**]).
Let $w \in I_{r,n}$. A maximal subsequence of consecutive integers in w will be referred to as a *block* of w. Clearly, w is the concatenation of its blocks. Let a_i be the length, k_i the last term of the i-th block in w. Then w determines a $2 \times m$ matrix

$$\begin{pmatrix} k_1 & \cdots & k_m \\ a_1 & \cdots & a_m \end{pmatrix},$$

where $0 < k_1 < \cdots < k_m \leq n$, $0 \leq a_i \leq k_i - k_{i-1}$, $k_0 = 0$, $\sum_i a_i = r$, and conversely such a $2 \times m$ matrix determines an unique element of $I_{r,n}$. In the sequel, we shall denote w by the associated $2 \times m$ matrix $\begin{pmatrix} k_1 & \cdots & k_m \\ a_1 & \cdots & a_m \end{pmatrix}$.

Note that if $G = SL(n)$, and w is $\begin{pmatrix} k_1 & \cdots & k_m \\ a_1 & \cdots & a_m \end{pmatrix}$, P_w is simply the parabolic subgroup with $S \setminus \{\alpha_{k_1}, \ldots, \alpha_{k_m}\}$ as the associated set of simple roots.

Next we describe the small resolution in terms of the $2 \times m$ matrix.

9.1.9. Schubert varieties in the Grassmannian. Let $w \in I_{r,n}$ so that $X(w)$ is a Schubert variety in the Grassmannian $G_{r,n}$. As above, let

$$w = \begin{pmatrix} k_1 & \cdots & k_m \\ a_1 & \cdots & a_m \end{pmatrix}.$$

Set $b_i = k_{i+1} - k_i - a_{i+1}$, $0 \le i < m$, $k_0 = 0$. Set $a_0 = b_m = \infty$. Choose $0 \le i < m$ so that $b_i \le a_i$, and $a_{i+1} \le b_{i+1}$; note that such an i exists since $b_0 < a_0 \, (= \infty)$ and $b_m \, (= \infty) > a_m$. Let

(9.1.10) $$w' = \begin{pmatrix} k_1 & \cdots & k_{i-1} & k_i + a_{i+1} & k_{i+2} & \cdots & k_m \\ a_1 & \cdots & a_{i-1} & a_i + a_{i+1} & a_{i+2} & \cdots & a_m. \end{pmatrix}$$

Then the number of blocks in w' is $m-1$ except for the case when $m = 1$, in which case $X(w')$ is a smooth variety. This choice of w' corresponds to the first step in the Bott–Samelson-type resolution as given by 9.1.7. Iterating this procedure, we obtain the Bott–Samelson-type resolution as given by 9.1.7.

9.1.11. Example. Let $w = (3, 4, 6, 9, 10) \in I_{5,10}$ so that it determines a Schubert variety in the Grassmannian $G_{5,10}$, of 5-planes in 10-space. In matrix notation,

$$w = \begin{pmatrix} 4 & 6 & 10 \\ 2 & 1 & 2 \end{pmatrix}.$$

We have, $(a_1, a_2, a_3) = (2, 1, 2)$ and $(b_0, b_1, b_2) = (2, 1, 2)$ (as above, we set $a_0 = b_3 = \infty$). Let $i = 1$ so that $a_i \ge b_i$ and $a_{i+1} \le b_{i+1}$. This leads to

$$w' = (3, 4, 5, 9, 10) = \begin{pmatrix} 5 & 10 \\ 3 & 2 \end{pmatrix}.$$

Then $X(w')$ is a subvariety of $X(w)$ such that $P_w \cdot X(w') = X(w)$. Here P_w denotes the stabilizer of $X(w)$. Again, for this w', one has $(a_1', a_2') = (3, 2)$, $(b_0', b_1') = (2, 3)$. Choosing $i = 1$ the conditions $a_i' \ge b_i'$ and $a_{i+1}' \le b_{i+1}'$ are satisfied. This choice of i leads to $w'' = (3, 4, 5, 6, 7) = \begin{pmatrix} 7 \\ 5 \end{pmatrix}$.

Note that $X(w'')$ is a smooth subvariety of $X(w')$ (cf. Corollary 9.3.3) and $P_{w'} \cdot X(w'') = X(w')$.

In the example, P_w is the parabolic with $S \setminus \{\alpha_4, \alpha_6\}$ as the associated set of simple roots. Similarly, $P_{w'}$, resp. $P_{w''}$, is the parabolic with $S \setminus \{\alpha_5\}$, resp. $S \setminus \{\alpha_7\}$, as the associated set of simple roots. Write $R = P_w \cap P_{w'}$, $R' = P_{w'} \cap P_{w''}$. Then

$$P_w \times^R P_{w'} \times^{R'} X(w'') \longrightarrow X(w)$$

is a small resolution of the Schubert variety $X(w)$.

9.1.12. Small resolutions for symplectic and orthogonal Grassmannians (Type C, D). Theorem 9.1.7 has been generalized by Sankaran and Vanchinathan to the symplectic Grassmannian and the orthogonal Grassmannian. In these cases, the theorems only apply to certain Schubert varieties. In §9.1.15, the Schubert varieties not admitting any small resolutions will be discussed.

Let $G = Sp(2n)$ and $P = P_n$, the maximal parabolic subgroup associated to the simple root α_n (the simple roots being indexed as on page 207). The set W^P of minimal representatives of W/W_P may be identified with $\{(a_1, \dots, a_n)\}$ such that

1. $1 \le a_1 < a_2 < \cdots < a_n \le 2n$.
2. If $k \in (a_1, \dots, a_n)$, then $2n + 1 - k \notin (a_1, \dots, a_n)$.

Note that for $1 \le k \le 2n$, precisely one of $\{k, 2n + 1 - k\}$ belongs to $(a_1, \dots, a_n)$. Thus, if r is such that $a_r \le n$, $a_{r+1} > n$, then $(a_1, \dots, a_n)$ is completely determined by $(a_1, \dots, a_r)$. Hence W^P may be identified with $\cup_{0 \le r \le n} I_{r,n}$, where $I_{r,n} = \{(i_1, \dots, i_r), 1 \le i_1 < i_2 < \cdots < i_r \le n\}$. For example, the element (n) considered as an element of W^P has the reduced expression $(s_{n-1}s_n)(s_{n-2}s_{n-1}s_n) \cdots (s_1 \cdots s_n)$.

9.1.13. Theorem. [143] *Let $X(w) \subset G/P_n$, $w \in I_{r,n}$. Let $w_r \le n - r$. There exist Schubert varieties*

$$X(w) = X(w^1) \supset X(w^2) \supset \cdots \supset X(w^s),$$

such that

$$P_{(1)} \times^{R_1} P_{(2)} \times \cdots \times^{R_{s-2}} P_{(s-1)} \times^{R_{s-1}} X(w^s) = \tilde{X}(w^1) \to X(w^1)$$

is a $P_{(1)}$-equivariant small resolution of $X(w^1)$, where $P_{(i)}$ is the stabilizer of $X(w^i)$, $1 \le i \le s$ and $R_i = P_{(i)} \cap P_{(i+1)}$, $1 \le i \le s - 1$.

Now let $G = SO(2n)$ and $P = P_n$, the maximal parabolic subgroup associated to the simple root α_n. Then as above, W^P may be identified with $\cup_{0 \le r \le n, \, n-r \text{ even}} I_{r,n}$.

9.1.14. Theorem. [143] *Let $X(w) \subset G/P_n$, $w \in I_{r,n}$, where $n - r$ is even. Further, let either $w_r < n - r$ or $r \ge 2$, $w_r = n$, $w_{r-1} \le n - r$. There exist Schubert varieties*

$$X(w) = X(w^1) \supset X(w^2) \supset \cdots \supset X(w^s),$$

such that

$$P_{(1)} \times^{R_1} P_{(2)} \times \cdots \times^{R_{s-2}} P_{(s-1)} \times^{R_{s-1}} X(w^r) = \tilde{X}(w^1) \to X(w^1)$$

is a $P_{(1)}$-equivariant small resolution of $X(w^1)$, where $P_{(i)}$ is the stabilizer of $X(w^i)$, $1 \le i \le s$ and $R_i = P_{(i)} \cap P_{(i+1)}$, $1 \le i \le s-1$.

We will describe the first iteration of the procedure to find $w^1, \ldots, w^s$ in Theorems 9.1.13 and 9.1.14. Let $G = Sp(2n)$ or $SO(2n)$. Let $X(w) \subset G/P_n$. Let $w = \begin{pmatrix} k_1 & \cdots & k_m \\ a_1 & \cdots & a_m \end{pmatrix}$ as in type A case in the previous subsection.

As above, let P_w denote the stabilizer of $X(w)$. If $G = Sp(2n)$, then P_w is the parabolic subgroup with $S \setminus \{\alpha_{k_1}, \ldots, \alpha_{k_m}\}$ as the associated set of simple roots. Also, if $G = SO(2n)$, $a_m \ge 2$ and $k_m = n$, then P_w is as above. However, if $G = SO(2n)$, $a_m = 1$, $k_m = n$, then P_w is the parabolic subgroup with $S \setminus \{\alpha_{k_1}, \ldots, \alpha_{k_{m-1}}\}$ as the associated set of simple roots.

Let w be either

$$\begin{pmatrix} k_1 & \cdots & k_m \\ a_1 & \cdots & a_m \end{pmatrix}, \ G = Sp(2n), \ SO(2n)$$

or

$$\begin{pmatrix} k_1 & \cdots & k_m & n \\ a_1 & \cdots & a_m & 1 \end{pmatrix}, \ G = SO(2n).$$

Let M denote the number of blocks in w; note that $M = m$ or $m+1$. Set $b_i = k_{i+1} - k_i - a_{i+1}$, $0 \le i < m$, $k_0 = 0$, $k_{m+1} = n$, $a_{m+1} = 1$, $b_m = N - k_m$, $a_0 = \infty$. Choose i, $0 \le i < m$. Let

$$w' = \begin{pmatrix} k_1 & \cdots & k_{i-1} & k_i + a_{i+1} & k_{i+2} & \cdots & k_M \\ a_1 & \cdots & a_{i-1} & a_i + a_{i+1} & a_{i+2} & \cdots & a_M \end{pmatrix}.$$

Then $X(w')$ is a subvariety of $X(w)$ such that $P_w \cdot X(w') = X(w)$ and this gives the first step in the resolution for Theorems 9.1.13 and 9.1.14.

9.1.15. Schubert varieties not admitting any small resolutions.
Suppose $p : \tilde{X} \to X$ is a resolution of a normal irreducible variety. Then by Zariski's main theorem (cf. [63]), the fiber $p^{-1}(x)$ over any singular point x of X has positive dimension. Hence, any normal irreducible variety X with codimension 2 singular locus cannot have any small resolution.

Using this fact, Sankaran–Vanchinathan construct a class of Schubert varieties in $Sp(2n)/B$ which do not admit any small resolution (by showing that these Schubert varieties have codimension 2 singular loci):

9.1.16. Theorem. [143] Let $G = Sp(2n)$. Let $w = (n)$ $(\in I_{1,n})$, $n \ge 3$.
Let Q be any parabolic subgroup contained in P_n, the maximal parabolic subgroup with $S \setminus \{\alpha_n\}$ as the associated set of simple roots. Let $X(\Lambda)$ be the inverse image of $X(w)$ $(\subset G/P_n)$ under the canonical projection $G/Q \to G/P_n$. Then $X(\Lambda)$ does not admit any small resolution.

9.1.17. Poincaré and Kazhdan–Lusztig polynomials. Using the notation from in the previous two subsections, let $w = \begin{pmatrix} k_1 & \cdots & k_m \\ a_1 & \cdots & a_m \end{pmatrix}$. Assume that the following conditions on w hold:

1. $k_m < N - a_m$.
2. $(a_m + a_{m-1} + \cdots + a_i) - (b_{m-1} + \cdots + b_i) < N - k_m$, $i \geq 1$.

Under the above assumptions on w, $X(w)$ admits a small resolution; indeed choose i, $0 \leq i < m$ such that $b_i \leq a_i$, $a_0 = \infty$, and $a_{i+1} \leq b_{i+1}$; note that such an i exists since $b_0 < a_0 (= \infty)$ and $b_m > a_m$. With w' as above, this corresponds to the first step in the Bott–Samelson-type resolution as given by Theorems 9.1.7, 9.1.13, and 9.1.14. Now the conditions (1) and (2) above hold on w' also, and iterating this procedure, we obtain the Bott–Samelson-type resolution as given by Theorems 9.1.13, 9.1.14.

9.1.18. Proposition. [143] *Let $w \in I_{r,n}$. Let $X(\tau)$ be a Schubert subvariety of $X(w)$.*

1. *Let $G = SL(n)$ or $Sp(2n)$. Then $X(\tau)$ is P_w-stable if and only if there exists a sequence $c(\tau, w) := (c_1, \ldots, c_m)$ of nonnegative integers such that $0 \leq a_i + c_i - c_{i-1} \leq k_i - k_{i-1}$, where $c_m = 0$ if $G = SL(n)$, and*

$$\tau = \begin{pmatrix} k_1 & k_2 & \cdots & k_m \\ a_1 + c_1 & a_2 + c_2 - c_1 & \cdots & a_m + c_m - c_{m-1} \end{pmatrix}.$$

2. *Let $G = SO(2n)$. Then $X(\tau)$ is P_w-stable if and only if there exists a sequence $c(\tau, w) := (c_1, \ldots, c_m)$ of nonnegative integers such that $0 \leq a_i + c_i - c_{i-1} \leq k_i - k_{i-1}$, where $c_m \equiv 0 \, (mod \, 2)$ if $k_m = n$, and $0 \leq a_m + c_m - c_{m-1} \leq 1$ if $(k_m, a_m) = (n, 1)$ such that*
   (a) *If $k_m \leq n - 2$, then*

$$\tau = \begin{pmatrix} k_1 & k_2 & \cdots & k_m & n \\ a_1 + c_1 & a_2 + c_2 - c_1 & \cdots & a_m + c_m - c_{m-1} & \epsilon \end{pmatrix},$$

   where $\epsilon = \frac{1}{2}(1 - (-1)^{c_m})$.
   (b) *If $k_m = n$, then*

$$\tau = \begin{pmatrix} k_1 & k_2 & \cdots & k_m & n \\ a_1 + c_1 & a_2 + c_2 - c_1 & \cdots & a_m + c_m - c_{m-1} & \epsilon \end{pmatrix},$$

   where $\epsilon = \frac{1}{2}(1 - (-1)^{c_m})$, and $a_m + c_m - c_{m-1} = 0$ if $\epsilon = 1$.
   (c) *If $k_m \leq n - 2$, then*

$$\tau = \begin{pmatrix} k_1 & k_2 & \cdots & k_m \\ a_1 + c_1 & a_2 + c_2 - c_1 & \cdots & a_m + c_m - c_{m-1} \end{pmatrix}.$$

9.1.19. Definition. The depth sequence $c(\tau, w)$ (cf. [119], [157]). With τ, w as in Proposition 9.1.18, the sequence $c(\tau, w) := (c_1, \ldots, c_m)$ is called the *depth* of τ in w.

For $a, b \in \mathbb{Z}$, let $\begin{bmatrix} a \\ b \end{bmatrix}$ denote the Gaussian binomial coefficient

$$\begin{bmatrix} a \\ b \end{bmatrix} = \frac{(q^a - 1) \cdots (q^{a-b+1} - 1)}{(q^b - 1) \cdots (q - 1)},$$

where $\begin{bmatrix} a \\ b \end{bmatrix}$ is to be understood as 0 if either $b < 0$ or $a < b$, and $\begin{bmatrix} 0 \\ 0 \end{bmatrix} = 1$.

9.1.20. Theorem. [144, 157]

1. *Let $X(w)$ be a Schubert variety in the Grassmannian $G_{r,n}$. For the small resolution $p : \tilde{X}(w) \to X(w)$ (cf. 9.1.7), we have the following inductive formula for the Poincaré polynomial of the fiber over e_ϕ:*

$$P_{\phi,w}(q) = \sum_d q^{(c_i - d)(c_{i+1} - d)} \begin{bmatrix} a_{i+1} - c_i + c_{i+1} \\ c_{i+1} - d \end{bmatrix} \begin{bmatrix} b_i + c_i - c_{i+1} \\ c_i - d \end{bmatrix} P_{\phi(d),w'}(q)$$

   where w' is as in (9.1.10), $(c_1, \ldots, c_m) = c(\phi, w)$, and $X(\phi(d))$ is the $P_{w'}$-stable subvariety of $X(w')$ with depth sequence $c(\phi(d), w') = (c_1, \ldots, c_{i-1}, d, c_{i+2}, \ldots, c_m)$.

2. *Let $X(w)$ be a Schubert variety in G/P_n, $G = Sp(2n), SO(2n)$. Further, let w satisfy the conditions in 9.1.17. For the small resolution $p : \tilde{X}(w) \to X(w)$ (cf. 9.1.13, 9.1.14), we have the following inductive formula for the Poincaré polynomial of the fiber over e_ϕ:*

$$P_{\phi,w}(q) = \sum_d q^{(c_i - d)(c_{i+1} - d)} \begin{bmatrix} a_{i+1} - c_i + c_{i+1} \\ c_{i+1} - d \end{bmatrix} \begin{bmatrix} b_i + c_i - c_{i+1} \\ c_i - d \end{bmatrix} P_{\phi(d),w'}(q)$$

   where w' is as in §9.1.10, $(c_1, \ldots, c_m) = c(\phi, w)$ and $X(\phi(d))$ is the $P_{w'}$-stable subvariety of $X(w')$ with depth sequence $c(\phi(d), w') = (c_1, \ldots, c_{i-1}, d, c_{i+2}, \ldots, c_M)$.

We can simplify the computation of $P_{\phi,w}(q)$. Let $G = Sp(2n), SO(2n)$. Let $w, \phi \in W/W_{P_n}$ be as in Theorem 9.1.20. Let w_0, ϕ_0 in W be the representatives of maximal length of w, ϕ, respectively. Let

$$\Lambda = \begin{pmatrix} k_1 & \cdots & k_m & N + c_m \\ a_1 & \cdots & a_m & c_m \end{pmatrix},$$

where $N = n$ or $n + 1$ according as $G = SO(2n)$ or $Sp(2n)$. Now Λ defines a Grassmannian Schubert variety $X^{Gr}(\Lambda) \subset SL(N + c_m)/P_{r+c_m}$, where $r = \sum_{1 \le i \le m} a_i$. Let $X^{Gr}(\Phi)$ be the P_Λ-stable Schubert variety of $X^{Gr}(\Lambda)$ with depth sequence $(c_1, c_2, \ldots, c_m, 0)$. Let $\Lambda_0, \Phi_0 \in W(SL(N + c_m))$ denote the representatives of maximal length of Λ, Φ, respectively.

Let $P^W_{y,w}$ be the Kazhdan–Lusztig polynomial of the Weyl group W, in order to distinguish different Weyl groups.

9.1.21. Theorem. [144] *Let $G = Sp(2n), SO(2n)$, and W the Weyl group of G. Let w be as in 9.1.20. Let $\theta \in W$ be any element such that*

$\theta \leq w_0$, w_0 being as above. Let $\bar{\theta} \in W/W_{P_n}$ denote its projection. Let $X(\phi) = P_w \cdot X(\bar{\theta}) \subset X(w)$. With notation as above, we have

$$P_{\theta,w_0}^W = P_{\phi_0,w_0}^W = P_{\Phi_0,\Lambda_0}^{W'}$$

where $W' = W(SL(N + c_m))$, the Weyl group of $SL(N + c_m)$.

9.1.22. Example. [144] Let $G = SO(24)$. Then $W = W(SO(24))$ is a subgroup of S_{24}. Let $w = \begin{pmatrix} 6 & 8 \\ 3 & 1 \end{pmatrix} \in W/W_{P_{12}}$. Note that w satisfies conditions (1) and (2) of 9.1.17 and hence admits a small resolution. Then $w_0 = (24, 23, 22, 18, 16, 15, 14, 13, 8, 6, 5, 4) \in W$; note that it suffices to give the first twelve entries in w_0 (cf. 9.1.12). Take $\theta = (5, 6, 8, 13, 21, 16, 14, 7, 1, 10, 3, 2)$. Then $\theta \leq w_0$. Sorting θ into increasing order, we have $\bar{\theta} = (1, 2, 3, 5, 6, 7, 8, 10, 13, 14, 16, 21)$. Now P_w, the stabilizer of $X(w)(\subset SO(24)/P_{12})$, is the maximal parabolic subgroup with $S \setminus \{\alpha_6, \alpha_8\}$ as the associated set of simple roots. The P_w-saturation of $X(\theta)$ is $X(\phi)$, where $\phi = \begin{pmatrix} 6 & 8 & 12 \\ 5 & 2 & 1 \end{pmatrix}$. Hence by Theorem 9.1.20, the Kazhdan–Lusztig polynomial $P_{\theta,w_0}(q) = P_t(p^{-1}(e_\phi))$, the Poincaré polynomial of the fiber over e_ϕ for the small resolution $p : \tilde{X}(w) \to X(w)$ (with $q = t^2$).

Now for $w = \begin{pmatrix} 6 & 8 \\ 3 & 1 \end{pmatrix}$, one has $(k_1, k_2) = (6, 8)$, $(a_1, a_2) = (3, 1)$, $(b_0, b_1) = (3, 1)$. Hence, the choice $i = 1$ leads to the first step of a small resolution of $X(w)$. When $i = 1$, we have $w' = \begin{pmatrix} 7 \\ 4 \end{pmatrix}$. One then takes $w'' = \begin{pmatrix} 4 \\ 4 \end{pmatrix}$. We have $X(w'') \simeq SO(16)/P_8$ and hence is smooth.

Now $c(\phi, w) = (c_1, c_2) = (2, 3)$. Since $i = 1$, the possible nonzero terms in the formula for $P_t(p^{-1}(e_\phi))$, as given by Theorem 9.1.20, are the terms corresponding to $d = 0, 1$, or 2. Also, we have, $b_2 - c_2 + c_1 = 0$, and hence the term corresponding to $d = c_1 = 2$ is the only nonzero term. Thus we obtain $P_{\phi,w}(q) = q^0 \begin{bmatrix} 1+3-2 \\ 1 \end{bmatrix} \cdot \begin{bmatrix} 0 \\ 0 \end{bmatrix} \cdot P_{\phi(2),w'}(q)$. Now $\phi(2) = \begin{pmatrix} 7 \\ 6 \end{pmatrix}$, and as above, using the formula given by Theorem 9.1.20, we obtain

$$P_{\phi(2),w'}(q) = q^0 \begin{bmatrix} 4+2-0 \\ 2 \end{bmatrix} \cdot 1 = \begin{bmatrix} 6 \\ 2 \end{bmatrix} = (1 + q^2 + q^4)(1 + q + q^2 + q^3 + q^4).$$

Hence we obtain

$$P_{\theta,w_0}(q) = P_{\phi,w}(q) = (1 + q)(1 + q^2 + q^4)(1 + q + q^2 + q^3 + q^4).$$

9.1.23. Remark. As mentioned above, Boe (cf.[17]) has obtained a description of all Kazhdan–Lusztig polynomials in the minuscule (and cominuscule) cases. On the other hand, as seen in §9.1.15, some of the Schubert varieties in the cominuscule cases do not admit a small resolution. This shows that there is some combinatorial insight in Boe's algorithm which applies not only to Schubert varieties admitting a small resolution, but also to Schubert varieties not admitting a small resolution!

9.2. Brion–Polo results

Given $w \in W$ and a parabolic subgroup Q, let L_Q denote the Levi subgroup of Q containing B. Let Q^- be the unique parabolic subgroup such that $Q \cap Q^- = L_Q$. Let U_Q, respectively U_Q^-, be the unipotent radical of Q, respectively Q^-. For a root $\alpha \in R$, let U_α denote the associated root subgroup. For a subset I of S, let P_I denote the parabolic subgroup with I as the associated subset of simple roots. If $Q = P_I$, then we shall denote $L_Q, W_Q, R_Q, R_Q^\pm$ also by $L_I, W_I, R_I, R_I^\pm$, respectively. We shall denote by $W_I^{\max}$ the set of maximal representatives in W of W/W_I (cf. §2.5.4). Let e_{wQ} denote the point wQ of G/Q; let $C_Q(w)$ denote the B-orbit Be_{wQ} through e_{wQ} so $X_Q(w) = \overline{C_Q(w)}$ is the Schubert variety in G/Q associated to w. Recall that if $Q = B$, we shall denote U_Q, W_Q, e_{wQ}, etc., by just U, W, e_w etc.

Let $y \le w$. In [30], Brion and Polo study the singularity of $X_Q(w)$ along $C_Q(y)$ by using a canonical T-stable transversal $\mathcal{N}_Q(y, w)$ to $C_Q(y)$ in $X_Q(w)$. It is shown that under certain hypotheses on y and w, the T-variety $\mathcal{N}_Q(y, w)$ is isomorphic to the orbit closure of a highest-weight vector in a certain Weyl module for a certain reductive subgroup containing T. As a consequence, Brion and Polo compute the Kazhdan–Lusztig polynomial $P_{y,w}$ and $\text{mult}_{yQ} X_Q(w)$ (the multiplicity of e_{yQ} on $X_Q(w)$) for such y, w's. The hypotheses on y and w are then verified for the case of Q being a minuscule parabolic subgroup and $X_Q(y)$ being an irreducible component of $X_Q(w) \backslash P \cdot e_{wQ}$, where P is the stabilizer in G of $X_Q(w)$; as a consequence, it is shown that $X_Q(w) \backslash P \cdot e_{wQ}$ is precisely $\text{Sing} X_Q(w)$, the singular locus of $X_Q(w)$. These results are then extended to Schubert varieties in the symplectic Grassmannian, in a smooth quadric, and in the variety of flags of type $(1, n)$ in K^{n+1}. We state below the main results of [30].

9.2.1. Definition. The Weyl module. Given a dominant character $\lambda \in X(T)$, as in Chapter 2, §2.11.10 we shall denote the associated Weyl module by $V_K(\lambda)$; note that $V_K(\lambda)^* = H^0(G/B, L(\lambda))$. More generally, for a subset I of S and a L_I-dominant character λ, i.e., λ is a character such that $(\lambda, \alpha^*) \ge 0$, $\forall \alpha \in I$, we shall denote by $V_I(\lambda)$, the Weyl module for L_I with highest weight λ; note that $V_I(\lambda)^* = H^0(P_I/B, L(\lambda))$.

9.2.2. Definition. The set $\mathcal{N}_Q(y, w)$. Given $y, w \in W$, $y \le w$, let $\mathcal{N}_Q(y, w) = ((yU_Q^- y^{-1}) \cap U^-)e_{yQ} \cap X_Q(w)$. This is a closed T-stable subvariety of $(yU_Q^- y^{-1})e_{yQ} \cap X_Q(w)$. In particular one has a T-equivariant isomorphism

$$(yU_Q^- y^{-1})e_Q \cap X_Q(w) \simeq C_Q(y) \times \mathcal{N}_Q(y, w).$$

We shall refer to $\mathcal{N}_Q(y, w)$ as a *transversal* to $C_Q(y)$ in $X_Q(w)$. If $Q = B$, we shall drop the suffix B and denote $X_B(w)$, $\mathcal{N}_B(y, w)$ etc., by just $X(w)$, $\mathcal{N}(y, w)$ etc.

9.2.3. Lemma. [30] *Let Q be a parabolic subgroup. Let $y, w \in W_Q^{\max}$, $y \leq w$. The canonical projection $\pi_Q : G/B \to G/Q$ induces an isomorphism $\mathcal{N}_B(y, w) \simeq \mathcal{N}_Q(y, w)$.*

9.2.4. Lemma. [30] *Let $y, w \in W, y < w$ and let $d = l(w) - l(y)$.*

1. *There exists a polynomial $K_{y,w}(q)$ of degree d such that for every prime number p and every integer $r \geq 1$, the number of $\mathbb{F}_{p^r}$-rational points of $\mathcal{N}(y, w) \setminus \{e_y\}$ equals $K_{y,w}(p^r)$.*
2. *If $\mathcal{N}(y, w) \setminus \{e_y\}$ is rationally smooth, then the Kazhdan–Lusztig polynomial $P_{y,w}(q) = (-K_{y,w})^{\leq (d-1)/2}(q)$, where, for any polynomial $P = \sum_i a_i q^i$ and any positive rational number t, $P^{\leq t}$ denotes the sum $\sum_{i \leq t} a_i q^i$.*

9.2.5. Closures of orbits of highest-weight vectors as transversals.

9.2.6. Definition. The cone $C(\lambda)$. Let λ be a dominant character and let P be the associated parabolic subgroup; note that P is the parabolic subgroup generated by B and $\{U_{-\alpha} \mid \alpha \in S , (\lambda, \alpha^*) = 0\}$. Then λ extends to a character of P and the associated line bundle $L_P(\lambda)$ on G/P is very ample. Furthermore, we have $V(\lambda)$, the Weyl module, is the dual of $H^0(G/P, L_P(\lambda))$. Hence we obtain an embedding $G/P \hookrightarrow \mathbb{P}(V(\lambda))$. Let $C(\lambda)$ denote the corresponding affine cone over G/P.

We have that $C(\lambda)$ is the G-orbit closure of a highest-weight vector. Further, $C(\lambda)$ is normal (cf.[**138**]). Denoting by K_λ the λ-weight space in $V(\lambda)$, we can identify the total space of the line bundle $L_P(\lambda)$ with $G \times^P K_\lambda$ (cf.Chapter 2, §2.8). We have (cf.[**30**]), the map $\phi : G \times^P K_\lambda \to C(\lambda)$, $(g, v) \mapsto gv$ is proper, birational and induces an isomorphism $G \times^P (K_\lambda \setminus \{0\}) \simeq C(\lambda) \setminus \{0\}$. Hence we obtain, by Zariski's main theorem (cf. [**63**]),

$$K[C(\lambda)] \simeq K[G \times^P K_\lambda] = \oplus_{n \geq 0} V(n\lambda)^*.$$

9.2.7. Definition. The multicone $C_I(\lambda_1, \ldots, \lambda_r)$. More generally, given dominant characters $\lambda_1, \ldots, \lambda_r$, let $P_1, \ldots, P_r$ be the associated parabolic subgroups, and let $Q = P_1 \cap \cdots \cap P_r$; let $I = S \setminus \{\alpha_1, \ldots, \alpha_r\}$ so that $Q = P_I$. Let $V = \oplus_{i=1}^r V(\lambda_i)$. Let E be the Q-submodule of V spanned by the highest-weight vectors. Let $C(\lambda_1, \ldots, \lambda_r) = GE$, which is closed since G/Q is complete. Then the line bundles $L_{P_i}(\lambda_i)$ define a closed immersion of G/Q into $\mathbb{P}(V(\lambda_1)) \times \cdots \times \mathbb{P}(V(\lambda_r))$ and the corresponding multicone gets identified with $C_I(\lambda_1, \ldots, \lambda_r)$. Also, $G \times^Q E$ is the total space of the vector bundle $\oplus_{i=1}^r L_Q(\lambda_i)$. Further, the map $\psi : G \times^Q E \to V$, $(g, v) \mapsto gv$ induces an isomorphism $G \times^Q E^\times \simeq GE^\times$, where $E^\times$ denotes the Q-stable, open subvariety of E consisting of those vectors whose projection onto $V(\lambda_i)$ is nonzero, for all $i = 1, \ldots, r$. Thus we obtain,

$$K[C_I(\lambda_1, \ldots, \lambda_r)] \simeq K[G \times^Q E] = \oplus_{n_1, \ldots, n_r \geq 0} V_I(n_1\lambda_1 + \cdots + n_r\lambda_r)^*.$$

9.2.8. Proposition. [30] *Let $I \subset S$, and $P = P_I, L = L_I$. Let $\beta \in S \setminus I$. Then $U_P^- e_P \cap \overline{Pe_{s_\beta P}}$ (which is an L-stable open neighborhood of $\overline{Pe_{s_\beta P}}$) is L-isomorphic to $C_I(-\beta)$, the orbit closure of a highest-weight vector in the Weyl module $V_I(-\beta)$.*

Let I, P, L, etc., be as in Proposition 9.2.8. Let $d = \dim \overline{Pe_{s_\beta P}}$, $I_0 = \{\alpha \in I \mid \langle \alpha, \beta \rangle = 0\}$. By Proposition 9.2.8 we have $d = 1 + \dim L/P_0 = 1 + \#(R_I^+ \setminus R_{I_0}^+)$. Note that if $d = 1$, then $\overline{Pe_{s_\beta P}} \simeq \mathbb{P}^1$. Let us now suppose that $d > 1$. For any subset A of W, let $H(A, q) = \sum_{w \in A} q^{l(w)}$. Let $\rho = \frac{1}{2} \sum_{\alpha \in R^+} \alpha$.

9.2.9. Corollary. *Proposition 9.2.8 implies the following:*

1. *The tangent space $T_{e_P}(\overline{Pe_{s_\beta P}})$ is L-isomorphic to $V_I(-\beta)$.*
2. *The multiplicity of $\overline{Pe_{s_\beta P}}$ at e_P equals*

$$(d - 1)! \prod_{\gamma \in R_I^+ \setminus R_{I_0}^+} \frac{\langle -\beta, \gamma \rangle}{\langle \rho, \gamma \rangle}.$$

3. *$\overline{Pe_{s_\beta P}}$ is smooth if and only if β is adjacent to a unique connected component J of I, J is of Type $\mathbf{A}_{d-1}$ or $\mathbf{C}_{d/2}$ (if d is even), and $J \cup \{\beta\}$ has no branch points and has β as a short root.*
4. *Let $y = w_I, w = w_I w_{I_0} s_\beta w_I$, w_I, respectively w_{I_0}, being the unique element of largest length in W_I, respectively W_{I_0}. Then*

$$P_{y,w}(q) = \left((1 - q) \frac{H(W_I, q)}{H(W_{I_0}, q)} \right)^{\leq (d-1)/2}.$$

9.2.10. Definition. The set $C_{[yQ, wQ]}$. Let Q be a parabolic subgroup of G containing B. Let $y, w \in W_Q^{\max}$, $y \leq w$. Let $C_{[yQ, wQ]}$ denote the union $\cup_{y \leq x \leq w} C_Q(x)$. This is a B-stable open subset of $X_Q(w)$ containing $C_Q(y)$ as the unique closed B-orbit.

9.2.11. Lemma. [30] *Let y, w be as above. Then $(y U_Q^- y^{-1}) e_{yQ} \cap X_Q(w)$ is the unique T-stable, open subset of $C_{[yQ, wQ]}$ containing e_{yQ}.*

Let Q be a parabolic subgroup of G containing B. It can be seen easily that the stabilizer in G of $C_Q(x)$ is the parabolic subgroup generated by B and $\{U_{-\alpha}, \alpha \in S \cap x(R_Q)\}$ and the stabilizer in G of $X_Q(x)$ is the parabolic subgroup generated by B and $\{U_{-\alpha}, \alpha \in S \cap x(R_Q \cup R^-)\}$.

9.2.12. Theorem. [30] *Let $y, w \in W_Q^{\max}$, $y \leq w$. Let I be a subset of $S \cap y(R_Q)$. Let $P = P_I, L = L_I$. Let $P_{yQ} = P \cap (yQy^{-1})$, the stabilizer in P of the point e_{yQ}. Let $X_Q(w) = PX_Q(s_\beta y)$, for some $\beta \in S \cap y(R^+ \setminus R_Q^+)$. Let $C_I(-\beta)$ denote the G-orbit closure of a highest-weight vector in $V_I(-\beta)$.*

1. *The morphism $\phi : \overline{Ps_\beta P}/P_{yQ} \to G/Q$, $gP_{yQ} \mapsto ge_{yQ}$ induces a P-equivariant isomorphism $\overline{Ps_\beta P}/P_{yQ} \simeq C_{[yQ,wQ]}$ and hence one has a locally trivial fibration $\pi : C_{[yQ,wQ]} \to \overline{Pe_{s_\beta}P}/P$ with fiber P/P_{yQ} $(\simeq C_Q(y))$.*

2. *We have an L-equivariant isomorphism $\psi : (yU_Q^- y^{-1})e_{yQ} \cap X_Q(w) \simeq C_I(-\beta) \times C_Q(y)$. In particular, we have $\mathcal{N}_Q(y, w) \simeq C_I(-\beta)$.*

9.2.13. Definition. The set $N_Q(y, w)$ Let y, w, I, β etc., be as in Theorem 9.2.12. Define $N_Q(y, w) = T_{yQ}\mathcal{N}_Q(y, w)$, the tangent space to $\mathcal{N}_Q(y, w)$ at e_{yQ}; it is an L-submodule of $T_{yQ}(G/Q)$, isomorphic to the normal space to $C_Q(y)$ in $X_Q(w)$ at e_{yQ}.

Let $I_0 = \{\alpha \in I \mid \langle \alpha, \beta \rangle = 0\}$. Let $d = l(w) - l(y) = \dim X_Q(w) - \dim X_Q(y) = 1 + \#(R_I^+ \setminus R_{I_0}^+)$. Let $\text{mult}_{yQ}X_Q(w)$ denote the multiplicity of $X_Q(w)$ at e_{yQ}. We have the following corollary to Theorem 9.2.12:

9.2.14. Corollary. *The following equations hold:*

$$(9.2.15) \qquad w = w_I w_{I_0} s_\beta y$$

$$(9.2.16) \qquad N_Q(y, w) \simeq V_I(-\beta)$$

$$(9.2.17) \qquad \text{mult}_{yQ} X_Q(w) = (d-1)! \prod_{\gamma \in R_I^+ \setminus R_{I_0}^+} \frac{\langle -\beta, \gamma \rangle}{\langle \rho, \gamma \rangle}$$

$$(9.2.18) \qquad P_{y,w}(q) = \left((1-q) \frac{H(W_I, q)}{H(W_{I_0}, q)} \right)^{\leq d - 1/2}.$$

9.2.19. Application to the minuscule case. In this section, we shall suppose that Q is a maximal parabolic subgroup associated to a minuscule fundamental weight ω, i.e., $(\omega, \beta^*) \leq 1$, for all $\beta \in R^+$. We shall further suppose that G is simply laced, which entails no loss of generality; for, if G is of type $\mathbf{B}_n$ or $\mathbf{C}_n$, then it is well-known that G/Q identifies with G'/Q', where G' is of type $\mathbf{D}_{n+1}$ or $\mathbf{A}_{2n-1}$, respectively, Q' is a maximal parabolic subgroup associated to a minuscule fundamental weight, and the Schubert varieties in G/Q get identified with the Schubert varieties in G'/Q'. For $w \in W_Q^{\max}$, we shall denote by $\text{Bd}(X_Q(w))$, the boundary of $X_Q(w)$, i.e.,

$\mathrm{Bd}(X_Q(w)) = (X_Q(w) \setminus P_J e_{wQ})$, where P_J denotes $\mathrm{Stab}(X_Q(w))$, the stabilizer in G of $X_Q(w)$. We shall denote by $\geq$ the usual partial order on $X(T)$, namely, $\lambda, \mu \in X(T), \lambda \geq \mu \Leftrightarrow \lambda - \mu \in \mathbb{N}R^+$.

9.2.20. Lemma. [30] *Let* $y, w \in W_Q^{\max}$, $y \leq w$.

1. *Suppose* $X_Q(y)$ *is an irreducible component of* $\mathrm{Bd}(X_Q(w))$. *Then there exists a unique simple root* β *such that* $X_Q(y) \subset X_Q(s_\beta y) \subseteq X_Q(w)$ *and we have* $X_Q(w) = PX_Q(s_\beta y)$, *where* $P = \mathrm{Stab}(X_Q(w)) \cap \mathrm{Stab}(C_Q(y))$.
2. *The irreducible components of* $\mathrm{Bd}(X_Q(w))$ *are exactly* $\{X_Q(s_\gamma w)\}$, γ *being a minimal element of* $\{\alpha \in R^+ \mid X_Q(s_\alpha w) \subseteq \mathrm{Bd}(X_Q(w))\}$.

Lemma 9.2.20 together with the results of §9.2.5 yields the following:

9.2.21. Theorem. [30] *Let* $y, w \in W_Q^{\max}$. *Let* $J = S \cap w(R^- \cup R_Q)$.

1. $\mathrm{Sing} X_Q(w) = \mathrm{Bd}(X_Q(w))$.
2. *Suppose* $X_Q(y)$ *is an irreducible component of* $\mathrm{Bd}(X_Q(w))$. *Let* β *be the unique simple root such that* $X_Q(y) \subset X_Q(s_\beta y) \subseteq X_Q(w)$ *and let* I *be the union of the connected components of* $J \cap y(R_Q)$ *to which* β *is adjacent. Then the normal space* $N_Q(y, w)$ *is isomorphic to the* L_I-*module* $V_I(-\beta)$ *and* $\mathcal{N}_Q(y, w)$ *identifies with the closure of the* L_I-*orbit of a highest-weight vector in* $V_I(-\beta)$.
3. *Thus* $\mathcal{N}_Q(y, w)$ *is determined by the pair* (I, I'), *where* $I' = I \cup \beta$ *and we have the following possibilities for* (I, I').
   (a) *I is of type* $\mathbf{A}_p \times \mathbf{A}_q$ *and* I' *is of type* $\mathbf{A}_{p+q+1}$. *Then* $\mathcal{N}_Q(y, w)$ *is isomorphic to the cone of decomposable vectors in* $K^{p+1} \otimes K^{q+1}$ *and has dimension* $p + q + 1$. *In this case* $\mathrm{mult}_{yQ} X_Q(w) = \binom{p+q}{p}$, $P_{y,w} = \sum_{i=0}^{\min\{p,q\}} t^i$.
   (b) *I is of type* $\mathbf{A}_n$ *and* I' *is of type* $\mathbf{D}_{n+1}$. *Then* $\mathcal{N}_Q(y, w)$ *is isomorphic to the cone of decomposable vectors in* $\Lambda^2 K^{n+1}$ *and has dimension* $2n - 1$. *We have* $\mathrm{mult}_{yQ} X_Q(w) = \frac{1}{n} \binom{2n-2}{n-1}$, $P_{y,w} = \sum_{i=0}^{[\frac{n-1}{2}]} t^{2i}$; *here, for a rational number* r, $[r]$ *denotes the integral part of* r.
   (c) *I is of type* $\mathbf{D}_n$ *and* I' *is of type* $\mathbf{D}_{n+1}$. *Then* $\mathcal{N}_Q(y, w)$ *is isomorphic to a nondegenerate quadratic cone in* K^{2n} *and has dimension* $2n-1$. *We have* $\mathrm{mult}_{yQ} X_Q(w) = 2$, $P_{y,w} = 1 + t^{n-1}$.
   (d) *I is of type* $\mathbf{D}_5$ *and* I' *is of type* $\mathbf{E}_6$. *Then* $N_Q(y, w)$ *identifies with* $V = K^{16}$, *a half-spin representation of* $Spin(10)$, *and* $\mathcal{N}_Q(y, w)$ *is isomorphic to the cone of pure half-spinors in* V *and has dimension* 11. *We have* $\mathrm{mult}_{yQ} X_Q(w) = 12$, $P_{y,w} = 1 + t^3$.
   (e) *I is of type* $\mathbf{E}_6$ *and* I' *is of type* $\mathbf{E}_7$. *Then* $N_Q(y, w)$ *identifies with* $V = K^{27}$, *a minuscule representation of* $\mathbf{E}_6$, *and* $\mathcal{N}_Q(y, w)$ *is isomorphic to the orbit closure of a highest-weight vector in*

V and has dimension 17. We have $\text{mult}_{yQ}X_Q(w) = 78$, $P_{y,w} = 1 + t^4 + t^8$.

9.2.22. A generalization to certain multi-cones.

The following theorem generalizes part of Theorem 9.2.12. For a subset J of R, we shall denote by $J^\perp$ the set of roots orthogonal to all of J.

9.2.23. Theorem. [30]

Let Q be a parabolic subgroup of G and let $y, w \in W_Q^{\max}$. Let $I = \{\alpha \in S \mid P_\alpha X_Q(w) = X_Q(w) \text{ and } P_\alpha C_Q(y) = C_Q(y)\}$. Suppose there exist linearly independent positive roots $\beta_1, \ldots, \beta_q$ satisfying the following conditions:

a. For all $\alpha \in I$, $i = 1, \ldots, q$ and $a > 0$, $-\beta_i + a\alpha$ is not a root.
b. $X_Q(y) \subset X_Q(s_{\beta}y) \subseteq X_Q(w)$, for all $i = 1, \ldots, q$.
c. $X_Q(w) = \overline{P_I U_{-\beta_1} \cdots U_{-\beta_q} X_Q(y)}$ and $\dim X_Q(w) = \dim X_Q(y) + q + \#(R_I^+ \setminus R_{I_0}^+)$, where $I_0 = I \cap \{\beta_1, \ldots, \beta_q\}^\perp$.

Then

1. $\mathcal{N}_Q(y, w)$ is L_I-isomorphic to $C_I(-\beta_1, \ldots, -\beta_q)$, the L_I-orbit closure of the sum of highest-weight vectors in the L_I-module $\oplus_{i=1}^q V_I(-\beta_i)$.
2. As a consequence, $N_Q(y, w)$ identifies with $\oplus_{i=1}^q V_I(-\beta_i)$.
3. If $C_I(-\beta_1, \ldots, -\beta_q) \setminus \{0\}$ is rationally smooth, then we have

$$P_{y,w}(q) = \left(- \sum_{J \subseteq \{\beta_1, \ldots, \beta_q\}, J \neq \emptyset} (q-1)^{|J|} \frac{H(W_I, q)}{H(W_{I \cap J^\perp}, q)}\right)^{\leq \frac{1}{2}(l(w) - l(y) - 1)}.$$

9.2.24. Remark. [30]

(i) The hypotheses of the theorem are satisfied, for instance, when $\beta_1, \ldots, \beta_q$ are pairwise orthogonal simple roots such that $X_Q(w) = P_I X_Q(s_{\beta_1} \cdots s_{\beta_q} y)$ and $X_Q(y) \subset X_Q(s_{\beta}y)$, $\beta = \beta_i$, $i = 1, \ldots, q$.
(ii) The hypothesis (c) can be weakened as $U_{-\beta_1} \cdots U_{-\beta_q} e_{yQ} \subset X_Q(w)$ and $\dim X_Q(w) \leq \dim X_Q(y) + q + \#(R_I^+ \setminus R_{I_0}^+)$.

9.2.25. One application to the symplectic Grassmannian.

Let $G = Sp(2n)$ and let Q be the maximal parabolic subgroup with $S \setminus \{\alpha_n\}$ as the associated set of simple roots. Note that G/Q is not minuscule, but cominuscule.

9.2.26. Lemma. [30]

Let $y, w \in W_Q^{\max}$. Suppose $X_Q(y)$ is an irreducible component of $\text{Bd}(X_Q(w))$. Then there exists a simple root β such that $X_Q(y) \subset X_Q(s_{\beta}y) \subset X_Q(w)$. Let I denote the union of the connected components of $S \cap w(R^- \cup R_Q) \cap y(R_Q)$ to which β is adjacent. Then precisely one of the following holds:

1. We have $X_Q(w) = P_I X_Q(s_{\beta}y)$, and either
   (a) I is of type $\mathbf{A}_r \times \mathbf{A}_t$ and $I \cup \{\beta\}$ is of type $\mathbf{A}_{r+t+1}$, or
   (b) I is of type $\mathbf{A}_r$ and $I \cup \{\beta\}$ is of type $\mathbf{C}_{r+1}$.

2. *We have, $\beta = \alpha_m$, $I = \{\alpha_{m-r}, \ldots, \alpha_{m-1}\} \cup \{\alpha_{m+1}, \ldots, \alpha_{n-1}\}$, for some $r < m < n$ and $X_Q(w) = P_{\{\alpha_n\}} P_I X_Q(s_m y)$. In this case $l(w) - l(y) = n - m + r + 1$.*

Preserving the notation of Lemma 9.2.26 we have

9.2.27. Proposition. *[30] $\mathrm{Sing} X_Q(w) = \mathrm{Bd}(X_Q(w))$. Indeed, if $X_Q(y)$ is an irreducible component of $\mathrm{Bd}(X_Q(w))$, then with notation as in Lemma 9.2.26 we have the following:*

1. *In 1(a) of Lemma 9.2.26, $\mathcal{N}_Q(y, w)$ is isomorphic to the cone of decomposable tensors in $K^{n-m} \otimes K^{r+1}$.*
2. *In 1(b) of Lemma 9.2.26, $N_Q(y, w) \simeq S^2 K^{r+1}$ and $\mathcal{N}_Q(y, w)$ is isomorphic to the cone over the 2-tuple embedding of $\mathbb{P}^r$ in $\mathbb{P}(S^2 K^{r+1})$. We have $\mathrm{mult}_{yQ} X_Q(w) = 2^r$, $P_{y,w} = 1$.*
3. *In (2) of Lemma 9.2.26, $\mathcal{N}_Q(y, w)$ is isomorphic to $\mathcal{C}$, the orbit closure of the sum of the highest weight vectors in the $GL(r + 1) \times GL(n - m)$-module $K^{r+1} \otimes K^{n-m} \oplus S^2 K^{r+1} = N_Q(y, w)$. We have $\mathrm{mult}_{yQ} X_Q(w) = \sum_{i=0}^{r} \binom{n-m+r}{i}$, $P_{y,w} = \sum_{i=0}^{\min\{r, n-m\}} t^i$. Further, $\mathcal{C}$ identifies with the contraction to a point of the zero section of the vector bundle $\mathcal{O}(-1) \otimes K^{n-m} \oplus \mathcal{O}(-2)$ over $\mathbb{P}^r$.*

9.2.28. The group $\mathrm{Spin}(2n + 1)$. Let $G = \mathrm{Spin}(2n + 1)$ and let Q be the maximal parabolic with ω_1 as the associated fundamental weight, the natural representation (cf. [30]). In this case G/Q is a smooth quadric hypersurface $\mathcal{Q} \subset \mathbb{P}(K^{2n+1})$. Further, each Schubert variety is the intersection of $\mathcal{Q}$ with a linear B-stable subspace. But the B-stable subspaces of K^{2n+1} are a flag of totally isotropic subspaces $V_1, \ldots, V_n$ (and their orthogonals) $V_{n+1}, \ldots, V_{2n}$, and $V_{2n+1} = K^{2n+1}$, the subspaces being indexed by their dimensions. From this it follows that the Schubert varieties in G/Q are the projective spaces $\mathbb{P}(V_1), \ldots, \mathbb{P}(V_n) = \mathcal{Q} \cap \mathbb{P}(V_{n+1})$, and the quadratic cones $\mathcal{Q} \cap \mathbb{P}(V_{n+2}), \ldots, \mathcal{Q} \cap \mathbb{P}(V_{2n+1})$. Denoting these by $X_0, \ldots, X_{2n-1}$ (indexed by their dimensions), we have $X_0, \ldots, X_{n-1}$ and X_{2n-1} are smooth (clearly); for $n \leq i \leq 2n-2$, X_i is singular along X_{2n-i-2} with a nondegenerate quadratic cone of dimension $2(i+1-n)$ as a transversal singularity. It follows that the multiplicity of X_i along X_{2n-i-2} is 2, where the corresponding Kazhdan–Lusztig polynomial is trivial.

9.2.29. The variety $F(1, n)$. [30] Let $G = SL(n + 1)$, $n \geq 3$. Consider the variety $F(1, n)$ of flags of type $(1, n)$ in K^{n+1}, i.e., the set of all pairs of vector spaces $\{V_1 \subset V_2\}$ such that $\dim(V_1) = 1$ and $\dim(V_n) = n$. Let $\{e_i, 1 \leq i \leq n + 1\}$ be the standard basis of K^{n+1}. For $i = 0, \ldots, n + 1$, let $E_i = K$-span of $\{e_q, q \leq i\}$. It can be seen that Schubert varieties in $F(1, n)$ are precisely

$$X_{i,j} = \{(l, H) \in \mathbb{P}^n \times (\mathbb{P}^n)^* \mid l \subset H, l \subseteq E_i, E_{j-1} \subseteq H\}$$

for $1 \leq i \neq j \leq n+1$. It is easily seen that $X_{i,j}$ is smooth if $i < j$ or $j = 1$ or $i = n + 1$. For $2 \leq j < i \leq n$, we have $X_{j-1,i+1} \subset X_{i,j}$. Further it can be seen that $X_{i,j}$ is smooth outside $X_{j-1,i+1}$ and that the transversal along $X_{j-1,i+1}$ is isomorphic to $\{(x,y) \in E_i/E_{j-1} \times (E_i/E_{j-1})^* \mid \langle x, y \rangle = 0\}$. This set is a nondegenerate quadratic cone in $K^{2(i-j+1)}$. The Kazhdan–Lusztig polynomial corresponding to this cone is $1 + q^{i-j}$ (cf. Theorem 9.2.21, (3c)).

9.3. Irreducible components of $\mathrm{Sing}X(w)$ in special cases

The irreducible components of $\mathrm{Sing}X(w)$ have been determined in [**112**] for $X(w)$ in G/P, for G classical and P certain maximal parabolic subgroup of minuscule type. We recall this result below.

TYPE A. Let $G = SL(n)$ and $P = P_d$, the maximal parabolic subgroup with associated set of simple roots being $S \setminus \{\alpha_d\}$. Then it is well-known (see Chapter 3) that G/P gets identified with the Grassmannian variety $G_{d,n} =$ the set of d-dimensional subspaces of K^n. As seen in Chapter 3, $W_{P_d}^{\min}$, the set of minimal representatives, may be identified as

$$W_{P_d}^{\min} = \{(a_1, \dots, a_d) \mid 1 \leq a_1 < a_2 < \cdots < a_d \leq n\}.$$

For the discussion below we follow the Young diagram representation of Schubert varieties in $G_{d,n}$; namely, given $(a_1 \cdots a_d) \in W_{P_d}$, we associate the partition $\mathbf{a} := (\mathbf{a}_1, \dots, \mathbf{a}_d)$, where $\mathbf{a}_i = a_{d-i+1} - (d - i + 1)$. For a partition $\mathbf{a} = (\mathbf{a}_1, \dots, \mathbf{a}_d)$, where $\mathbf{a}_d \leq n - d$, we shall denote by $X_{\mathbf{a}}$ the Schubert variety corresponding to $(a_1 \cdots a_d)$. Then recall (cf. Chapter 3) that $\dim X_{\mathbf{a}} = |\mathbf{a}| = \mathbf{a}_1 + \cdots + \mathbf{a}_d$. Let

$$\mathbf{a} = (p_1^{q_1}, \dots, p_r^{q_r}) = (\underbrace{p_1, \dots, p_1}_{q_1 \text{ times}}, \dots, \underbrace{p_r, \dots, p_r}_{q_r \text{ times}})$$

(we say that $\mathbf{a}$ consists of r rectangles: $p_1 \times q_1, \dots, p_r \times q_r$).

9.3.1. Theorem. [**112, 113, 152**] *Let $\mathbf{a}$ consist of r rectangles. Then $\mathrm{Sing}X_{\mathbf{a}}$ has $r - 1$ components $X_{\mathbf{a}'_1}, \dots, X_{\mathbf{a}'_{r-1}}$, where*

$$(9.3.2) \qquad \mathbf{a}'_i = (p_1^{q_1}, \dots, p_{i-1}^{q_{i-1}}, p_i^{q_i-1}, (p_{i+1}-1)^{q_{i+1}+1}, p_{i+2}^{q_{i+2}}, \dots, p_r^{q_r}),$$

for $1 \leq i \leq r - 1$ and $p_1 > p_2 > \cdots > p_r$.

Let $\mathbf{a}/\mathbf{a}'_i$ denote the set-theoretic difference $\mathbf{a} \setminus \mathbf{a}'_i$. Note that $\mathbf{a}/\mathbf{a}'_i$, $1 \leq i \leq r - 1$, are simply the hooks in the Young diagram $\mathbf{a}$. For example, if $\mathbf{a} = (17, 5^4, 3^2, 1^3)$, then $\mathbf{a}'_2 = (17, 5^3, 2^3, 1^3)$.

9.3.3. Corollary. $X_{\mathbf{a}}$ *is smooth if and only if $\mathbf{a}$ consists of one rectangle.*

TYPE B. Let $V = K^{2n+1}$ together with a nondegenerate, symmetric bilinear form. Taking the matrix of the form (with respect to the standard basis $\{e_1, \dots, e_{2n+1}\}$ of V) to be the $2n + 1 \times 2n + 1$, anti-diagonal matrix with 1s all along the anti-diagonal except at the $n + 1 \times n + 1$-th place where the entry is 2, the associated quadratic form Q on V is given by $Q(x_1, x_2, \ldots x_n, z, y_1, y_2, \ldots y_n) = x_1 y_n + x_2 y_{n-1} + \cdots + x_n y_1 + z^2$. We identify G as the subgroup of $SL(2n + 1)$ leaving Q invariant.

I. Let $P = P_1$, the maximal parabolic, the associated set of simple roots being $S \setminus \{\alpha_1\}$ (see figure on page 207). Then G/P_1 can be identified with the quadric $Q = 0$ in $\mathbb{P}(V)$. Further, all the Schubert varieties in G/P_1 are given by $(Y_i)_{\mathrm{red}}$ for $Y_i, i = 0, 1, \ldots, 2n, i \neq n + 1$ defined below.

$Y_0 = \{Q_n = x_1 y_n + x_2 y_{n-1} + \cdots + x_n y_1 + z^2 = 0 \text{ in } \mathbb{P} = \mathbb{P}(V)\}$

$Y_1 = \{Q_{n-1} = x_2 y_{n-1} + \cdots + x_n y_1 + z^2 = 0, \text{ and } x_1 = 0 \text{ in } \mathbb{P}\}$

$\vdots$

$Y_i = \{Q_{n-i} = x_{i+1} y_{n-i} + \cdots + x_n y_1 + z^2 = 0, \text{ and } x_1 = \cdots = x_i = 0 \text{ in } \mathbb{P}\}$

$\vdots$

$Y_{n-1} = \{Q_1 = x_n y_1 + z^2 = 0, \text{ and } x_1 = \cdots = x_{n-1} = 0 \text{ in } \mathbb{P}\}$

$Y_n = \{Q_0 = z^2 = 0, \text{ and } x_1 = \cdots = x_n = 0 \text{ in } \mathbb{P}\}$

$Y_{n+2} = \{x_1 = \cdots = x_n = z = y_1 = 0 \text{ in } \mathbb{P}\}$

$\vdots$

$Y_{n+i} = \{x_1 = \cdots = x_n = z = y_1 = \cdots = y_{i-1} = 0 \text{ in } \mathbb{P}\}$

$\vdots$

$Y_{2n} = \{x_1 = \cdots = x_n = z = y_1 = \cdots = y_{n-1} = 0 \text{ in } \mathbb{P}\}$
$\phantom{Y_{2n}} = \text{ the point } \{(0, \ldots, 0, 1)\}.$

The Bruhat order on G/P_1 is just a chain:

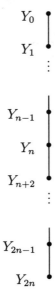

$$Y_0$$
$$Y_1$$
$$\vdots$$
$$Y_{n-1}$$
$$Y_n$$
$$Y_{n+2}$$
$$\vdots$$
$$Y_{2n-1}$$
$$Y_{2n}$$

Note that codimension of Y_i in G/P_1 is i, resp. $i-1$, according to whether $i \leq n$ or $i > n$. We have

- $Y_0, (Y_n)_{\text{red}}$ and $Y_{n+i}, 2 \leq i \leq n$ are nonsingular (clearly).
- $Y_i, 1 \leq i \leq n-1$ are singular, and the singular locus of Y_i is $Y_{2n+1-i}, 1 \leq i \leq n-1$. These are obtained by computing $X_{-\beta}q_{\text{id}}, \beta \in R^+$. See [**102**] for details.

II. Let $P = P_n$, the maximal parabolic, the associated set of simple roots being $S \setminus \{\alpha_n\}$. Then G/P can be identified with the isotropic Grassmannian of n spaces in the $2n+1$-dimensional space with a nondegenerate symmetric bilinear form $(,)$. We can identify $W_{P_n}^{\min}$ as

$$W_{P_n}^{\min} = \left\{ (a_1 \cdots a_n) \left| \begin{array}{ll} (1) & 1 \leq a_1 < a_2 < \cdots < a_n \leq 2n, \\ & a_i \neq n+1, 1 \leq i \leq n \\ (2) & \text{for } 1 \leq i \leq 2n, \text{ if } i \in \{a_1, \ldots, a_n\} \\ & \text{then } 2n+2-i \notin \{a_1, \ldots, a_n\} \end{array} \right. \right\}.$$

To $(a_1, \ldots, a_n) \in W_{P_n}^{\min}$, we associate the partition $\mathbf{a} := (\lambda_1, \ldots, \lambda_n)$, where

(9.3.4)
$$\lambda_{n+1-i} = \begin{cases} a_i - i & \text{if } a_i \leq n \\ a_i - i - 1 & \text{if } a_i > n. \end{cases}$$

For example, the partition associated with $(1\,3\,7\,8\,10) \in B_5$ is $(4, 3, 3, 1, 0)$. The conditions on the a_i's imply that the partition λ is a *self-dual* partition contained in an $n \times n$ square. For a partition $\lambda = (\lambda_1, \ldots, \lambda_n)$, we shall denote by X_λ the Schubert variety corresponding to $(a_1, \ldots, a_n)$. Thus here again Schubert varieties in G/P are indexed by self-dual partitions contained in an $n \times n$ grid.

9.3.5. Theorem. [112] *Let λ be a self-dual partition. Then we have* $\mathrm{Sing}X_\lambda = \cup X_\mu$, *where $\mu \subset \lambda$, and either λ/μ is a disjoint sum of two hooks that are dual to each other, or $\lambda/\mu = (r+i, r^{r-1}, 1^i) / ((r-1)^{r-1})$ for some r, i with $i > 0$ (the sum of two hooks dual to each other connected at one box), or $\lambda/\mu = (r^2, 2^{r-2}) /(0^r)$ for some $r > 2$ (self-dual double hook).*

TYPE C. Let $V = K^{2n}$ together with a nondegenerate, skew-symmetric bilinear form $(,)$. Let $H = SL(V)$ and $G = Sp(V) = \{A \in SL(V) \mid A$ leaves the form $(,)$ invariant$\}$. Taking the matrix of the form with respect to the standard basis $\{e_1, \ldots, e_{2n}\}$ of V to be

$$E = \begin{pmatrix} 0 & J \\ -J & 0 \end{pmatrix}$$

where J is the anti-diagonal $(1, \ldots, 1)$ of size $n \times n$, we may realize $G = Sp(V)$ as the subgroup of $SL(2n)$ leaving $(,)$ invariant.

I. Let $P = P_1$ be the maximal parabolic, the associated set of simple roots being $S \setminus \{\alpha_1\}$. Then G/P_1 can be identified with $\mathbb{P}(V)$. Further, fixing a coordinate system $(x_1, x_2, \ldots, x_{2n})$, the Schubert varieties in G/P_1 can be identified with the $2n$ linear subspaces in $\mathbb{P}(V)$ given by

$$x_1 = 0; \; x_1 = x_2 = 0; \; x_1 = \ldots x_{2n-1} = 0.$$

In particular, all the Schubert varieties in G/P_1 are smooth.

II. Let $P = P_n$ be the maximal parabolic, the associated set of simple roots being $S \setminus \{\alpha_n\}$. Then G/P can be identified with the isotropic Grassmannian of n-spaces in a $2n$-dimensional space with a nondegenerate skew-symmetric bilinear form $(,)$. We can identify $W_{P_n}^{\min}$ as

$$W_{P_n}^{\min} = \left\{ (a_1 \cdots a_n) \; \middle| \; \begin{array}{ll} (1) & 1 \le a_1 < a_2 < \cdots < a_n \le 2n, \; 1 \le i \le n \\ (2) & \text{for } 1 \le i \le 2n, \text{ if } i \in \{a_1, \ldots, a_n\} \\ & \text{then } 2n+1-i \notin \{a_1, \ldots, a_n\} \end{array} \right\}.$$

To $(a_1, \ldots, a_n) \in W_{P_n}^{\min}$, we associate the partition $\mathbf{a} := (\lambda_1, \ldots, \lambda_n)$, where $\lambda_{n+1-i} = a_i - i$. For example, the partition associated with $(13679) \in \mathfrak{C}_5$ is $(4, 3, 3, 1, 0)$. The conditions on the a_i's imply that the partition λ is a *self-dual* partition contained in an $n \times n$ square. For a partition $\lambda = (\lambda_1, \ldots, \lambda_n)$, we shall denote by X_λ the Schubert variety corresponding to $(a_1, \ldots, a_n)$. Thus Schubert varieties in G/P are indexed by self-dual partitions contained in $n \times n$.

9.3.6. Theorem. [112] *Let λ be a self-dual partition. Then we have* $\mathrm{Sing} X_\lambda = \cup X_\mu$, *where $\mu \subset \lambda$, and either λ/μ is a sum of two hooks that are dual to each other, or λ/μ is a self-dual hook (different from a single box).*

TYPE D. Let $V = K^{2n}$ together with a nondegenerate, symmetric bilinear form. Taking the matrix of the form (with respect to the standard basis $\{e_1, \dots, e_{2n}\}$ of V) to be the $2n \times 2n$, anti-diagonal matrix with 1s all along the anti-diagonal, the associated quadratic form Q on V is given by

$$Q(x_1, x_2, \dots, x_n, y_1, y_2, \dots, y_n) = x_1 y_n + x_2 y_{n-1} + \cdots + x_n y_1.$$

We identify G as the subgroup of $SL(2n)$ leaving Q invariant.

I. Let $P = P_1$ be the maximal parabolic, the associated set of simple roots being $S \setminus \{\alpha_1\}$. Then G/P_1 can be identified with the quadric $Q = 0$ in $\mathbb{P}(V)$. Further, all the Schubert varieties in G/P_1 are given by $Y_0, \dots, Y_{n-1}, Y'_n, Y_n, \dots, Y_{2n-2}$ defined below. We have

$$Y_0 = \{Q_n = x_1 y_n + x_2 y_{n-1} + \cdots + x_n y_1 = 0 \text{ in } \mathbb{P} = \mathbb{P}(V)\}$$
$$Y_1 = \{Q_{n-1} = x_2 y_{n-1} + \cdots + x_n y_1 = 0, \text{ and } x_1 = 0 \text{ in } \mathbb{P}\}$$

$$\vdots$$

$$Y_i = \{Q_{n-i} = x_{i+1} y_{n-i} + \cdots + x_n y_1 = 0, \text{ and } x_1 = \cdots = x_i = 0 \text{ in } \mathbb{P}\}$$

$$\vdots$$

$$Y_{n-2} = \{Q_2 = x_{n-1} y_2 + x_n y_1 = 0, \text{ and } x_1 = \cdots = x_{n-2} = 0 \text{ in } \mathbb{P}\}$$
$$Y_{n-1} = \{x_1 = \cdots = x_n = 0\} \text{ in } \mathbb{P}\}$$
$$Y'_{n-1} = \{x_1 = \cdots = x_{n-1} = y_1 = 0\}$$
$$Y_n = Y_{n-1} \cup Y'_{n-1} = \{x_1 = \cdots = x_n = y_1 = 0\} \text{ in } \mathbb{P}\}$$
$$Y_{n+1} = \{x_1 = \cdots = x_n = y_1 = y_2 = 0 \text{ in } \mathbb{P}\}$$

$$\vdots$$

$$Y_{n+i} = \{x_1 = \cdots = x_n = z = y_1 = \cdots = y_{i+1} = 0 \text{ in } \mathbb{P}\}$$

$$\vdots$$

$$Y_{2n-2} = \{x_1 = \cdots = x_n = y_1 = \cdots = y_{n-1} = 0 \text{ in } \mathbb{P}\}$$
$$= \text{ the point } \{(0, \dots, 0, 1)\}.$$

The Bruhat order on G/P_1 is almost a chain:

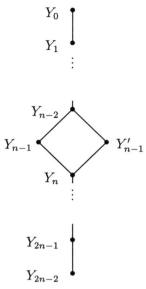

Note that here the subscript gives the codimension in G/P_1 of the corresponding Schubert variety.

We have

1. Y_0, Y'_{n-1} and $Y_{n+i}, -1 \le i \le n-2$ are nonsingular (clearly).
2. $Y_i, 1 \le i \le n-2$ are singular, and the singular locus of Y_i is $Y_{2n-1-i}, 1 \le i \le n-2$. These are obtained by computing $X_{-\beta}q_{\mathrm{id}}, \beta \in R^+$. See [**102**] for details.

II. Let $P = P_n$ be the maximal parabolic, the associated set of simple roots being $S \setminus \{\alpha_n\}$. Then G/P can be identified with the isotropic Grassmannian of n spaces. We can identify $W_{P_n}^{\min}$ as

$$
W_{P_n}^{\min} = \left\{ (a_1 \cdots a_n) \,\middle|\,
\begin{array}{ll}
(1) & 1 \le a_1 < a_2 < \cdots < a_n \le 2n,\ 1 \le i \le n \\[4pt]
(2) & \#\{i, 1 \le i \le n \mid a_i > n\} \text{ is even} \\[4pt]
(3) & \text{for } 1 \le i \le 2n, \text{ if } i \in \{a_1, \ldots, a_n\} \\
& \text{then } 2n+1-i \notin \{a_1, \ldots, a_n\}
\end{array}
\right\}.
$$

Note that for $w_1 = (a_1, \ldots, a_n), w_2 = (b_1, \ldots, b_n)$ in $W_{P_n}^{\min}$, we have that $w_1 \ge w_2$, if and only if $a_i \ge b_i,\ 1 \le i \le n$. Let $P = P_n, Q = P_{n-1}$. Consider the map $\delta : W_P \to W_Q, \delta(a_1, \ldots, a_n) = (b_1, \ldots, b_{n-1})$, where $(b_1, \ldots, b_{n-1})$ is obtained from $(a_1, \ldots, a_n)$ by replacing n by $n'(= n+1)$, resp. n' by n, if n, resp. n', is present in $\{a_1, \ldots, a_n\}$. Note that if $a_n > n$, then precisely one of $\{n, n'\}$ is present in $(a_1, \ldots, a_{n-1})$; if $a_n = n$, then $(a_1, \ldots, a_n) = (1, \ldots, n)$, and $\delta(a_1, \ldots, a_n) = (1, \ldots, n-1)$. It is easily seen that δ is a bijection preserving the Bruhat order. In fact δ is induced by the isomorphism of the varieties $G/P \to G/Q$.

Let us denote $W' = W(SO(2n-1))$, and define $\theta : W'_{P_{n-1}} \to W_P$ as $\theta(a_1, \dots, a_{n-1}) = (a_1, \dots, a_n)$, where $a_n = n$ or n' and the choice is made so that $\#\{i, 1 \le i \le n \mid a_i > n\}$ is even (the i' in $(a_1, \dots, a_{n-1})$ (resp. $\theta(a_1, \dots, a_{n-1})$) should be understood as $2n-i$ (resp. $2n+1-i$)). Then it is easily seen that θ is a bijection preserving the Bruhat order. In fact θ is induced by the isomorphism of the varieties $SO(2n-1)/P_{n-1} \to SO(2n)/P$.

In view of the isomorphisms θ, and δ, we have results for Schubert varieties in $G/P, G/Q$ (G being $SO(2n)$) similar to Theorem 9.3.5.

In the case of P being a minuscule (maximal) parabolic subgroup, as seen in §7.2, we have the following result.

9.3.7. Theorem. [30] *Let $X(w)$ be a Schubert variety in a minuscule G/P. Let Q be the parabolic subgroup maximal for the property that $X(w)$ is stable for multiplication on the left by Q. Then $\operatorname{Sing}X(w) = X(w)\backslash Qe_w$, e_w being the point in G/P corresponding to the coset wP.*

9.4. Multiplicity at a singular point

In this section, we discuss the multiplicity at a singular point on a Schubert variety. We review the results of [112] which gives a recursive formula for the multiplicity and the Hilbert polynomial at a singular point on Schubert varieties in a minuscule G/P. We have also included two closed formulas due to Kreiman–Lakshmibai ([89]), Rosenthal–Zelevinsky [142] for the multiplicity at a singular point for Schubert varieties in the Grassmannian. We have also included a closed formula for the Hilbert polynomial at a singular point due to Kreiman–Lakshmibai ([89]).

9.4.1. Multiplicity at a point P on a Schubert variety X. Let X be a Schubert variety in G/Q, Q being a parabolic subgroup. Recall the definition of $\operatorname{mult}_P X$ for $P \in X$ from Chapter 4 (cf. §4.7), namely, $\operatorname{mult}_P X = e_A$, where $A = \mathcal{O}_{X,P}$, the stalk at P (notation being as in Chapter 4) and e_A is given by $e_A = r!a$, with a, r being respectively, the leading coefficient and the degree of $P_A(x)$, the Hilbert polynomial of A. Given $P \in X$, let e_τ be the T-fixed point of the B-orbit through P. We have $\operatorname{mult}_P X = \operatorname{mult}_{e_\tau} X$. Thus, it suffices to compute $\operatorname{mult}_{e_\tau} X$ for all T-fixed points e_τ in X. In [112], $\operatorname{mult}_{e_\tau} X$ has been computed for all Schubert varieties in G/P, P being a maximal parabolic subgroup of minuscule type, i.e., the associated fundamental weight ω satisfies $(\omega, \beta^*) \le 1$, for all $\beta \in R^+$ and also for Schubert varieties in the symplectic Grassmannian $Sp(2n)/P_n$, P_n being the maximal parabolic subgroup with $S \backslash \{\alpha_n\}$ as the associated set of simple roots (see also Chapter 7 for other related results). We describe these results below. For $\theta \in W$, and $P = P_d$ (the maximal parabolic subgroup with $S \setminus \{\alpha_d\}$ as the associated set of simple roots), we shall denote by $\theta^{(d)}$, the element in W^{P_d} representing θW_{P_d}. For the rest of this section, we shall suppose that P is minuscule.

Let $P = P_d$ and $\tau, w \in W^P, w \geq \tau$. Set $R_P(w, \tau) = \{\beta \in \tau(R^+ \setminus R_P^+) \mid w \geq (s_\beta \tau)^{(d)}\}$.

9.4.2. Theorem. [112] $T(w, \tau)$, *the tangent space to* $X_P(w)$ *at* e_τ *is spanned by* $\{X_{-\beta}, \beta \in R_P(w, \tau)\}$.

Let U_τ^- be the unipotent subgroup of G generated by the root subgroups $U_{-\beta}, \beta \in \tau(R^+ \setminus R_P^+)$. We have

$$U_{-\beta} \simeq \mathbb{G}_a, \ U_\tau^- \simeq \prod_{\beta \in \tau(R^+ - R_P^+)} U_{-\beta}.$$

We shall denote the coordinate system on $U_\tau^- e_\tau$ induced by the above identification by $\{x_{-\beta}, \ \beta \in \tau(R^+ - R_P^+)\}$. We shall denote by A_τ the polynomial algebra $K[x_{-\beta}, \ \beta \in \tau(R^+ - R_P^+)]$. Let $A_{\tau,w} = A_\tau / \mathfrak{I}_P(\tau, w)$, where $\mathfrak{I}_P(\tau, w)$ is the ideal of elements of A_τ that vanish on $X_P(w) \cap U_\tau^- e_\tau$. Then $A_{\tau,w}$ is generated as an algebra by $\{x_{-\beta}, \ \beta \in R_P(w, \tau)\}$, in view of Theorem 9.4.2.

9.4.3. Homogeneity of $\mathfrak{I}_P(\tau, w)$. Let ω be the fundamental weight associated to P (P being minuscule). We shall denote the extremal weight vectors in $H^0(G/P, L_\omega)$ by $\{p_\tau, \ \tau \in W^P\}$ (§2.11.13). We have (cf. §2.11.15) that $\{p_\tau, \ \tau \in W^P\}$ is a basis for $H^0(G/P, L_\omega)$. Given $y \in \mathfrak{g}$, we identify y with the corresponding right invariant vector field D_y on G. Considering $H^0(G/P, L_\omega)$ as a $\mathfrak{g}$-module, we have

$$D_y f = yf, \ f \in H^0(G/P, L_\omega).$$

Further, the evaluations of $\frac{\partial f}{\partial x_{-\beta}}$ and X_β at e_τ coincide for $\beta \in \tau(R^+ \setminus R_P^+)$; recall that for a root γ, X_γ denotes the element in the Chevalley basis of $\mathfrak{g}$ associated to γ. We take a total order on $\tau(R^+ \setminus R_P^+)$ and denote its elements as $\{\beta_1, \ldots, \beta_N\}$, where $N = \#(R^+ \setminus R_P^+) \ (= \dim G/P)$.

Given $f \in H^0(G/P, L_\omega)$, let

$$\mathcal{D}_f = \{D = X_{-\beta_s} \cdots X_{-\beta_1} \mid Df = a_D p_\tau, \ a_D \in K^*\}.$$

Here, $X_{-\beta_s} \cdots X_{-\beta_1}$ is considered as an element of $U(\mathfrak{g})$, the universal enveloping algebra of $\mathfrak{g}$. If $\mathcal{D}_f \neq \emptyset$, then f is a weight vector for the action of T, and in A_τ, f gets identified with $\sum_{D \in \mathcal{D}_f} a_D x_D$, where if $D = X_{-\beta_s} \cdots X_{-\beta_1}$, then x_D is the monomial $x_{-\beta_s} \cdots x_{-\beta_1}$. If $D = X_{-\beta_s} \cdots X_{-\beta_1}$ we say ord $D = s$.

9.4.4. Lemma. [112] *Let* $\tau, \theta \in W^P$. *Suppose* $D_1 = X_{-\beta_s} \cdots X_{-\beta_1}$, $D_2 = X_{-\beta_t'} \cdots X_{-\beta_1'}$ *are such that* $D_i p_\theta = p_\tau a_{D_i}$, $a_{D_i} \in K^*$, $i = 1, 2$. *Then* $s = t$, *i.e.,* $\mathrm{ord}D_1 = \mathrm{ord}D_2$.

Hence we obtain that if for $f \in H^0(G/P, L_\omega)$, $\mathcal{D}_f \neq \emptyset$, then for $D_1, D_2 \in \mathcal{D}_f$, ord $D_1 = $ ord D_2; thus in A_τ, f gets identified with a homogeneous polynomial. Now $I_P(X(w))$, the ideal of $X_P(w)$ in G/P, is generated by $\{p_\theta \mid w \not\geq \theta\}$ (cf. [146]). Hence we obtain that $\mathfrak{J}_P(\tau, w)$ ($=$ the ideal of elements of A_τ that vanish on $X_P(w) \cap U_\tau^- e_\tau$) is homogeneous. Hence denoting by $M_{\tau,w}$ the maximal ideal in $A_{\tau,w}$ that corresponds to the point e_τ, and $B_{\tau,w}$ the localization of $A_{\tau,w}$ at $M_{\tau,w}$, we obtain

$$\mathrm{gr}(B_{\tau,w}, M_{\tau,w} B_{\tau,w}) = A_{\tau,w}$$

(recall (cf. [45]) that if R is a graded K-algebra generated in degree 1, and $\mathfrak{m} = R_+$, then $\mathrm{gr}(R_\mathfrak{m}, \mathfrak{m} R_\mathfrak{m}) \cong R$).

9.4.5. A basis for $(M_{\tau,w})^r/(M_{\tau,w})^{r+1}$. Let us fix $\tau, w \in W^P$, $w \geq \tau$. Let $\theta \in W^P$, $\theta \leq w$ be such that θ and τ are comparable, i.e., either $\tau \geq \theta$ or $\theta \geq \tau$.

9.4.6. Lemma. [112] *Let w, τ, θ be as above. Then there exists a $D = X_{-\beta_s} \cdots X_{-\beta_1}$, $\beta_i \in R_P(w, \tau), 1 \leq i \leq s$ such that $Dp_\theta = a_D p_\tau$, for some $a_D \in K^*$.*

For θ as above, set

$$N_P(\theta) = \{D = X_{-\beta_s} \cdots X_{-\beta_1}, \beta_i \in R_P(w, \tau), 1 \leq i \leq s$$
$$\mid Dp_\theta = a_D p_\tau, a_D \in K^*\}.$$

We have, in view of Lemma 9.4.6, $N_P(\theta) \neq \emptyset$.

9.4.7. Definition. With notation as above, define $d_\theta :=$ ord D where $D \in N_P(\theta)$; note that the right-hand side is independent of D (cf. Lemma 9.4.4), so the definition makes sense.

9.4.8. Remark. If $\theta = w$, then we refer to d_w as the *degree* of $X_P(w)$ at e_τ and denote it by $\deg_\tau w$.

9.4.9. Definition. With notation as above, define $x_\theta := \sum_{D \in N_P(\theta)} a_D x_D$, where for $D = X_{-\beta_s} \cdots X_{-\beta_1}$, we set $x_D = x_{-\beta_s} \cdots x_{-\beta_1}$.

9.4.10. Theorem. [112] *With notation as above, the set $\{(x_{\tau_1} \cdots x_{\tau_m})\}$ satisfying*

(1) $w \geq \tau_1 \geq \cdots \geq \tau_m$
(2)   *for $1 \leq i \leq m$, either $\tau_i \geq \tau$ or $\tau_i \leq \tau$*
(3) $\sum d_{\tau_i} = r$

is a basis for $(M_{\tau,w})^r/(M_{\tau,w})^{r+1}$.

9.4.11. Remark. With the above notation, we have that p_θ, as an element of $A_{\tau,w}$, gets identified with x_θ.

9.4.12. A recursive formula for the multiplicity and the Hilbert polynomial. As above, let P be a maximal parabolic subgroup of minuscule type. As a consequence of Theorem 9.4.10, we obtain

9.4.13. Theorem. [112] *Denoting* $\text{mult}_{e_\tau} X(w)$ *by* $m_\tau(w)$, *we have*

$$m_\tau(w) \deg_\tau w = \sum m_\tau(w')$$

where the sum on the right-hand side runs over all Schubert divisors $X_P(w')$ *in* $X_P(w)$ *such that* $e_\tau \in X_P(w')$.

9.4.14. Corollary. [112] *The Hilbert polynomial* $F_{\tau,w}$ *of* $X_P(w)$ *at* e_τ *is determined inductively by the formula*

$$F_{\tau,w}(r + d) = F_{\tau,w}(r) + F_{\tau,H}(r + d)$$

where H *is the union of all Schubert divisors* $X_P(w')$ *in* $X_P(w)$ *such that* $e_\tau \in X_P(w')$ *and* $d = \deg_\tau w$.

Here, $F_{\tau,H}$ is computed by induction on the number of components in H, say $H = X_P(w') \cup H'$. Then

$$F_{\tau,H}(r) = F_{\tau,H'}(r) + F_{\tau,w'}(r) - F_{\tau,\ H' \cap X_P(w')}(r),$$

where note that $H' \cap X_P(w')$ is a union of Schubert varieties of smaller dimension.

9.4.15. The case $G_{d,n}$. Let $G = GL(n)$, $P = P_d$, any maximal parabolic subgroup so that $G/P_d = G_{d,n}$ is the Grassmannian. Let $w, \tau \in I_{d,n}$, $w \geq \tau$. Let $\mathcal{O}_\tau = U_\tau^- e_\tau$. To get $\deg_\tau w$, we have to express $p_w \,|\, \mathcal{O}_\tau$ (p_w being the Plücker coordinate corresponding to w) in terms of the coordinates $\{x_{-\beta}, \ \beta \in \tau(R^+ - R_P^+)\}$ and get the corresponding degree. For example, let us take $n = 6, d = 3, \tau = \text{id}$. Then we have an identification

$$\mathcal{O}_\tau = \left\{ \begin{pmatrix} 1 & 0 & 0 \\ 0 & 1 & 0 \\ 0 & 0 & 1 \\ x_{41} & x_{42} & x_{43} \\ x_{51} & x_{52} & x_{53} \\ x_{61} & x_{62} & x_{63} \end{pmatrix} \right\}.$$

If $w = (1, 4, 5)$, then $p_w = x_{42}x_{53} - x_{52}x_{43}$ and hence $\deg_{\text{id}}(1, 4, 5) = 2$; if $w = (1, 2, 6)$, then $p_w = x_{63}$ and hence $\deg_{\text{id}}(1, 4, 5) = 1$. In particular, note that for $w = (i_1, \dots, i_d)$,

$$\deg_{\text{id}} w = d - \#\{i_k \,|\, i_k \in \{1, \dots, d\}\}.$$

More generally, let $w = (i_1, \dots, i_d)$, $\tau = (j_1, \dots, j_d)$, where we suppose that $\tau \leq w$. Then expressing $p_w \,|\, \mathcal{O}_\tau$ in terms of the coordinates $\{x_{-\beta}, \ \beta \in \tau(R^+ - R_P^+)\}$, we obtain

$$\deg_\tau w = d - \#\{i_k \,|\, i_k \in \{j_1, \dots, j_d\}\}.$$

Observe that $m_\tau(w)$ may be interpreted in terms of maximal weighted chains in the Hasse diagram of $[\tau, w]$, where the maximal chains are given weights, as follows:

To each element $\phi \in [\tau, w]$, assign the number $\frac{1}{\deg_\tau \phi}$ or 1 according to whether $\phi >$ or $= \tau$. To a maximal chain $\underline{c} := \{\phi_0 = w > \phi_1 > \cdots > \phi_r = \tau\}$, where $r = l(w) - l(\tau)$, the codimension of $X(\tau)$ in $X(w)$, assign the weight $m(\underline{c}) := \frac{1}{\prod_{t=0}^r \deg_\tau \phi_t}$. Then $m_\tau(w)$ is simply $\sum_{\underline{c}} m(\underline{c})$, where the summation runs over all the maximal chains $\underline{c}$ in $[\tau, w]$.

9.4.16. Example. Consider $X(356) \subset G_{3,6}$. Take $\tau = (235)$. We have the following Hasse diagram:

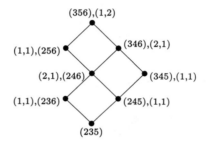

where at each vertex ϕ, the pair (a, b) stands for $(\deg\phi, m_\tau(\phi))$. We have the following five maximal chains:

$$\underline{c}_1 := \{(356) > (256) > (246) > (236) > (235)\}, \quad m(\underline{c}_1) = \frac{1}{2}$$

$$\underline{c}_2 := \{(356) > (256) > (246) > (245) > (235)\}, \quad m(\underline{c}_2) = \frac{1}{2}$$

$$\underline{c}_3 := \{(356) > (346) > (246) > (236) > (235)\}, \quad m(\underline{c}_3) = \frac{1}{4}$$

$$\underline{c}_3 := \{(356) > (346) > (246) > (245) > (235)\}, \quad m(\underline{c}_1) = \frac{1}{4}$$

$$\underline{c}_3 := \{(356) > (346) > (345) > (245) > (235)\}, \quad m(\underline{c}_1) = \frac{1}{2}.$$

Hence we obtain $\sum_{i=1}^5 m(\underline{c}_i) = 2$ which is in fact equal to $m_\tau(w)$, as computed from Theorem 9.4.13.

9.4.17. A closed formula for the multiplicity and the Hilbert polynomial. In this section, we give closed formulas for the multiplicities and the Hilbert polynomials at singular points on Schubert varieties in the Grassmannian.

Let $X = X(w)$ be a Schubert variety in the Grassmannian $G_{d,n}$ ($\cong SL(n)/P_d$). Let us denote P_d by just P. Consider a $\tau \in W^P$ such that $\tau \leq w$.

9.4.18. Evaluation of Plücker coordinates on $U_{\mathrm{id}}^- e_{\mathrm{id}}$. Let us first consider the case $\tau = \mathrm{id}$. We identify $U^- e_{\mathrm{id}}$ with

$$
\left\{
\begin{pmatrix}
\mathrm{Id}_{d \times d} \\
\begin{matrix} x_{d+1\,1} & \cdots & x_{d+1\,d} \\ \vdots & & \vdots \\ x_{n\,1} & \cdots & x_{n\,d} \end{matrix}
\end{pmatrix},
\quad x_{ij} \in K, \quad d+1 \le i \le n, 1 \le j \le d
\right\}.
$$

Let A be the affine algebra of $U^- e_{\mathrm{id}}$. Let us identify A with the polynomial algebra $K[x_{-\beta}, \beta \in R^+ \setminus R_P^+]$. To be very precise, we have $R^+ \setminus R_P^+ = \{\epsilon_j - \epsilon_i, 1 \le j \le d, d+1 \le i \le n\}$; given $\beta \in R^+ \setminus R_P^+$, say $\beta = \epsilon_j - \epsilon_i$, we identify $x_{-\beta}$ with x_{ij}. As seen in §9.4.3. we have that the expression for $p_\theta|_{U^- e_{\mathrm{id}}}$ in the local coordinates $x_{-\beta}$'s is homogeneous.

9.4.19. Example. Consider $G_{2,4}$. Then

$$
U^- e_{\mathrm{id}} =
\left\{
\begin{pmatrix}
1 & 0 \\
0 & 1 \\
x_{31} & x_{32} \\
x_{41} & x_{42}
\end{pmatrix},
\quad x_{ij} \in K
\right\}.
$$

On $U^- e_{\mathrm{id}}$, we have $p_{12} = 1$, $p_{13} = x_{32}$, $p_{14} = x_{42}$, $p_{23} = x_{31}$, $p_{24} = x_{41}$, $p_{34} = x_{31} x_{42} - x_{41} x_{32}$.

Let $Y_P(w) = X_P(w) \cap U^- e_{\mathrm{id}}$. As in section §9.4.3, let $\mathfrak{I}_P(w)$ ($=$ the ideal of elements of A that vanish on $Y_P(w)$) is homogeneous, and is generated by $\{p_\theta|_{U^- e_{\mathrm{id}}}|\ w \not\ge \theta\}$. Let $A_w = A/\mathfrak{I}_P(w)$.

9.4.20. A Gröbner basis for $\mathfrak{I}_P(w)$.

9.4.21. Definition. Let $K[X] := K[x_1, \cdots, x_n]$. A *monomial order* $\prec$ on the set of monomials in $K[X]$ is a total order such that given monomials $m, m_1, m_2, m_1 \prec m_2, m \ne 1$, we have, $m_1 \prec m m_1$, and $m m_1 \prec m m_2$. The largest monomial (with respect to $\prec$) present in a polynomial $f \in K[X]$ is called the *initial term* of f, and is denoted by $in(f)$.

9.4.22. Definition. Given an ideal $I \subset K[X]$, denote by $in(I)$ the ideal generated by the initial terms of the elements in I. A finite set $G \subset I$ is called a *Gröbner basis* of I (with respect to the monomial order $\prec$), if $in(I)$ is generated by the initial terms of the elements in G.

9.4.23. Remark. Note that a Gröbner basis of I generates I as an ideal.

Consider $K[X] := K[x_{d+1\,1}, x_{d+1\,2}, \cdots x_{d+1\,d}, \cdots x_{n\,1}, x_{n\,2}, \cdots x_{n\,d}]$. We introduce a total order on the variables as follows:

$$
x_{n1} > x_{n2} > \cdots > x_{nd} > x_{(n-1)1} > x_{(n-1)2}
$$
$$
> \cdots > x_{(n-1)d} > \cdots > x_{d+1\,1} > x_{d+1\,2} > \cdots > x_{d+1\,d}.
$$

9.4.24. Remark. Let

$$Z = \{(i,j), 1 \le j \le d, d+1 \le i \le n\}.$$

Note that the above order extends the partial order on Z given by $(i,j) \succeq (k,l) \Longleftrightarrow j \le l < k \le i$. Note also that if s_{ij} denotes the transposition in S_n corresponding to (i,j), then the latter condition is equivalent to the condition that $s_{ij} \ge s_{kl}$ (as elements of S_n; note that for $(i,j) \in Z, s_j s_{j+1} \cdots s_{i-1}$ gives a reduced expression for s_{ij}).

The above total order induces a total order, namely the lexicographic order, on the set of monomials in $K[X]$; denote it by $\prec$. Note that $\prec$ is a monomial order. Note also that the initial term (with respect to $\prec$) of a minor of X is equal to the product of its elements on the skew diagonal.

9.4.25. Proposition. [89] *For* $\theta \in W^P(= I_{d,n})$, *let* $f_\theta = p_\theta|_{U^- e_{\mathrm{id}}}$. *Then* $\{f_\theta \mid w \not\succeq \theta\}$ *is a Gröbner basis for* $\mathfrak{I}_P(w)$.

9.4.26. Definition. Given a graded ring $R = \oplus_{m \in \mathbb{Z}^+} R_m$,

$$H_R(t) := \sum_{m \in \mathbb{Z}^+} (\dim_K R_m) t^m$$

is called the *Hilbert series* of R.

Let us denote $in(\mathfrak{I}_P(w))$ by $\mathfrak{J}_P(w)$; set $C_w = A/\mathfrak{J}_P(w)$. Note that $\mathfrak{J}_P(w)$ is generated by square-free monomials. The following theorem gives an explicit closed expression for the Hilbert series of the graded ring C_w. Let Z_w be the set of all $J \subseteq Z$ such that for every subset $\{(i_1,j_1),(i_2,j_2),\ldots,(i_r,j_r)\} \subset J$ with $j_1 < j_2 < \ldots j_r < i_r < \ldots < i_1$ we have $w \ge s_{i_1 j_1} \cdots s_{i_r j_r}$ in W^P. Note in particular $(i,j) \in Z_w \Longleftrightarrow w \ge s_{ij}(\text{in } W^P)$, i.e, if and only if $X_{-\beta}$ belongs to $T(w,\mathrm{id})$, β being the root $\epsilon_j - \epsilon_i \in R^+ \setminus R_P^+$ (cf. Theorem 9.4.2). Note also that a subset $\{(i_1,j_1),(i_2,j_2),\ldots(i_r,j_r),j_1 < j_2 < \ldots j_r < i_r < \ldots < i_1\}$ of Z corresponds to a collection of mutually commuting reflections $s_{\gamma_1},\ldots,s_{\gamma_r}$ with $s_{\gamma_1} > \ldots > s_{\gamma_r}$.

9.4.27. Theorem. [89] $H_{C_w}(t) = \sum_{m \in \mathbb{Z}^+}(\sum_{J \in Z_w} \binom{m-1}{\#J-1} t^m)$.

9.4.28. Corollary. *Let* r_w *be the maximum in* $\{\#J, J \in Z_w\}$. *Then* $H_{C_w}(m)$, *the Hilbert polynomial of* $C_w = \sum_{\{J \in Z_w, \#J=r_w\}} \binom{m-1}{\#J-1}$. *In particular, the degree of* $H_{C_w}(m) = r_w - 1$.

9.4.29. Corollary. e_{C_w}, *the multiplicity of* $C_w = \#\{J|\#J = r_w\}$.

9.4.30. Theorem. [89] *The multiplicity,* m_w, *of* $X(w)$ *at the point* e_{id} *equals* $\#\{J|\#J = r_w\}$.

PROOF. The result follows from the general theory of Gröbner bases (cf. [45]). Since $A_w = A/\mathfrak{I}_P(w)$, $C_w = A/\mathfrak{J}_P(w)$, and $\mathfrak{J}_P(w)$ equals $in(\mathfrak{I}_P(w))$, there exists a flat family over $\mathbb{A}^1 (= \operatorname{Spec} K[t])$ with generic fiber (u invertible) Spec A_w and special fiber ($u = 0$) Spec C_w. Now A_w and C_w being graded, we have in addition that A_w and C_w have the same Hilbert polynomial and hence the same multiplicity. The required result now follows since m_w equals the multiplicity of A_w.   □

9.4.31. Example. Consider $G_{2,4}$ (cf. Example 9.4.19), and $w = (2413)$. We have $K[X] = K[x_{31}, x_{32}, x_{41}, x_{42}]$. We have, $J \in Z_w$ if and only if $J \not\supset \{(3,2),(4,1)\}$. Hence we get, $Z_w = Z_1 \cup Z_2 \cup Z_3$, where

$$Z_1 = \{\{(3,1)\}, \{(3,2)\}, \{(4,1)\}, \{(4,2)\}\}$$
$$Z_2 = \{\{(3,1),(3,2)\}, \{(3,1),(4,1)\}, \{(3,1),(4,2)\},$$
$$\{(4,1),(4,2)\}, \{(3,2),(4,2)\}\}$$
$$Z_3 = \{\{(3,1),(4,1),(4,2)\}, \{(3,1),(3,2),(4,2)\}\}.$$

We have, $r_w = 3, m_w = 2$.

9.4.32. Evaluation of Plücker coordinates on $U_\tau^- e_\tau$. Let now τ be any other element in W^P, say $\tau = (a_1, \ldots, a_n)$. Then $U_\tau^- e_\tau$ consists of $\{N_{d,n}\}$, where $N_{d,n}$ is obtained from $\begin{pmatrix} \text{Id} \\ X \end{pmatrix}_{n \times d}$ (with notations as above) by permuting the rows by τ^{-1}. Note that $U_\tau^- e_\tau = \tau U^- e_{\text{id}}$.

9.4.33. Example. Consider $G_{2,4}$, and let $\tau = (2314)$. Then $\tau^{-1} = (3124)$, and

$$U_\tau^- e_\tau = \left\{ \begin{pmatrix} x_{31} & x_{32} \\ 1 & 0 \\ 0 & 1 \\ x_{41} & x_{42} \end{pmatrix}, \quad x_{ij} \in k \right\}.$$

Note that the $\text{Id}_{d \times d}$ matrix is in the position determined by the entries in τ considered as a d-tuple; in the above example, the 2-tuple corresponding to τ is $(2,3)$. We have on $U_\tau^- e_\tau$, $p_{12} = -x_{32}$, $p_{13} = x_{31}$, $p_{14} = x_{31}x_{42} - x_{41}x_{32}$, $p_{23} = 1$, $p_{24} = x_{42}$, $p_{34} = -x_{41}$.

As in the case $\tau = \text{id}$, we have that for $\theta \in W^P$, $p_\theta|_{U_\tau^- e_\tau}$ is homogeneous in local coordinates.

9.4.34. Proposition. *Let $\theta \in W^P$. We have a natural isomorphism*

$$K[x_{-\beta}, \beta \in R^+ \setminus R_P^+] \cong K[x_{-\tau(\beta)}, \beta \in R^+ \setminus R_P^+],$$

given by

$$p_\theta \mapsto p_{\tau\theta},$$

here we have denoted $p_\theta|_{U^- e_{\text{id}}}$, $p_{\tau\theta}|_{U_\tau^- e_\tau}$ by just p_θ, $p_{\tau\theta}$, respectively.

We have similar results at e_τ to those at e_{id} which we state below. Let $A_\tau = K[x_{-\tau(\beta)}, \beta \in R^+ \setminus R_P^+]$. Let $Y_P(\tau, w) = X_P(w) \cap U_\tau^- e_\tau$, $\mathfrak{I}_P(\tau, w)$ (=

the ideal of elements of A_τ that vanish on $Y_P(\tau, w)$). Then $\mathfrak{I}_P(\tau, w)$ is homogeneous, and is generated by $\{p_\theta|_{U_\tau^- e_\tau} \mid w \not\geq \theta\}$. Let $A_{\tau, w} = A_\tau / \mathfrak{I}_P(\tau, w)$. We take the same monomial order as before.

9.4.35. Proposition. [89] *For $\theta \in W^{P_d}(= I_{d,n})$, let $f_{\tau, \theta} = p_\theta|_{U_\tau^- e_\tau}$. Then $\{f_{\tau, \theta} \mid w \not\geq \theta\}$ is a Gröbner basis for $\mathfrak{I}_P(\tau, w)$.*

Let us denote $in(\mathfrak{I}_P(\tau, w))$ by $\mathfrak{J}_P(\tau, w)$; set $C_{\tau, w} = A_\tau / \mathfrak{J}_P(\tau, w)$. Let

$$Z_\tau = \{\tau(i, j), (i, j) \in Z\},$$

where recall that $Z = \{(i, j), 1 \leq j \leq d, d + 1 \leq i \leq n\}$; note that Z_τ may be identified with $\tau(R^+ \setminus R_P^+)$.

Let $Z_{\tau, w}$ be the set of all $J \subseteq Z_\tau$ such that for every subset of the form $\{\tau(i_1, j_1), \tau(i_2, j_2), \ldots \tau(i_r, j_r)\} \subset J$ with $j_1 < j_2 < \ldots j_r < i_r < \ldots < i_1$, we have $w \geq \tau s_{i_1 j_1} \cdots s_{i_r j_r}$ in W^P. Note in particular $(i, j) \in Z_{\tau, w} \iff w \geq \tau s_{ij}$ (in W^P), i.e, if and only if $X_{\tau(-\beta)}$ belongs to $T(w, \tau)$, β being the root $\epsilon_j - \epsilon_i \in R^+ \setminus R_P^+$ (cf. Theorem 9.4.2).

9.4.36. Theorem. [89] $H_{C_{\tau, w}}(t) = \sum_{m \in \mathbb{Z}^+}(\sum_{J \in Z_{\tau, w}} \binom{m-1}{\#J-1} t^m)$.

9.4.37. Corollary. *Let $r_{\tau, w}$ be the maximum in $\{\#J, J \in Z_{\tau, w}\}$. Then $H_{C_{\tau, w}}(m)$, the Hilbert polynomial of $C_{\tau, w} = \sum_{\{J \in Z_{\tau, w}, \#J = r_{\tau, w}\}} \binom{m-1}{\#J-1}$. In particular, the degree of $H_{C_{\tau, w}}(m) = r_{\tau, w} - 1$.*

9.4.38. Corollary. *$e_{C_{\tau, w}}$, the multiplicity of $C_{\tau, w} = \#\{J \in Z_{\tau, w} | \#J = r_{\tau, w}\}$.*

9.4.39. Theorem. [89] *The multiplicity, $m_\tau(w)$, of $X(w)$ at the point e_τ equals $\#\{J \in Z_{\tau, w} | \#J = r_{\tau, w}\}$.*

9.4.40. Other determinantal formulas. We list below some determinantal formulas for $m_{id}(w)$ due to Herzog–Trung [**64**], Lakshmibai–Weyman [**112**], Rosenthal–Zelevinsky [**142**]. Rosenthal and Zelevinsky have in fact given determinantal formulas for $m(\tau, w), \tau \leq w$.

Historically, Abhyankar's results on the Hilbert series of the determinantal varieties gave the first formulas for the multiplicities at singular points on a Schubert variety [**3**]. Recall that a determinantal variety is the "opposite cell" in a suitable Schubert variety in a suitable Grassmannian. The Herzog–Trung formula below is a generalization of Abhyankar's formula. See also [**51, 88, 90**] for related results on determinantal varieties.

Herzog–Trung formula. Given $w \in I_{d,n}$, say $w = \{i_1, \ldots, i_d\}$. Let i_r be the largest entry $\leq d$. Let $s = d - r$. Let $\{b_1, \ldots, b_s\}$ be the complement of $\{j_1, \ldots, j_r\}$ in $\{1, 2, \ldots, d\}$, and $\{a_1, \ldots, a_s\} = \{n + 1 - i_d, n + 1 - i_{d-1}, \ldots, n + 1 - i_{r+1}\}$. We have

9.4.42. Theorem. [**64**] $m_{id}(w) = det\left(\binom{n-a_i-b_j}{n-a_i-d}\right)_{1 \leq i,j \leq s}$.

9.4.43. Example. Consider $w = (246)$ in $G_{3,6}$. We have, $d = 3, r = 1, s = 2, \{a_1, a_2\} = \{1, 3\} = \{b_1, b_2\}$, and $m_{\mathrm{id}}(w) = 5$.

Lakshmibai–Weyman formula. In order to describe the Lakshmibai–Weyman formula, we need to use the Frobenius notation for a partition which we describe first.

Given a partition $\lambda = (\lambda_1 \geq \lambda_2 \geq \ldots \lambda_r)$, represent λ as a diagram consisting of the points (i, j) of $\mathbb{Z}^2$ such that $1 \leq j \leq \lambda_i$. For example $\lambda = (5, 4, 3, 1)$ has the diagram

Here, we have adopted the convention, as with matrices, that the first coordinate i (the row index) increases as one goes downwards, and the second coordinate j (the column index) increases as one goes from left to right.

Let $\lambda' = (\lambda'_1 \geq \lambda'_2 \geq \ldots \lambda'_s)$ denote the conjugate of λ, namely the partition whose diagram is the transpose of λ, i.e., the diagram obtained by reflection in the main diagonal. In the above example, $\lambda' = (4, 3, 3, 2, 1)$. Let r be the number of nodes on the main diagonal (note that r is the largest integer t such that (t, t) belongs to the diagram of λ; note also that r equals $\#\{i | \lambda_i \geq i\}$). For $1 \leq i \leq r$, let $c_i := \lambda_i - i$, the number of nodes in the i-th row of λ to the right of (i, i), and $d_i := \lambda'_i - i$, the number of nodes in the i-th column of λ below (i, i). In the above example, we have $r = 3, (c_1, c_2, c_3) = (4, 2, 0), (d_1, d_2, d_3) = (3, 1, 0)$. The number r is called the *Durfee square size of* λ. One denotes λ by $(c_1, \ldots, c_r | d_1, \ldots, d_r)$ and calls it the *Frobenius notation* for λ. Note that

$$\lambda = (c_1+1, c_2+2, \ldots, c_r+r, (r)^{d_r}, (r-1)^{d_{r-1}-d_r-1}, \ldots, (2)^{d_2-d_3-1}, (1)^{d_1-d_2-1}).$$

Let $w = (a_1, \ldots, a_d) \in I_{d,n}$. Let $\lambda_w := (\lambda_1 \geq \lambda_2 \geq \ldots \geq \lambda_d)$ be the partition $(i_d - d, \ldots, i_2 - 2, i_1 - 1)$; note that $\lambda_i = a_{d+1-i} - (d+1-i)$. Let $\lambda_w = (c_1, \ldots, c_r | d_1, \ldots, d_r)$ be the Frobenius notation for λ_w. Note that $\lambda_i \geq i \Longleftrightarrow a_{d+1-i} \geq d+1$. Hence the Durfee square size r is just $\#\{i | \alpha_i \geq d+1\}$, which is just $\deg_{\mathrm{id}} w$ (cf. Remarks 9.4.8 and 9.4.15).

9.4.45. Theorem. [112] *Let notations be as above. Then*

$$m_{\mathrm{id}}(w) = \det\left(\binom{c_i + d_j}{c_i}\right)_{1 \leq i, j \leq r}.$$

9.4.46. Example. Consider $w = (236)$ in $G_{3,6}$. We have, $d = 3, r = 1, c_1 = 2 = d_1$, and $m_{\mathrm{id}}(w) = 6$.

Rosenthal–Zelevinsky formula. Rosenthal and Zelevinsky [142] have given a determinantal formula for $m_\tau(w)$ which we describe below.

For any d-tuple $\mathbf{i} = (i_1, \ldots, i_d)$, set $|\mathbf{i}| = i_1 + \cdots + i_d$. Let $w, \tau \in I_{d,n}$ be such that $w \geq \tau$ with $w = (i_1, \ldots, i_d)$ and $\tau = (j_1, \ldots, j_d)$. Then by Theorem 9.4.13, $m_\tau(w)$ is the unique solution of the partial difference equation

$$(9.4.48) \qquad m_\tau(w) = \frac{1}{\deg(\tau, w)} \sum_{\substack{|\mathbf{k}| = |w| - 1 \\ \tau \leq \mathbf{k} < w}} m_\tau(\mathbf{k})$$

subject to the initial condition $m_\tau(\tau) = 1$ and subject to the boundary condition $m_\tau(\mathbf{k}) = 0$ for any d-tuple $\mathbf{k}$ having the property that some of the indices $1 \leq k_1 \leq \ldots \leq k_d \leq n$ are equal.

9.4.49. Theorem. [142] *The multiplicity $m_\tau(w)$ is given by*

$$(9.4.50) \quad m_\tau(w) = (-1)^{s_1 + \cdots + s_d} \det \begin{bmatrix} \binom{i_1}{-s_1} & \cdots & \cdots & \binom{i_d}{-s_d} \\ \binom{i_1}{1-s_1} & \cdots & \cdots & \binom{i_d}{1-s_d} \\ \vdots & & & \vdots \\ \binom{i_1}{d-1-s_1} & \cdots & \cdots & \binom{i_d}{d-1-s_d} \end{bmatrix},$$

where

$$(9.4.51) \qquad s_q := \#\{j_p \mid i_q < j_p\}.$$

Example. Assume the indices w, τ satisfy $j_d \leq i_1$. In this situation the numbers $s_1, \ldots, s_d$ attain the smallest possible value: $s_1 = \ldots = s_d = 0$. Then the (p, q)-entry of the determinant in (9.4.50) has the form $P_p(i_q)$, where $P_p(t)$ is a polynomial with the leading term $t^{p-1}/(p-1)!$. It follows that

$$m_\tau(w) = \frac{1}{1! \cdots (d-1)!} V(w) = \frac{1}{1! \cdots (d-1)!} \prod_{p > q} (i_p - i_q),$$

where $V(w)$ is the Vandermonde determinant.

Example. Consider $w = (245), \tau = (124)$ in $G_{3,6}$. We have, $s_1 = 1, s_2 = 0 = s_3$, and $m_\tau(w) = 2$.

The above theorem is proved using the lemma below. Before stating the lemma, we need to introduce some notation. To any nonnegative integer vector $\mathbf{s} = (s_1, \ldots, s_d)$ we associate a polynomial $P_\mathbf{s}(\mathbf{t}) \in \mathbb{Q}[\mathbf{t}] = \mathbb{Q}[t_1, \ldots, t_d]$ defined by

$$(9.4.52) \qquad P_\mathbf{s}(\mathbf{t}) = (-1)^{|\mathbf{s}|} \det \begin{bmatrix} \binom{t_1}{-s_1} & \cdots & \cdots & \binom{t_d}{-s_d} \\ \binom{t_1}{1-s_1} & \cdots & \cdots & \binom{t_d}{1-s_d} \\ \vdots & & & \vdots \\ \binom{t_1}{d-1-s_1} & \cdots & \cdots & \binom{t_d}{d-1-s_d} \end{bmatrix};$$

here $\binom{t}{s}$ is the polynomial $t(t-1)\cdots(t-s+1)/s!$ for $s \geq 0$, and $\binom{t}{s} = 0$ for $s < 0$.

For $q = 1, \ldots, d$, let $\Delta_q : \mathbb{Q}[\mathbf{t}] \to \mathbb{Q}[\mathbf{t}]$ denote the partial difference operator $\Delta_q P(\mathbf{t}) = P(\mathbf{t}) - P(\mathbf{t} - e_q)$, where $e_1, \ldots, e_d$ are the unit vectors in $\mathbb{Q}^d$.

9.4.53. Lemma. *For any nonnegative integer vector* $\mathbf{s}$, *the corresponding polynomial* $P_{\mathbf{s}}(\mathbf{t})$ *satisfies the partial difference equation*

$$(9.4.54) \qquad (\Delta_1 + \ldots + \Delta_d)P = 0 .$$

9.4.55. Remark. The space of all polynomial solutions of the partial difference equation (9.4.54) can be described as follows. Let $\mathbf{y} = (y_1, \ldots, y_d)$ be an auxiliary set of variables, and let $\varphi : \mathbb{Q}[\mathbf{y}] \to \mathbb{Q}[\mathbf{t}]$ be the isomorphism of vectors spaces that sends each monomial $\prod_{q=1}^{d} y_q^{n_q}$ to $\prod_{q=1}^{d} t_q(t_q + 1) \cdots (t_q + n_q - 1)$. The map φ intertwines each Δ_q with the partial derivative $\frac{\partial}{\partial y_q}$. It follows that the space of solutions of (9.4.54) is the image under φ of the $\mathbb{Q}$-subalgebra in $\mathbb{Q}[\mathbf{y}]$ generated by all differences $y_p - y_q$.

9.4.56. Remark. The formula for $m_\tau(w)$ as given by Theorems 9.4.30, 9.4.39 has the added advantage over the formulas as given by Theorems 9.4.42, 9.4.45, 9.4.49 in that it does not involve any negative expressions; note that the latter formulas being determinantal formulas involve many negative terms.

9.5. The symplectic Grassmannian $Sp(2n)/P_n$

Let $G = Sp(2n)$ and P_n be the maximal parabolic subgroup with $S \setminus \{\alpha_n\}$ as the associated set of simple roots. We shall denote P_n by P. Recall from Chapter 3 that

$$W^P = \left\{ (a_1, \ldots, a_n) \middle| \begin{array}{ll} (1) & 1 \leq a_1 < \ldots < a_n \leq 2n \\ (2) & \text{for } 1 \leq i \leq n, \text{ precisely one of } \{i, i'\} \text{ is in} \\ & \{a_1, \ldots, a_n\} \end{array} \right\},$$

where, $i' = 2n + 1 - i$.

9.5.1. Description of $T(w, \tau)$. Let $w, \tau \in W^P$, $w \geq \tau$. Let $N_P(w, \tau) = \{\beta \in \tau(R^+ \setminus R_P^+) \mid X_{-\beta} \in T(w, \tau)\}$. As above, for $\theta \in W$, we shall denote by $\theta^{(n)}$ the element in W^P representing θW_P. Recall (cf. Theorem 5.4.5)

9.5.2. Theorem. *Let* $\beta = \tau(\alpha)$, *where* $\alpha \in R^+$.

1. *Let* $\alpha = \epsilon_j - \epsilon_n, 1 \leq j < n$, *or* $2\epsilon_j, 1 \leq j \leq n$. *Then* $\beta \in N_P(w, \tau)$ *if and only if* $w \geq (s_\beta \tau)^{(n)}$.
2. *Let* $\alpha = \epsilon_j + \epsilon_k, 1 \leq j < k \leq n$.
   (a) *Let* $\tau > (\tau s_\alpha)^{(n)} (= (s_\beta \tau)^{(n)})$. *Then* $\beta \in N_P(w, \tau)$ *necessarily*.
   (b) *Let* $\tau < (\tau s_\alpha)^{(n)}$.

(i) Let $\tau > (\tau s_{2\epsilon_j})^{(n)}$ or $(\tau s_{2\epsilon_k})^{(n)}$. Then $\beta \in N_P(w, \tau)$ if and only if $w \geq (s_\beta \tau)^{(n)}$.

(ii) Let $\tau < (\tau s_{2\epsilon_j})^{(n)}$ and $(\tau s_{2\epsilon_k})^{(n)}$.

    (A) If $\tau < (\tau s_{\epsilon_j - \epsilon_k})^{(n)}$, then $\beta \in N_P(w, \tau)$ if and only if $w \geq (s_\beta \tau)^{(n)} (= (\tau s_{\epsilon_j + \epsilon_k})^{(n)})$ or $(\tau s_{2\epsilon_j})^{(n)}$.

    (B) If $\tau > (\tau s_{\epsilon_j - \epsilon_k})^{(n)}$, then $\beta \in N_P(w, \tau)$ if and only if $w \geq (\tau s_{\epsilon_j - \epsilon_k} s_{2\epsilon_j})^{(n)}$.

As in Chapter 3, let $V = K^{2n}$ together with a nondegenerate skew symmetric bilinear form $(,)$. Then G/P can be identified with the set of all maximally isotropic subspaces ($= n$-dimensional subspaces $U \mid (u_1, u_2) = 0$, $u_1, u_2 \in U\}$).

We have a canonical inclusion

$$G/P \hookrightarrow G_{n,2n},$$

$G_{n,2n}$ being the Grassmannian of n-dimensional subspaces in $V = K^{2n}$. Let us consider the Plücker embedding of $G_{n,2n}$:

$$G/P \hookrightarrow G_{n,2n} \hookrightarrow \mathbb{P}(\wedge^n V).$$

Let L be the tautological line bundle on $\mathbb{P}(\wedge^n V)$. We shall denote the restrictions of L to $G_{n,2n}$ and G/P also by L. We have a canonical surjective map (induced by restriction)

$$H^0(G_{n,2n}, L) \to H^0(G/P, L).$$

For $\theta = (\theta_1, \ldots, \theta_n)$, $1 \leq \theta_1 < \ldots < \theta_n \leq 2n$, let us denote the associated Plücker coordinate on $G_{n,2n}$ by f_θ and the restriction of f_θ to G/P by f'_θ. Let $\theta = (a_1, \ldots, a_r, y_1, y'_1, y_2, y'_2, \ldots, y_s, y'_s)$, where $y_j \leq n$, $1 \leq j \leq s$ and $a'_i (= 2n + 1 - a_i) \notin \theta, 1 \leq i \leq r$. Let $\{x_1, \ldots, x_t\}$ be the complement of $\{|a_1|, \ldots, |a_r|, y_1, y_2, \ldots, y_s\}$ in $\{1, \ldots, n\}$ (where $|q| = \min\{q, 2n + 1 - q, 1 \leq q \leq 2n\}$). Let $\theta' = (a_1, \ldots, a_r, x_1, x'_1, x_2, x'_2, \ldots, x_t, x'_t)$. Then we have (cf. [108])

1. $f'_\theta = f'_{\theta'}$.
2. $\{f'_\theta : \theta \in I_{n,2n}\}$ is a basis of $H^0(G/P, L)$ (note that if $\theta = \theta'$ — in which case $s = 0$ — then $f'_\theta = p_\theta$, the extremal weight vector in $H^0(G/B, L)$ associated to θ).

Now we also have a basis $\{p_{\delta,\phi}\}$ for $H^0(G/P, L)$ indexed by admissible pairs in W^P (cf. Chapter 6 or [110]). The admissible pairs in W^P are simply pairs $\delta = (a_1, \ldots, a_n)$, $\phi = (b_1, \ldots, b_n)$ such that

$$(a_1, \ldots, a_n) \geq (b_1, \ldots, b_n), \quad \#\{i \mid a_i > n\} = \#\{i \mid b_i > n\}.$$

(If $\delta = \phi$, then we shall denote $p_{\delta,\delta}$ by just p_δ.)

Further, the two bases $\{f'_\theta\}$ and $\{p_{\delta,\phi}\}$ are one and the same. To be very precise, if $\theta \in W^P$, then $f'_\theta = \pm p_\theta$; if $\theta = (a_1, \ldots, a_r, y_1, y'_1, y_2, y'_2, \ldots, y_s,$

$y'_s)$, $\theta' = (a_1, \ldots, a_r, x_1, x'_1, x_2, x'_2, \ldots, x_t, x'_t)$, then up to ± 1, $f'_\theta = f'_{\theta'} = p_{\delta,\phi}$, where $\delta =$ bigger of $\{\rho, \xi\}$, $\phi =$ smaller of $\{\rho, \xi\}$, and ρ and ξ are given by

$$\rho = (a_1, \ldots, a_r, y_1, y_2, \ldots, y_s, x'_1, \ldots, x'_s),$$

$$\xi = (a_1, \ldots, a_r, y'_1, y'_2, \ldots, y'_s, x_1, \ldots, x_s).$$

9.5.3. Remark. With notation as above, we have for $\tau \in W^P$, $\tau \geq \theta$, resp. $\tau \leq \theta$, if and only if $\tau \geq \theta'$, resp. $\tau \leq \theta'$.

In view of the above facts, we can derive the results for G/P from those for $G_{n,2n}$, as discussed in §9.4.3.

Let $U^-_\tau, A_\tau, A_{\tau,w}, M_{\tau,w}$, etc., be defined as in §9.4.3. We have that $A_{\tau,w}$ is generated as an algebra by $\{x_{-\beta}, \ \beta \in N_P(w, \tau)\}$ in view of Theorem 9.5.2. For $f \in H^0(G/P, L)$, $\mathcal{D}_f$ is defined as in §9.4.3. If $\mathcal{D}_f \neq \emptyset$, then in A_τ, f is identified with $\sum_{D \in \mathcal{D}_f} a_D x_D$, where if $D = X_{-\beta_s} \cdots X_{-\beta_1}$, then x_D is the monomial $x_{-\beta_s} \cdots x_{-\beta_1}$. Hence we obtain that if for $f \in H^0(G/P, L)$, $\mathcal{D}_f \neq \emptyset$, then in A_τ, f gets identified with a homogeneous polynomial. Now $I(X_P(w))$, the ideal of $X_P(w)$ in G/P is generated by $\{p_{\delta,\phi} \mid w \not\geq \delta\}$ (cf. [**110**]). Hence we obtain that $\mathfrak{I}_P(\tau, w)$ ($=$ the ideal of elements of A_τ that vanish on $X_P(w) \cap U^-_\tau e_\tau$) is homogeneous. Hence denoting by $M_{\tau,w}$ the maximal ideal in $A_{\tau,w}$ that corresponds to the point e_τ, and $B_{\tau,w}$ the localization of $A_{\tau,w}$ at $M_{\tau,w}$, we obtain

$$\mathrm{gr}(B_{\tau,w}, M_{\tau,w} B_{\tau,w}) = A_{\tau,w}.$$

9.5.4. Lemma. [**112**] *Let $w, \tau \in W^P$, $w \geq \tau$. Let δ, ϕ, θ be such that $p_{\delta,\phi} = f'_\theta$. Further, let θ and τ be comparable as n-tuples. Then there exists an operator $D = X_{-\beta_s} \cdots X_{-\beta_1}$, $\beta_i \in R_P(w, \tau)$, $1 \leq i \leq s$, where $R_P(w, \tau) = \{\beta \in \tau(R^+ - R^+_P) \mid w \geq s_\beta \tau\}$ such that $D p_{\delta,\phi} = p_\tau a_D$, $a_D \in K^*$.*

9.5.5. Lemma. [**112**] *Let $w, \tau \in W^P$, $w \geq \tau$. Let δ, ϕ, θ be such that $p_{\delta,\phi} = f'_\theta$. Further, let θ and τ be comparable as n-tuples. Suppose $D_1 = X_{-\beta_s} \cdots X_{-\beta_1}$, $D_2 = X_{-\beta'_t} \cdots X_{-\beta'_1}$ are such that $D_i p_{\delta,\phi} = p_\tau a_{D_i}$, $a_{D_i} \in K^*$, $i = 1, 2$. Then $s = t$, i.e., $\mathrm{ord}\, D_1 = \mathrm{ord}\, D_2$.*

With notation as in Lemma 9.5.5, set

$$N_P(\delta, \phi) = \{D = X_{-\beta_s} \cdots X_{-\beta_1}, \ \beta_i \in N_P(w, \tau),$$
$$1 \leq i \leq s \mid D p_{\delta,\phi} = a_D p_\tau, \ a_D \in K^*\}.$$

Lemma 9.5.4 together with the fact that $R_P(w, \tau) \subseteq N_P(w, \tau)$ implies that $N_P(\delta, \phi) \neq \emptyset$.

9.5.6. Definition. Notation being as above, define $d_{\delta,\phi} := \mathrm{ord}\, D$ where $D \in N_P(\delta, \phi)$; note that the right-hand side is independent of D (cf. Lemma 9.5.5), so the definition makes sense. If $\delta = \phi$, then we shall denote $d_{\delta,\delta}$ by just d_δ. For $\delta = \phi = w$, we refer to d_w as the *degree* of $X_P(w)$ at e_τ and denote it by $\deg_\tau w$.

9.5.7. Definition. With notation as above, define

$$x_{\delta,\phi} := \sum_{D \in N_P(\delta,\phi)} a_D x_D,$$

where for $D = X_{-\beta_s} \cdots X_{-\beta_1}$, we set $x_D = x_{-\beta_s} \cdots x_{-\beta_1}$.

Note that from our discussion above, we have that in $A_{\tau,w}$, $p_{\delta,\phi}$ gets identified with $x_{\delta,\phi}$.

9.5.8. Theorem. *(cf. [112]) With notation as above,*

$$\left\{ x_{\delta_1,\phi_1} \cdots x_{\delta_m,\phi_m} \,\middle|\, \begin{array}{cc} (1) & w \geq \delta_1 \geq \phi_1 \geq \delta_2 \geq \ldots \geq \phi_m \\ (2) & for\ 1 \leq i \leq m, \theta_i\ and\ \tau\ are\ comparable \\ (3) & \sum d_{\delta_i,\phi_i} = r \end{array} \right\}$$

is a basis for $(M_{\tau,w})^r / (M_{\tau,w})^{r+1}$.

Let $X_P(w')$ be a Schubert divisor in $X_P(w)$, say $w' = w s_\beta$, for some $\beta \in R^+$. Denote $m(w, w') = (\omega_n, \beta^*)$, the Chevalley multiplicity of $X_P(w')$ in $X_P(w)$ (cf. §4.8.1). As a consequence of Theorem 9.5.8, we have

9.5.9. Theorem. [112] *Denoting* $\mathrm{mult}_{e_\tau} X$ *by* $m_\tau(w)$, *we have*

$$m_\tau(w) \deg_\tau w = \sum m_\tau(w') m(w, w')$$

where the sum on the right-hand side runs over all Schubert divisors $X_P(w')$ *in* $X_P(w)$ *such that* $e_\tau \in X_P(w')$.

9.5.10. Corollary. [112] *The Hilbert polynomial* $F_{\tau,w}$ *of* $X_P(w)$ *at* e_τ *is determined inductively by the formula*

$$F_{\tau,w}(r + d + d') = F_{\tau,w}(r + d') + \sum F_{\tau,\delta_i}(r + d + d'_i) + F_{\tau,H}(r + d + d')$$

where H *is the union of all Schubert divisors* $X_P(w')$ *in* $X_P(w)$ *such that* $e_\tau \in X_P(w')$, $d = \deg_\tau w$, $d' = \sum d_{w,\delta_i}, d'_i = d' - d_{w,\delta_i}$ *and* $\{\delta_i, 1 \leq i \leq m\}$ *are all the elements of* W^P *such that*

1. $w \geq \delta_i$ *and* (w, δ_i) *is an admissible pair.*
2. *If* θ_i *is the* n-*tuple such that* $f'_{\theta_i} = p_{w,\delta_i}$, *then* τ *and* θ_i *are comparable (note in particular that* τ *and* δ_i *are comparable).*

CHAPTER 10

Rank Two Results

Let G be simple of rank 2. In this chapter we describe $\operatorname{Sing} X(w)$, for all Schubert varieties $X(w)$ in G/B as well as G/P, P being a maximal parabolic subgroup. We demonstrate the techniques of Chapters 5 and 7.

10.1. Kumar's method

In order to apply Kumar's criteria (Theorem 7.2.1) for smoothness and rational smoothness we simply need to compute the expansion of any x_w in terms of the basis δ_v and note whether or not the coefficient has the appropriate form.

For any group G and any pair of simple reflections $s_1, s_2 \in W$, we compute all nonzero products of x_{s_1} and x_{s_2} directly from the definitions in Chapter 7.

$$x_{s_1} x_{s_2} = \frac{1}{\alpha_1} \left(\frac{1}{\alpha_2} (\delta_{\mathrm{id}} - \delta_{s_2}) - \frac{1}{s_1 \alpha_2} (\delta_{s_1} - \delta_{s_1 s_2}) \right)$$

$$x_{s_1} x_{s_2} x_{s_1} = \frac{1}{\alpha_1} \left(\frac{\alpha_2(\alpha_1^\vee)}{\alpha_2(s_1 \alpha_2)} (\delta_e - \delta_{s_1}) + \frac{1}{\alpha_2(s_2 \alpha_1)} (\delta_{s_2} - \delta_{s_2 s_1}) \right.$$
$$\left. - \frac{1}{(s_1 \alpha_2)(s_1 s_2 \alpha_1)} (\delta_{s_1 s_2} - \delta_{s_1 s_2 s_1}) \right)$$

$$x_{s_1} x_{s_2} x_{s_1} x_{s_2} = \frac{1}{\alpha_1} \left(\frac{m-1}{\alpha_2(s_1 \alpha_2)(s_2 \alpha_1)} (\delta_{\mathrm{id}} - \delta_{s_2}) - \frac{m-1}{\alpha_2(s_1 \alpha_2)(s_1 s_2 \alpha_1)} (\delta_{s_1} - \delta_{s_1 s_2}) \right.$$
$$+ \frac{1}{\alpha_2(s_2 \alpha_1)(s_2 s_1 \alpha_2)} (\delta_{s_2 s_1} - \delta_{s_2 s_1 s_2})$$
$$\left. - \frac{1}{(s_1 \alpha_2)(s_1 s_2 \alpha_1)(s_1 s_2 s_1 \alpha_2)} (\delta_{s_1 s_2 s_1} - \delta_{s_1 s_2 s_1 s_2}) \right)$$

$$x_{s_1}x_{s_2}x_{s_1}x_{s_2}x_{s_1} = \frac{1}{\alpha_1}\left(\frac{-2}{\alpha_2(s_1\alpha_2)(s_2\alpha_1)(s_1s_2\alpha_1)}(\delta_{\mathrm{id}} - \delta_{s_1})\right.$$

$$+ \frac{-\alpha_2(\alpha_1^\vee)}{\alpha_2(s_1\alpha_2)(s_2\alpha_1)(s_2s_1\alpha_2)}(\delta_{s_2} - \delta_{s_2s_1})$$

$$+ \frac{\alpha_2(\alpha_1^\vee)}{\alpha_2(s_1\alpha_2)(s_1s_2\alpha_1)(s_1s_2s_1\alpha_2)}(\delta_{s_1s_2} - \delta_{s_1s_2s_1})$$

$$+ \frac{1}{\alpha_2(s_2\alpha_1)(s_2s_1\alpha_2)(s_2s_1s_2\alpha_1)}(\delta_{s_2s_1s_2} - \delta_{s_2s_1s_2s_1})$$

$$- \frac{1}{(s_1\alpha_2)(s_1s_2\alpha_1)(s_1s_2s_1\alpha_2)(s_1s_2s_1s_2\alpha_1)}$$

$$\left. \cdot (\delta_{s_1s_2s_1s_2} - \delta_{s_1s_2s_1s_2s_1})\right)$$

Here $m = \alpha_1(\alpha_2^\vee)\alpha_2(\alpha_1^\vee)$ which is 2 for types B_2 and C_2 and 3 for type G_2. In the last equation we have assumed $\alpha_1(\alpha_2^\vee)\alpha_2(\alpha_1^\vee) = 3$ since this is the only case for which $x_{s_1}x_{s_2}x_{s_1}x_{s_2}x_{s_1}$ is not zero.

One can now easily find the singular locus of a Schubert variety indexed by any Weyl group element of the form $s_is_js_i\ldots$ by seeing which coefficients above are different from ± 1.

10.1.1. Theorem. [91] *The following is a complete description of the singular locus of the Schubert varieties in the case of rank two groups:*

1. *Type A_2: All six Schubert varieties are smooth.*
2. *Type B_2: Of the eight Schubert varieties, only $X_{s_2s_1s_2}$ is singular and its singular locus is X_{s_2}.*
3. *Type C_2: Of the eight Schubert varieties, only $X_{s_1s_2s_1}$ is singular and its singular locus is X_{s_1}.*
4. *Type G_2: Of the twelve Schubert varieties, there are five which are singular. The following is the complete list of singular ones and their singular loci:*

$$\begin{aligned}
\mathrm{Sing}(X_{s_1s_2s_1}) &= X_{s_1} \\
\mathrm{Sing}(X_{s_1s_2s_1s_2}) &= X_{s_1s_2} \\
\mathrm{Sing}(X_{s_2s_1s_2s_1}) &= X_{s_2s_1} \\
\mathrm{Sing}(X_{s_1s_2s_1s_2s_1}) &= X_{s_1s_2s_1} \\
\mathrm{Sing}(X_{s_2s_1s_2s_1s_2}) &= X_{s_2}\,.
\end{aligned}$$

Upon further examination of the equations above, one sees that the coefficients $c_{w,v}$ are all integers divided by the appropriate product of roots, which proves the following theorem.

10.1.2. Theorem. *All Schubert varieties in rank 2 groups are rationally smooth.*

10.2. Tangent space computations

If $G = SL(3)$, i.e., G is of Type A_2, then every Schubert variety in G/B as well as G/P_i, $i = 1, 2$ is smooth.

10.2.1. Type B_2, C_2. From the Dynkin diagram (on page 207) it is clear that the root systems of type B_2 and C_2 are the same; so we will only demonstrate these computations in type C_2.

Let G be $Sp(4)$ so its corresponding root system is of type C_2. We have $R^+ = \{\alpha_1, \alpha_2, \alpha_1 + \alpha_2, 2\alpha_1 + \alpha_2\}$. We shall denote these as $\{\tilde{\alpha}_1, \tilde{\alpha}_2, \tilde{\alpha}_3, \tilde{\alpha}_4\}$. We shall denote the elements of W as

$$\tau_0 = \phi_0 = id \quad \tau_1 = s_1 \quad \tau_2 = s_2 s_1 \qquad \tau_3 = s_1 s_2 s_1$$
$$\phi_1 = s_2 \quad \phi_2 = s_1 s_2 \quad \phi_3 = s_2 s_1 s_2 \quad w_0 = s_1 s_2 s_1 s_2 (= s_2 s_1 s_2 s_1).$$

Note that w_0 is the unique element of largest length in W. The poset of the Bruhat order is given in the picture below.

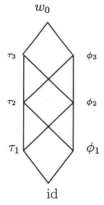

With $\mathcal{B}_i$, $i = 1, 2$ as in § 5.2, we have

$$\mathcal{B}_1 = \{q_{\tau_i}, 0 \le i \le 3\},$$
$$\mathcal{B}_2 = \{q_{\phi_i}, 0 \le i \le 3, \ q_{\phi_2, \phi_1}\}.$$

In the first set, the trivial pairs (τ, τ) are the only admissible pairs in W^{P_1} and we have denoted $q_{\tau, \tau}$ by just q_τ. In the second set, (ϕ_2, ϕ_1) is the only nontrivial admissible pair in W^{P_2}.

Let q_{id} be the highest-weight vector in $V_{\mathbb{Z}, \omega_i}$, $i = 1, 2$. By weight considerations, we have (up to ± 1),

$$X_{-\tilde{\alpha}_1} q_{id} = \begin{cases} q_{\tau_1}, & i = 1, \\ 0, & i = 2. \end{cases}$$

$$X_{-\tilde{\alpha}_2} q_{id} = \begin{cases} 0, & i = 1, \\ q_{\phi_1}, & i = 2. \end{cases}$$

$$X_{-\tilde{\alpha}_3} q_{\mathrm{id}} = \begin{cases} q_{\tau_2}, & i = 1, \\ q_{\phi_2, \phi_1}, & i = 2. \end{cases}$$

$$X_{-\tilde{\alpha}_4} q_{\mathrm{id}} = \begin{cases} q_{\tau_3}, & i = 1, \\ q_{\phi_2}, & i = 2. \end{cases}$$

Hence by consideration of $\dim T(w, e)$, we obtain that $X(\tau_3)$ is the only singular Schubert variety. We determine $\mathrm{Sing}\, X(w)$ by computing $X_{-\beta} q_\tau$, $\beta \in \tau(R^+)$, $\tau \le w$ and using Theorem 5.1.1. In view of the fact that Schubert varieties are nonsingular in codimension 1, it suffices to compute $X_{-\beta} q_\tau$ for $\tau = \tau_1, \phi_1$. We describe these computations below. We shall denote by $q_{\tau,i}$ the extremal-weight vector in $V_{\mathbb{Z}, \omega_i}$ of weight $\tau(\omega_i)$, $i = 1, 2$. Further for $1 \le j \le 4$, we shall denote $\tau(\tilde{\alpha}_j)$ by β_j.

Case 1. $\tau = \phi_1$. We have (up to ± 1),

$$X_{-\beta_1} q_{\tau,i} = \begin{cases} q_{\tau_2}, & i = 1, \\ 0, & i = 2. \end{cases}$$

$$X_{-\beta_2} q_{\tau,i} = \begin{cases} 0, & i = 1, \\ q_{\mathrm{id}}, & i = 2. \end{cases}$$

$$X_{-\beta_3} q_{\tau,i} = \begin{cases} q_{\tau_1}, & i = 1, \\ q_{\phi_2, \phi_1}, & i = 2. \end{cases}$$

$$X_{-\beta_4} q_{\tau,i} = \begin{cases} q_{\tau_3}, & i = 1, \\ q_{\phi_3}, & i = 2. \end{cases}$$

Hence we obtain (by consideration of $\dim T(w, \tau)$, $w = \tau_3$) that e_{ϕ_1} is smooth on $X(w)$.

Case 2. $\tau = \tau_1$. We have (up to ± 1),

$$X_{-\beta_1} q_{\tau,i} = \begin{cases} q_{\mathrm{id}}, & i = 1, \\ 0, & i = 2. \end{cases}$$

$$X_{-\beta_2} q_{\tau,i} = \begin{cases} 0, & i = 1, \\ q_{\phi_2}, & i = 2. \end{cases}$$

$$X_{-\beta_3} q_{\tau,i} = \begin{cases} q_{\tau_3}, & i = 1, \\ q_{\phi_2, \phi_1}, & i = 2. \end{cases}$$

$$X_{-\beta_4} q_{\tau,i} = \begin{cases} q_{\tau_2}, & i = 1, \\ q_{\phi_1}, & i = 2. \end{cases}$$

Hence we obtain that e_{τ_1} is singular on $X(\tau_3)$. Hence we obtain

10.2.4. Theorem. *With notation as above, we have*

1. $X(\tau_3)$ *is the only singular Schubert variety in G/B and $\operatorname{Sing}X(\tau_3) = X(\tau_1)$.*
2. *All Schubert varieties in G/P_1 are smooth, while $X_P(\phi_2)$ is the only singular Schubert variety in G/P_2 with e_{id} as the only singular point.*
3. *All Schubert varieties in G/B as well as G/P_i, $i = 1, 2$, are rationally smooth.*

10.2.5. The group G_2. Let G be simple of type G_2. We have $R^+ = \{\alpha_1, \alpha_2, \alpha_1 + \alpha_2, 2\alpha_1 + \alpha_2, 3\alpha_1 + \alpha_2, 3\alpha_1 + 2\alpha_2\}$. We shall denote these as $\{\tilde{\alpha}_1, \tilde{\alpha}_2, \tilde{\alpha}_3, \tilde{\alpha}_4, \tilde{\alpha}_5, \tilde{\alpha}_6\}$. We shall denote the elements of W as

$$\phi_1 = s_1 \quad \phi_2 = s_2 s_1 \quad \phi_3 = s_1 s_2 s_1 \quad \phi_4 = s_2 s_1 s_2 s_1 \quad \phi_5 = s_1 s_2 s_1 s_2 s_1$$
$$\tau_1 = s_2 \quad \tau_2 = s_1 s_2 \quad \tau_3 = s_2 s_1 s_2 \quad \tau_4 = s_1 s_2 s_1 s_2 \quad \tau_5 = s_2 s_1 s_2 s_1 s_2$$

$$\tau_0 = \phi_0 = \mathrm{id} \quad w_0 = (s_1 s_2)^3 = (s_2 s_1)^3$$

The Bruhat–Chevalley order of W is given by the following poset:

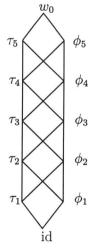

We have that (ϕ_3, ϕ_2) is the only nontrivial admissible pair in W^{P_1}, and hence q_{ϕ_3, ϕ_2} is the only non-extremal-weight vector in $\mathcal{B}_1$. Further, $V_{\mathbb{Z}, \omega_2}$ is simply $\mathfrak{g}_{\mathbb{Z}}$, and $\mathcal{B}_2$ is simply the Chevalley basis. To be very precise, we have (cf. [**93**]) the extremal-weight vectors $q_{\tau_i}, 0 \le i \le 5$, are $X_{\pm\beta}$, β being a long root. The non-extremal-weight vectors are given by
$$X_{2\alpha_1 + \alpha_2} = e_{\tau_2, \tau_1}, \quad X_{\alpha_1 + \alpha_2} = f_{\tau_2, \tau_1},$$
$$X_{-(2\alpha_1 + \alpha_2)} = f_{\tau_4, \tau_3}, \quad X_{-(\alpha_1 + \alpha_2)} = e_{\tau_4, \tau_3},$$
$$X_{\alpha_1} = q_{\tau_3, \tau_1}, \quad X_{-\alpha_1} = q_{\tau_4, \tau_2}, \quad H_{\alpha_1} = q_{\tau_4, \tau_1}, \quad H_{\alpha_2} = q_{\tau_3, \tau_2}.$$

Thus we have

$$\mathcal{B}_1 = \{q_{\phi_i}, 0 \le i \le 5,\ q_{\phi_3,\phi_2}\},$$
$$\mathcal{B}_2 = \{q_{\tau_i}, 0 \le i \le 5,\ q_{\tau,\phi},\ \tau = \tau_4, \tau_3,\ \phi = \tau_2, \tau_1, e_{\tau,\phi}, f_{\tau,\phi},$$
$$(\tau, \phi) = (\tau_4, \tau_3), (\tau_2, \tau_1)\}.$$

10.2.6. Computation of $X_{-\beta}q_{\mathrm{id}}$. For $i = 1, 2$, we shall denote the highest-weight vector in $V_{\mathbb{Z},\omega_i}$ by just q_{id}. We have (up to ± 1),

$$X_{-\tilde{\alpha}_1} q_{\mathrm{id}} = \begin{cases} q_{\phi_1}, & i = 1, \\ 0, & i = 2. \end{cases}$$

$$X_{-\tilde{\alpha}_2} q_{\mathrm{id}} = \begin{cases} 0, & i = 1, \\ q_{\tau_1}, & i = 2. \end{cases}$$

$$X_{-\tilde{\alpha}_3} q_{\mathrm{id}} = \begin{cases} q_{\phi_2}, & i = 1, \\ e_{\tau_2,\tau_1}, & i = 2. \end{cases}$$

$$X_{-\tilde{\alpha}_4} q_{\mathrm{id}} = \begin{cases} q_{\phi_3,\phi_2}, & i = 1, \\ f_{\tau_2,\tau_1}, & i = 2. \end{cases}$$

$$X_{-\tilde{\alpha}_5} q_{\mathrm{id}} = \begin{cases} q_{\phi_3}, & i = 1, \\ q_{\tau_2}, & i = 2. \end{cases}$$

$$X_{-\tilde{\alpha}_6} q_{\mathrm{id}} = \begin{cases} q_{\phi_4}, & i = 1, \\ q_{\tau_4,\tau_1} + 2q_{\tau_3,\tau_2}, & i = 2. \end{cases}$$

Hence by consideration of dim $T(w, e)$, we obtain that the singular Schubert varieties are given by $X(w)$, $w = \phi_i$, $i = 3, 4, 5$, and $w = \tau_j$, $j = 4, 5$. We determine $\mathrm{Sing}\, X(w)$ by computing $X_{-\beta}q_\tau$, $\beta \in \tau(R^+), \tau \le w$ and using Theorem 5.1.1. In view of the fact that Schubert varieties are non-singular in codimension 1, it suffices to compute $X_{-\beta}q_\tau$, for $\tau = \tau_i, \phi_i$, $i = 1, 2, 3$. We describe these computations below. We shall denote by $q_{\tau,i}$ the extremal-weight vector in $V_{\mathbb{Z},\omega_i}$ of weight $\tau(\omega_i)$, $i = 1, 2$. Further for $1 \le j \le 6$, we shall denote $\tau(\tilde{\alpha}_j)$ by β_j.

Case 1. $\tau = \phi_1$. We have (up to ± 1),

$$X_{-\beta_1} q_{\tau,i} = \begin{cases} q_{\mathrm{id}}, & i = 1, \\ 0, & i = 2. \end{cases}$$

$$X_{-\beta_2} q_{\tau,i} = \begin{cases} 0, & i = 1, \\ q_{\tau_2}, & i = 2. \end{cases}$$

$$X_{-\beta_3}q_{\tau,i} = \begin{cases} q_{\phi_3}, & i = 1, \\ f_{\tau_2,\tau_1}, & i = 2. \end{cases}$$

$$X_{-\beta_4}q_{\tau,i} = \begin{cases} q_{\phi_3,\phi_2}, & i = 1, \\ e_{\tau_2,\tau_1}, & i = 2. \end{cases}$$

$$X_{-\beta_5}q_{\tau,i} = \begin{cases} q_{\phi_2}, & i = 1, \\ q_{\tau_1}, & i = 2. \end{cases}$$

$$X_{-\beta_6}q_{\tau,i} = \begin{cases} q_{\phi_5}, & i = 1, \\ q_{\tau_4,\tau_1} + 2q_{\tau_3,\tau_2}, & i = 2. \end{cases}$$

Hence we obtain, by consideration of dim $T(w,\tau)$, that e_{ϕ_1} is singular on $X(\phi_i)$, $i = 3, 4, 5$, and $X(\tau_4)$.

Case 2. $\tau = \tau_1$. We have (up to ± 1),

$$X_{-\beta_1}q_{\tau,i} = \begin{cases} q_{\phi_2}, & i = 1, \\ 0, & i = 2. \end{cases}$$

$$X_{-\beta_2}q_{\tau,i} = \begin{cases} 0, & i = 1, \\ q_{\mathrm{id}}, & i = 2. \end{cases}$$

$$X_{-\beta_3}q_{\tau,i} = \begin{cases} q_{\phi_1}, & i = 1, \\ e_{\tau_2,\tau_1}, & i = 2. \end{cases}$$

$$X_{-\beta_4}q_{\tau,i} = \begin{cases} q_{\phi_3,\phi_2}, & i = 1, \\ q_{\tau_3,\tau_1}, & i = 2. \end{cases}$$

$$X_{-\beta_5}q_{\tau,i} = \begin{cases} q_{\phi_4}, & i = 1, \\ q_{\tau_3}, & i = 2. \end{cases}$$

$$X_{-\beta_6}q_{\tau,i} = \begin{cases} q_{\phi_3}, & i = 1, \\ q_{\tau_4,\tau_2}, & i = 2. \end{cases}$$

B. Hence we obtain that e_{τ_1} is singular on $X(\phi_i)$, $X(\tau_i)$, $i = 4, 5$.

Case 3. $\tau = \phi_2$. We have (up to ± 1),

$$X_{-\beta_1}q_{\tau,i} = \begin{cases} q_{\mathrm{id}}, & i = 1, \\ 0, & i = 2. \end{cases}$$

$$X_{-\beta_2}q_{\tau,i} = \begin{cases} 0, & i = 1, \\ q_{\tau_3}, & i = 2. \end{cases}$$

$$X_{-\beta_3}q_{\tau,i} = \begin{cases} q_{\phi_4}, & i = 1, \\ q_{\tau_3,\tau_1}, & i = 2. \end{cases}$$

$$X_{-\beta_4}q_{\tau,i} = \begin{cases} q_{\phi_3,\phi_2}, & i = 1, \\ e_{\tau_2,\tau_1}, & i = 2. \end{cases}$$

$$X_{-\beta_5}q_{\tau,i} = \begin{cases} q_{\phi_1}, & i = 1, \\ q_{\mathrm{id}}, & i = 2. \end{cases}$$

$$X_{-\beta_6}q_{\tau,i} = \begin{cases} q_{\phi_5}, & i = 1, \\ q_{\tau_3}, & i = 2. \end{cases}$$

C. Hence we obtain that e_{ϕ_2} is singular on $X(\phi_i)$, $i = 4, 5$.

Case 4. $\tau = \tau_2$. We have (up to ± 1),

$$X_{-\beta_1}q_{\tau,i} = \begin{cases} q_{\phi_3}, & i = 1, \\ 0, & i = 2. \end{cases}$$

$$X_{-\beta_2}q_{\tau,i} = \begin{cases} 0, & i = 1, \\ q_{\mathrm{id}}, & i = 2. \end{cases}$$

$$X_{-\beta_3}q_{\tau,i} = \begin{cases} q_{\mathrm{id}}, & i = 1, \\ f_{\tau_2,\tau_1}, & i = 2. \end{cases}$$

$$X_{-\beta_4}q_{\tau,i} = \begin{cases} q_{\phi_3,\phi_2}, & i = 1, \\ q_{\tau_4,\tau_2}, & i = 2. \end{cases}$$

$$X_{-\beta_5}q_{\tau,i} = \begin{cases} q_{\phi_5}, & i = 1, \\ q_{\tau_4}, & i = 2. \end{cases}$$

$$X_{-\beta_6}q_{\tau,i} = \begin{cases} q_{\phi_2}, & i = 1, \\ q_{\tau_3,\tau_2}, & i = 2. \end{cases}$$

Hence we obtain that e_{τ_2} is singular on $X(\phi_5)$ and $X(\tau_4)$.

Case 5. $\tau = \phi_3$. We have (up to ± 1),

$$X_{-\beta_1} q_{\tau,i} = \begin{cases} q_{\phi_1}, & i = 1, \\ 0, & i = 2. \end{cases}$$

$$X_{-\beta_2} q_{\tau,i} = \begin{cases} 0, & i = 1, \\ q_{\tau_4}, & i = 2. \end{cases}$$

$$X_{-\beta_3} q_{\tau,i} = \begin{cases} q_{\phi_5}, & i = 1, \\ q_{\tau_4,\tau_2}, & i = 2. \end{cases}$$

$$X_{-\beta_4} q_{\tau,i} = \begin{cases} q_{\phi_3,\phi_2}, & i = 1, \\ f_{\tau_2,\tau_1}, & i = 2. \end{cases}$$

$$X_{-\beta_5} q_{\tau,i} = \begin{cases} q_{\mathrm{id}}, & i = 1, \\ q_{\mathrm{id}}, & i = 2. \end{cases}$$

$$X_{-\beta_6} q_{\tau,i} = \begin{cases} q_{\phi_4}, & i = 1, \\ q_{\tau_3,\tau_2}, & i = 2. \end{cases}$$

Hence we obtain that e_{ϕ_3} is singular on $X(\phi_5)$.

Case 6. : $\tau = \tau_3$. We have (up to ± 1),

$$X_{-\beta_1} q_{\tau,i} = \begin{cases} q_{\phi_4}, & i = 1, \\ 0, & i = 2. \end{cases}$$

$$X_{-\beta_2} q_{\tau,i} = \begin{cases} 0, & i = 1, \\ q_{\tau_1}, & i = 2. \end{cases}$$

$$X_{-\beta_3} q_{\tau,i} = \begin{cases} q_{\mathrm{id}}, & i = 1, \\ q_{\tau_3,\tau_1}, & i = 2. \end{cases}$$

$$X_{-\beta_4} q_{\tau,i} = \begin{cases} q_{\phi_3,\phi_2}, & i = 1, \\ e_{\tau_4,\tau_3}, & i = 2. \end{cases}$$

$$X_{-\beta_5} q_{\tau,i} = \begin{cases} q_{\phi_5}, & i = 1, \\ q_{\tau_5}, & i = 2. \end{cases}$$

$$X_{-\beta_6} q_{\tau,i} = \begin{cases} q_{\phi_1}, & i = 1, \\ q_{\tau_3,\tau_2}, & i = 2. \end{cases}$$

Hence we obtain that e_{τ_3} is smooth on all $X(w)$, $w \geq \tau_3$.

Using Cases 1 through 6 above and the fact that Schubert varieties are nonsingular in codimension one, we obtain the following:

10.2.13. Theorem. *Let G be a simple group of type G_2.*

1. *The singular Schubert varieties varieties in G/B are given by $X(w)$, $w = \phi_i$, $i = 3, 4, 5$ and $w = \tau_j$, $j = 4, 5$. Further, $\mathrm{Sing} X(\phi_i) = X(\phi_{i-2})$, $i = 3, 4, 5$, $\mathrm{Sing} X(\tau_4) = X(\tau_2)$, and Sing $X(\tau_5) = X(\tau_1)$.*

2. *The singular Schubert varieties in G/P_1 are $X(\phi_i)$, $i = 3, 4$. Further, $\mathrm{Sing} X(\phi_3) = X(\phi_1)$, and $\mathrm{Sing} X(\phi_4) = e_{\mathrm{id}}$.*

3. *The singular Schubert varieties in G/P_2 are $X(\tau_i)$, $i = 2, 3, 4$. Further, $\mathrm{Sing} X(\tau_i) = X(\tau_{i-2})$.*

4. *All Schubert varieties in G/B as well as G/P_i, $i = 1, 2$, are rationally smooth.*

CHAPTER 11

Related Combinatorial Results

In this chapter we present some results relating smoothness and the factoring of the Poincaré polynomial of a Schubert variety. We also present results on counting the number of smooth Schubert varieties for $G = SL(n)$.

11.1. Factoring the Poincaré polynomial of a Schubert variety

Recall that if the Poincaré polynomial of a Schubert variety is symmetric, then $X(w)$ is rationally smooth by Theorem 6.2.4. It turns out that these Poincaré polynomials are not just symmetric, but in fact have nice factorizations as a product of symmetric factors. The first theorem below is a generalization of the Kostant–Macdonald formula (originally proved by Kostant [83] and then Macdonald [122]). The factoring rules given in Theorem 11.2.5 can be used as a recursive criteria for rational smoothness. While other factoring rules also hold for the Poincaré polynomials, these are sufficient to completely factor the Poincaré polynomials of all rationally smooth Schubert varieties of classical type.

11.1.1. Theorem. [33] *Let W be any Weyl group. If $X(w)$ is smooth then the Poincaré polynomial factors as*

$$(11.1.2) \qquad\qquad P_w(t) = \prod_{\substack{\alpha > 0 \\ s_\alpha \leq w}} \frac{1 - t^{\mathrm{ht}\alpha + 1}}{1 - t^{\mathrm{ht}\alpha}}.$$

11.1.3. Remark. Note that the converse to Theorem 11.1.1 need not be true. For example, in Type C_3, $X(263415)$, $X(513462)$, and $X(153426)$ are not smooth but their Poincaré polynomials do factor as in (11.1.2). Thus, the existence of such a factorization is only a necessary condition for smoothness.

11.1.4. Remark. The theorem does not extend to all rationally smooth Schubert varieties. The following Schubert varieties are rationally smooth in Type C_3, but (11.1.2) does not hold: $X(362514)$, $X(531642)$, and $X(632541)$. Nevertheless, the existence of such a factorization is a sufficient condition for rational smoothness [32].

In type A, there is a second factoring formula for the Poincaré polynomials of smooth Schubert varieties. This formula will be generalized in the next section.

11.1.5. Theorem. [52, 116] *Let* $w \in S_n$, *and assume* $w_d = n$ *and* $w_n = e$. *The Poincaré polynomial of* w *factors in the form*

$$P_w(t) = (1 + t^1 + \cdots + t^\mu) P_{w'}(t).$$

under the following circumstances:

1. *If* $n = w_d > w_{d+1} > \cdots > w_n$, *then* $P_w(t)$ *factors with* $w' = ws_d \cdots s_{n-1}$ *and* $\mu = n - d$.
2. *If* w *contains a consecutive sequence ending in* $w_n = e$, *then* P_w *factors with* $w' = s_{n-1} \cdots s_{e+1} s_e w$ *and* $\mu = n - e$. *Note* w *has a consecutive sequence if and only if* w^{-1} *has Property 1.*

Furthermore, $P_w(t)$ *factors completely using the above algorithm if and only if* $X(w)$ *is smooth.*

11.1.6. Remark. Lascoux [116] actually gives the corresponding factorization of the Kazhdan–Lusztig basis element $\tilde{C}_w$. Note, this is not exactly the basis given in 6.1.6. This factorization is complete if and only if $X(w)$ is smooth (in the type A case). Theorem 11.1.5 follows as a corollary.

Apply Theorem 11.1.5 to the element $w_0 = n \ldots 1$ (in one-line notation) to get the well-known formula for the Poincaré polynomial of $\mathcal{S}_n$ [69, 3.7], namely

$$(11.1.7) \qquad\qquad P_{w_0}(t) = \prod_{k=1}^{n-1} (1 + t + \ldots + t^k).$$

It is interesting to compare the factoring formulas in Theorems 11.1.1 and 11.1.5 in the type A case. In this case the degrees of the factors of the Poincaré polynomial form a partition, and the sequence counting the positive roots of heights 1,2, ... form the conjugate partition. More precisely, we have the following theorem.

11.1.8. Theorem. [13] *Fix* $w \in S_n$. *Let* h_i *be the number of positive roots in the set* $\{\beta \in R_+ : \mathrm{ht}(\beta) = i \text{ and } s_\beta \leq w\}$. *If* $P_w(t) = (1 + t + \cdots + t^{\mu_1})(1 + t + \cdots + t^{\mu_2}) \ldots (1 + t + \cdots + t^{\mu_k})$ *and* $\mu_1 \geq \mu_2 \geq \cdots \geq \mu_k$, *then the partition* $\mu = \mu_1 \ldots \mu_k$ *is conjugate to the partition* $h_1 \ldots h_k$ *and* $h_i = 0$ *for all* $i > k$.

At this time no analog of Theorem 11.1.8 is known for other types.

11.2. Structure of Bruhat intervals

In this section, we show that the factoring theorems in §11.1 reflect structure on the intervals in the Bruhat order.

11.2.1. Definition. The element $m(w, P)$. For every $w \in W$ and every parabolic subgroup W_P, it was shown in [14] that there exists a unique maximal element $m(w, P) \in W_P \cap \{v \leq w\}$. If $\mathbf{a} = a_1 \ldots a_p$ is a reduced word for w, let $b_1 \ldots b_k$ be the subword obtained by removing all elements which are not in W_P. The element $m(w, P)$ is defined to be $m(b_1 \ldots b_k)$ which can be obtained recursively as follows:

(11.2.2)

$$m(b_1 \ldots b_k) = \begin{cases} m(b_1 \ldots b_{k-1}) & \text{if } m(b_1 \ldots b_{k-1}) > m(b_1 \ldots b_{k-1})s_{b_k} \\ m(b_1 \ldots b_{k-1})s_{b_k} & \text{if } m(b_1 \ldots b_{k-1}) < m(b_1 \ldots b_{k-1})s_{b_k}. \end{cases}$$

For any Weyl group W, any parabolic subgroup P, and any $w \in W$, let $w = uv$ be the unique decomposition of w such that $u \in W_P$ and v is a minimal coset representative for $W_P w$ (cf. §2.5).

11.2.3. Theorem. *Let $w \in W$ and let α_i be a simple root which corresponds to an end node of the Dynkin diagram. Let P_i be the corresponding maximal parabolic subgroup. If w (or w^{-1}) has the minimal coset decomposition $w = m(w, P) \cdot \gamma$ and the Bruhat interval below γ in $W/\{s_\alpha\}$ is a chain, then the Bruhat interval $B_w = \{v \in W : v \leq w\}$ contains the poset $B_{m(w,P)} \times \{0, 1, \ldots, d\}$ as a subposet on all vertices i.e., some edges are removed.*

PROOF. Assume the chain below γ in $W/\{s_\alpha\}$ is labeled by id, $s_1, s_1 s_2$, $\ldots, s_1 s_2 \ldots s_p$ where $p = l(\gamma)$. Given any $x \leq w$, say x has minimal coset decomposition $x = yz$ with respect to P, then $y \leq m(w, P)$ since $m(w, P)$ is the unique maximal element below w and in W_P. Furthermore, $z \leq \gamma$ by Proposition 11.2.7 below, so by assumption $z = s_1 s_2 \ldots s_k$ for some k. Therefore, we can define a map

(11.2.4) $$\phi : B_w \to B_v \times \{0, 1, \ldots, d\}$$

by mapping x to $(y, l(z))$. Note that this map is injective and rank preserving since $l(x) = l(y) + l(z)$.

Conversely, given any $y \in P$ such that $y \leq m(w, P)$ and any $k \in \{0, 1, \ldots, d\}$, we have $\phi(as_1 \ldots s_k)$ equals (a, k). Then clearly $as_1 \ldots s_k \leq w$ since $as_1 \ldots s_k$ can be written as a subword of the reduced word for w which is the concatenation of a reduced word for $m(w, P)$ and a reduced word for γ. Hence ϕ is surjective. $\square$

11.2.5. Corollary. *Using the notation in Theorem 11.2.3, we get a factoring formula for the Poincaré polynomial as well:*

(11.2.6) $$P_w(t) = (1 + t^1 + \cdots + t^{l(\gamma)})P_{m(w,P)}(t).$$

11.2.7. Proposition. [**137**, *Lemma 3.2*] *For any Weyl group W, any parabolic subgroup P, and any $w \in W$, let $w = uv$ be the unique decomposition of w such that $u \in W_P$ and v is a minimal coset representative for $W_P w$. If $c \in W$ is also a minimal length element in the coset $W_P c$, then $c \leq w$ if and only if $c \leq v$.*

11.2.8. Conjecture. *Let W be any Weyl group. For $w \in W$, if X_w is a rationally smooth Schubert variety, then the interval in the Bruhat poset weakly below w contains a product of chains as a subposet. The converse is true by Theorem 6.2.4.*

11.3. Generating function for smooth permutations

There is a surprising generating function for the number of permutations corresponding to smooth Schubert varieties. This formula was first proved by M. Haiman (preprint dated 1992) and also appears in the work of M. Bona [19] using results of Stankova [150].

11.3.1. Theorem. [60] *Let v_n be the number of $w \in S_n$ for which $X(w)$ is smooth in $SL(n)/B$. Then the generating function $V(t) = \sum_n v_n t^n$ is given by*

$$(11.3.2) \qquad V(t) = \frac{1 - 5t + 3t^2 + t^2\sqrt{1-4t}}{1 - 6t + 8t^2 - 4t^3}$$

$$(11.3.3) \qquad = 1 + t + 2t^2 + 6t^3 + 22t^4 + 88t^5 + 366t^6$$

$$(11.3.4) \qquad + 1552t^7 + 6652t^8 + 28696t^9 + O(t^{10}).$$

The following data shows the number of smooth Schubert varieties in types A, B, C, and D:

	A	B	C	D
$n = 1$	1	2	2	1
$n = 2$	2	7	7	4
$n = 3$	6	28	28	22
$n = 4$	22	116	114	108
$n = 5$	88	490	472	490
$n = 6$	366	2094	1988	2164
$n = 7$	1552	9014	8480	9474

(11.3.5)

11.3.6. Remark. The problem of finding a generating function for the number of (rationally) smooth Schubert varieties is still open for all types except type A.

11.4. Bona's results

Recall from Chapter 8 that we say a permutation $w = (w_1 \ldots w_n) \in S_n$ *avoids* a pattern $(a_1 \ldots a_k) \in S_k$ if no subsequence of w has the same relative order as $(a_1 \ldots a_k)$. Permutations indexing smooth Schubert varieties in type A avoid the pair of patterns 3412 and 4231 (Th. 8.1.1). Let $S_n(u, v)$ be the number of $w \in S_n$ such that w avoids both u and v. M. Bona (cf. [19]) has completely classified all pairs of permutations in S_4 such that $S_n(u, v) = S_n(3412, 4231)$ i.e., equinumerous to the smooth permutations.

Two pairs of patterns are *equivalent* if one can be transformed into the other by some combination of reversing, complementing, and inverse-taking.

11.4.1. Theorem. [19] *The following five inequivalent pairs of permutation classes are equinumerous:* $S_n(3412, 4231) = S_n(1342, 2431) = S_n(1342, 3241) = S_n(1342, 2314) = S_n(1324, 2413)$. *There are no more inequivalent pairs* (u, v) *for which the values of* $S_n(u, v)$ *equals* $S_n(3412, 4231)$ *for all* n.

CHAPTER 12

Related Varieties

In this chapter we present results on the classical determinantal varieties from the aspect of their relationship to Schubert varieties. We then review the results of [56], [103] wherein two classes of affine varieties — a certain class of ladder determinantal varieties, a certain class of quiver varieties respectively — are shown to be normal, Cohen–Macaulay by identifying them with opposite cells in certain Schubert varieties. For the ladder determinantal varieties and certain of the quiver varieties, we also have a description of the singular locus (cf. [56], [101]) which gives information on the irreducible components of the associated Schubert variety and verifies the conjecture in §8.2 for these Schubert varieties.

We begin this section with a review of the opposite cells in the Schubert varieties in $SL(n)$. These results will be needed to relate the above mentioned varieties with Schubert varieties.

12.1. Opposite cells in Schubert varieties in $SL(n)/B$

Let $G = SL(n)$, the special linear group of rank $n - 1$. Let T be the maximal torus consisting of all the diagonal matrices in G, and B the Borel subgroup consisting of all the upper triangular matrices in G.

12.1.1. The opposite big cell in G/Q. Fix a parabolic subgroup Q. The sequence $a_1 < \cdots < a_t$ is defined by $Q = \cap_{t=1}^{k} P_{a_t}$, where P_{a_i} denotes the maximal parabolic subgroup with $S \setminus \{\alpha_{a_i}\}$ as the associated set of simple roots. Let $a = n - a_k$, and Q be the parabolic subgroup consisting of all the elements in G of the form

$$
\begin{pmatrix}
A_1 & * & * & \cdots & * & * \\
0 & A_2 & * & \cdots & * & * \\
\vdots & \vdots & \vdots & & \vdots & \vdots \\
0 & 0 & 0 & \cdots & A_k & * \\
0 & 0 & 0 & \cdots & 0 & A
\end{pmatrix},
$$

where each A_t is a matrix of size $c_t \times c_t$, $c_t = a_t - a_{t-1}$, $1 \le t \le k$, with $a_0 = 0$, A is a matrix of size $a \times a$, and $x_{ml} = 0$, $m > a_t$, $l \le a_t$, $1 \le t \le k$. Denote by O^- the subgroup of G generated by $\{U_\alpha \mid \alpha \in R^- \setminus R_Q^-\}$. Then

O^- consists of the elements of G of the form

$$\begin{pmatrix} I_1 & 0 & 0 & \cdots & 0 & 0 \\ * & I_2 & 0 & \cdots & 0 & 0 \\ \vdots & \vdots & \vdots & & \vdots & \vdots \\ * & * & * & \cdots & I_k & 0 \\ * & * & * & \cdots & * & I_a \end{pmatrix},$$

where I_t is the $c_t \times c_t$ identity matrix, $1 \le t \le k$, I_a is the $a \times a$ identity matrix, and if $x_{ml} \ne 0$, with $m \ne l$, then $m > a_t$, $l \le a_t$ for some t, $1 \le t \le k$. Further, the restriction of the canonical morphism $f : G \to G/Q$ to O^- is an open immersion, and $f(O^-) \cong B^- e_{\mathrm{id},Q}$. Thus the *opposite cell* $B^- e_{\mathrm{id},Q}$ gets identified with O^-.

12.1.2. Evaluation of Plücker coordinates on the opposite big cell in the Grassmannian.
Let P_d be the maximal parabolic subgroup of G with $S \setminus \{\alpha_d\}$ as the associated set of simple roots. Recall from §3.1.6 that

$$P_d = \left\{ M \in G \,\middle|\, M = \begin{pmatrix} * & * \\ 0_{(n-d) \times d} & * \end{pmatrix} \right\},$$

$$G_{d,n} \cong G/P_d.$$

Consider the morphism $\phi_d : G \to \mathbb{P}(\wedge^d V)$, where $\phi_d = f_d \circ \theta_d$, $f_d : G_{d,n} \hookrightarrow \mathbb{P}(\wedge^d(K^n))$, $\theta_d : G \to G/P_d$. Then $p_j(\phi_d(g))$ is simply the minor of g consisting of the first d columns and the rows with indices $j_1, \ldots, j_d$. Now, denote by Z_d the unipotent subgroup of G generated by $\{U_\alpha \mid \alpha \in R^- \setminus R^-_{P_d}\}$. We have as in §12.1.1,

$$Z_d = \left\{ \begin{pmatrix} I_d & 0_{d \times (n-d)} \\ A_{(n-d) \times d} & I_{n-d} \end{pmatrix} \in G \right\}.$$

As in §12.1.1, we identify Z_d with the opposite big cell in G/P_d. Then, given $z \in Z_d$, the Plücker coordinate p_j evaluated at z is simply a certain minor of A, which may be explicitly described as follows. Let $\underline{j} = (j_1, \ldots, j_d)$, and let j_r be the largest entry $\le d$. Let $\{k_1, \ldots, k_{d-r}\}$ be the complement of $\{j_1, \ldots, j_r\}$ in $\{1, \ldots, d\}$. Then this minor of A is given by column indices $k_1, \ldots k_{d-r}$, and row indices $j_{r+1}, \ldots, j_d$ (here the rows of A are indexed as $d+1, \ldots, n$). Conversely, given a minor of A, say with column indices $b_1, \ldots, b_s$, and row indices $i_{d-s+1}, \ldots, i_d$, it is the evaluation of the Plücker coordinate $p_{\underline{i}}$ at z, where $\underline{i} = (i_1, \ldots, i_d)$ may be described as follows: $\{i_1, \ldots, i_{d-s}\}$ is the complement of $\{b_1, \ldots, b_s\}$ in $\{1, \ldots, d\}$, and $i_{d-s+1}, \ldots, i_d$ are simply the row indices (again, the rows of A are indexed as $d+1, \ldots, n$).

By convention, if $\underline{j} = (1, \ldots, d)$, then $p_{\underline{j}}$ evaluated at z is 1. We shall consider the element 1 (in $K[Z_d]$) as the minor of A with row indices (and column indices) given by the empty set.

12.1.3. Evaluation of the Plücker coordinates on the opposite big cell in G/Q. Consider

$$f : G \to G/Q \hookrightarrow G/P_{a_1} \times \cdots \times G/P_{a_k} \hookrightarrow \mathbf{P}_1 \times \cdots \times \mathbf{P}_k,$$

where $\mathbf{P}_t = \mathbb{P}(\wedge^{a_t} V)$. Denoting the restriction of f to O^- also by just f, we obtain an embedding $f : O^- \hookrightarrow \mathbf{P}_1 \times \cdots \times \mathbf{P}_k$, O^- having been identified with the opposite big cell in G/Q. For $z \in O^-$, the multi-Plücker coordinates of $f(z)$ are simply all the $a_t \times a_t$ minors of z with column indices $\{1, \ldots, a_t\}$, $1 \leq t \leq k$.

12.1.4. Equations defining the cones over Schubert varieties in $G_{d,n}$. Let $Q = P_d$. Given a d-tuple $\underline{i} = (i_1, \ldots, i_d) \in I_{d,n}$, let us denote the associated element of $W_{P_d}^{\min}$ by $\theta_{\underline{i}}$. For simplicity of notation, let us denote P_d by just P, and $\theta_{\underline{i}}$ by just θ. Recall (cf. Chapter 2, §2.10) that the homogeneous co-ordinate ring A, resp. $A_P(\theta)$, of $G/P = (G_{d,n})$, resp. $X_P(\theta)$, for the Plücker embedding is given by

$$A = \bigoplus_{n \in \mathbb{Z}^+} H^0(G/P, L^n)$$

$$A_P(\theta) = \bigoplus_{n \in \mathbb{Z}^+} H^0(X_P(\theta), L^n),$$

where L is the ample generator of $\mathrm{Pic}\, G/P$. From §2.10, we have that the restriction map $A \to A_P(\theta)$ is surjective, and the kernel is generated as an ideal by $\{p_{\underline{j}} \mid \underline{i} \not\geq \underline{j}\}$.

12.1.5. Equations defining multicones over Schubert varieties in G/Q. Let Q be as in §12.1.1. Let $X_Q(w) \subset G/Q$. Recall (cf. Chapter 2, §2.10) that the multi-homogeneous coordinate ring C, resp. $C_Q(w)$, of G/Q, resp. $X_Q(w)$, is given by

$$C = \bigoplus_{\underline{a}} H^0(G/Q, \bigotimes_i L_i^{a_i})$$

$$C_Q(w) = \bigoplus_{\underline{a}} H^0(X_Q(w), \bigotimes_i L_i^{a_i}),$$

notation being as in Chapter 2, §2.10. We have that the kernel of the restriction map $B \to B_w$ is generated by the kernel of $R_1 \to (R(w))_1$; but now, this kernel is the span of

$$\{p_{\underline{i}} \mid \underline{i} \in I_{d,n}, d \in \{a_1, \ldots, a_k\}, \ w^{(d)} \not\geq \underline{i}\},$$

where $w^{(d)}$ is the d-tuple corresponding to the Schubert variety that is the image of $X_Q(w)$ under the projection $G/Q \to G/P_d$, $a_1 \leq d \leq a_k$.

12.1.6. Definition. Ideal of the opposite cell in $X_Q(w)$. Let Q be as in §12.1.1, $Y_Q(w) = B^- e_{\mathrm{id},Q} \cap X_Q(w)$, the opposite cell in $X_Q(w)$ (cf. Chapter 4, §4.4). Considering $Y_Q(w)$ as a closed subvariety of $\mathcal{O}^-$, we obtain (in view of §12.1.5) that the ideal defining $Y_Q(w)$ in $\mathcal{O}^-$ is generated by

$$\{p_{\underline{i}}|_{\mathcal{O}^-} \mid \underline{i} \in I_{d,n}, d \in \{a_1, \ldots, a_k\},\ w^{(d)} \not\geq \underline{i}\}.$$

The following two lemmas relate to the evaluation of Plücker coordinates on the opposite cell of a Schubert variety in G/Q. Let $G = SL(n)$, $1 \leq a_1 < \cdots < a_h \leq n$, $Q = P_{a_1} \cap \cdots \cap P_{a_h}$. Let $\mathcal{O}^-$ be the opposite big cell in G/Q. Let $X = (x_{ba})$, $1 \leq b, a \leq n$ be a generic $n \times n$ matrix and H the one-sided ladder in X defined by the outside corners $(a_i + 1, a_i)$, $1 \leq i \leq h$. Clearly, $\mathbb{A}(H) \simeq \mathcal{O}^-$. Let $X^- = (x_{ba}^-)$, $1 \leq b, a \leq n$, where

$$x_{ba}^- = \begin{cases} x_{ba}, & \text{if } (b,a) \in H \\ 1, & \text{if } b = a \\ 0, & \text{otherwise.} \end{cases}$$

Note that, given $\tau \in W^{a_i}$, for some i, $1 \leq i \leq h$, the function $p_\tau|_{\mathcal{O}^-}$ represents the determinant of the $a_i \times a_i$ submatrix T of X^- whose row indices are $\{\tau(1), \ldots, \tau(a_i)\}$, and column indices are $\{1, \ldots, a_i\}$. Let

$$H_i = \{x_{ba} \mid a_i + 1 \leq b \leq n, 1 \leq a \leq a_i\}\, 1 \leq i \leq h.$$

12.1.7. Lemma. *Let M be a $t \times t$ matrix contained in H_i, for some i, $1 \leq i \leq h$, with row indices $r_1 < \cdots < r_t$. Then $\det M$ belongs to the ideal of $K[H]$ generated by $p_\phi|_{\mathcal{O}^-}$, with $\phi \in W^{a_i}$ such that $\{r_1, \ldots, r_t\} = \{\phi(1), \ldots, \phi(a_i)\} \cap \{a_i + 1, \ldots, n\}$.*

PROOF. Denote by $c_1 < \cdots < c_t$ the column indices of M. Let $\tau = (\{1, \ldots, a_i\} \setminus \{c_1, \ldots, c_t\}) \cup \{r_1, \ldots, r_t\}$. Then $\tau \in W^{a_i}$, and $p_\tau|_{\mathcal{O}^-} = \det T$, where T is the $a_i \times a_i$ submatrix of X^- with row indices $\{\tau(1), \ldots, \tau(a_i)\}$ and column indices $\{1, \ldots, a_i\}$. Using Laplace expansion with respect to the last t rows of T, we obtain

$$(*) \qquad \det T = \sum \pm \det N_{c_1', \ldots, c_t'} \det M_{c_1', \ldots, c_t'},$$

the sum being taken over all subsets with t elements $\{c_1', \ldots, c_t'\}$ of $\{1, \ldots, a_i\}$ where $N_{c_1', \ldots, c_t'}$ is the $(a_i - t) \times (a_i - t)$ submatrix of X^- with row indices $\{1, \ldots, a_i\} \setminus \{c_1, \ldots, c_t\}$ and column indices $\{1, \ldots, a_i\} \setminus \{c_1', \ldots, c_t'\}$, and $M_{c_1', \ldots, c_t'}$ is the $t \times t$ submatrix of X^- with row indices $\{r_1, \ldots, r_t\}$ and column indices $\{c_1', \ldots, c_t'\}$. Note that $M_{c_1, \ldots, c_t} = M$, and $N_{c_1, \ldots, c_t}$ is a lower triangular matrix, with all diagonal entries equal to 1, and hence $\det M$ appears in $(*)$, and its coefficient is ± 1. Also note that $N_{c_1', \ldots, c_t'}$ is obtained from $N_{c_1, \ldots, c_t}$ by replacing the columns with indices $c_1', \ldots, c_t'$ by the columns with indices $c_1, \ldots, c_t$.

Let $\geq$ denote the partial order on I_{t,a_i}, namely $(d_1, \ldots, d_t) \geq (c_1, \ldots, c_t)$ if $d_j \geq c_j$ for all $1 \leq j \leq t$. We prove the lemma by decreasing induction with respect to the order $\geq$ on the t-tuple $(c_1, \ldots, c_t)$ consisting of the column indices of M.

If $c_j > a_{i-1}$ for all $1 \leq j \leq t$, then for $\{c'_1, \ldots, c'_t\} \neq \{c_1, \ldots, c_t\}$ we have $\det N_{c'_1, \ldots, c'_t} = 0$, since at least one of $c_1, \ldots, c_t$ is an index for a column in $N_{c'_1, \ldots, c'_t}$, and all entries of this column are 0. Thus, in this case (∗) reduces to $\det T = \pm \det M$, i.e., $\det M = \pm p_\tau|_{O^-}$, with $\tau \in W^{a_i}$ such that $\{\tau(1), \ldots, \tau(a_i)\} \cap \{a_i + 1, \ldots, n\} = \{r_1, \ldots, r_t\}$.

Assume now that the assertion is true for all matrices with row indices $r_1 < \cdots < r_t$ and column indices $d_1 < \cdots < d_t$ such that $(d_1, \ldots, d_t) > (c_1, \ldots, c_t)$, i.e., such that $d_j \geq c_j$ for all $1 \leq j \leq t$ and $(d_1, \ldots, d_t) \neq (c_1, \ldots, c_t)$. We shall now prove it for the matrix M with row indices $r_1 < \cdots < r_t$ and column indices $c_1 < \cdots < c_t$. Consider a typical $N_{c'_1, \ldots, c'_t}$ in (∗). If there exists a j such that $c'_j < c_j$, then the column with index c_j is replacing the column with index c'_j, while obtaining $N_{c'_1, \ldots, c'_t}$ from $N_{c_1, \ldots, c_t}$; hence $N_{c'_1, \ldots, c'_t}$ is still lower triangular, but the diagonal entry in the column with index c_j is 0, which implies that $\det N_{c'_1, \ldots, c'_t} = 0$. Consequently, we obtain

$$\det T = \pm \det M + \sum \pm \det N_{c'_1, \ldots, c'_t} \det M_{c'_1, \ldots, c'_t},$$

and hence

$$\det M = \pm p_\tau|_{O^-} + \sum \pm \det N_{c'_1, \ldots, c'_t} \det M_{c'_1, \ldots, c'_t},$$

the sum being taken over all $(c'_1, \ldots c'_t) \in I_{t,a_i}$ such that $(c'_1, \ldots, c'_t) > (c_1, \ldots, c_t)$. The required result now follows by induction hypothesis.   □

12.1.8. Lemma. *Let $1 \leq t \leq a \leq a_i$, $1 \leq s \leq n$ and $\tau \in W^{a_i}$ such that $\tau(a - t + 1) \geq s$. Then $p_\tau|_{O^-}$ belongs to the ideal of $K[H]$ generated by t-minors in X^- with row indices $\geq s$ and column indices $\leq a$.*

PROOF. Let T be the $a_i \times a_i$ submatrix of X^- with row indices $\{\tau(1), \ldots \tau(a_i)\}$ and column indices $\{1, \ldots, a_i\}$. Then $p_\tau|_{O^-} = \det T$. Using Laplace expansion with respect to the first a columns, we have $\det T = \sum_p \det A_p \det B_p$, where A_p, resp. B_p, is an $a \times a$, resp. $(a_i - a) \times (a_i - a)$, matrix. Clearly, all the column indices of a typical A_p are $\leq a$, and since $\tau(a - t + 1) \geq s$, at least t of the row indices of A_p are $\geq s$. Using Laplace expansion for A_p with respect to t rows with indices $\geq s$, we obtain $\det A_p = \sum_q \det C_q \det D_q$, where C_q, resp. D_q, is a $t \times t$, resp. $(a - t) \times (a - t)$, matrix, the row indices of C_q are $\geq s$, and column indices of C_q are $\leq a$. The required result follows from this.   □

12.2. Determinantal varieties

Even though determinantal varieties are classically well known, their relationship to Schubert varieties has been developed only recently (cf. [108], [132]). In this section we present this aspect of determinantal varieties. For the sake of completeness of the treatment of this aspect, we have included proofs for all of the statements.

Let $Z = M_{m,n}$, the space of all $m \times n$ matrices with entries in K. We shall identify Z with $\mathbb{A}^{mn}$. We have $K[Z] = K[x_{i,j}, 1 \leq i \leq m, 1 \leq j \leq n]$.

12.2.1. Definition. The variety D_t. Let $X = (x_{ij})$, $1 \leq i \leq m$, $1 \leq j \leq n$ be an $m \times n$ matrix of indeterminates. Let $A \subset \{1, \ldots, m\}$, $B \subset \{1, \ldots, n\}$ be two subsets of the same size s, where $s \leq \min \{m, n\}$. We shall denote by $[A|B]$ the s-minor of X with row indices given by A, and column indices given by B. For $1 \leq t \leq \min \{m, n\}$, let $I_t(X)$ be the ideal generated by $\{[A|B], A \subset \{1, \ldots, m\}, B \subset \{1, \ldots, n\}, \#A = \#B = t\}$. Let $D_t \subset Z$ be the variety with $\sqrt{I_t(X)}$ as the defining ideal. D_t is called a *determinantal variety* since its defining equations are certain minors in a matrix.

We shall see among other things that $\sqrt{I_t(X)} = I_t(X)$.

12.2.2. Definition. The partial order among minors. As above, let $X = (x_{ba})$, $1 \leq b \leq m$, $1 \leq a \leq n$, be a $m \times n$ matrix of indeterminates. We introduce a partial order on the set of all minors of X as follows: $[i_1, \ldots, i_r | j_1, \ldots, j_r] \leq [i'_1, \ldots, i'_s | j'_1, \ldots, j'_s]$ if

1. $r \geq s$
2. $i_r \geq i'_s, i_{r-1} \geq i'_{s-1}, \ldots, i_{r-s+1} \geq i'_1$
3. $j_1 \leq j'_1, j_2 \leq j'_2, \ldots, j_s \leq j'_s$.

We say that an ideal I of $K[X]$ is *cogenerated* by a given minor M if I is generated by $\{M' \mid M'$ a minor of X and $M' \not\geq M\}$.

12.2.3. Lemma. *Let $1 \leq t \leq \min \{m, n\}$. Let M be the $(t-1)$-minor $[A|B]$ of X, where $A = \{m+2-t, m+3-t, \ldots, m\}$, $B = \{1, 2, \ldots, t-1\}$; note that M is the $(t-1)$-minor of X consisting of the last $(t-1)$ rows of X, and the first $(t-1)$ columns of X. Then $I_t(X)$ is cogenerated by M.*

PROOF. Let us write $A = \{i_1, \ldots, i_{t-1}\}$, $B = \{j_1, \ldots, j_{t-1}\}$ as $t-1$-tuples. Let I be the ideal cogenerated by M. Then I is generated by $\mathcal{G}_t$, where $\mathcal{G}_t = \{$all minors $M' \mid M' \not\geq M\}$. Let $N = [i'_1, \ldots, i'_s | j'_1, \ldots, j'_s]$ be a minor of X of size s, $N \neq M$. If $s \geq t$, then clearly $N \not\geq M$, and hence $N \in \mathcal{G}_t$. Let now $s \leq t-1$. We have clearly $N \geq M$, since $i'_s \leq m (= i_{t-1}), i'_{s-1} \leq m-1 (= i_{t-2}), \ldots, i'_1 \leq m+1-s (= i_{t-s})$, and $j'_l \geq j_l (= l)$, $1 \leq l \leq s$; hence $N \notin \mathcal{G}_t$. From this it follows that $\mathcal{G}_t = \{M' \mid M'$ is a minor of X of size $\geq t\}$. But then the ideal generated by $\mathcal{G}_t$ is precisely $I_t(X)$. Thus we obtain $I = I_t(X)$ and the result follows. $\square$

12.2.4. The monomial order $\prec$ and Gröbner bases. We introduce a total order on the variables as follows:

$$x_{m1} > x_{m2} > \cdots > x_{mn} > x_{(m-1)1} > x_{(m-1)2}$$
$$> \cdots > x_{(m-1)n} > \cdots > x_{11} > x_{12} > \cdots > x_{1n}.$$

This induces a total order, namely the lexicographic order, on the set of monomials in $K[X] = K[x_{11}, \ldots, x_{mn}]$, denoted by $\prec$. Recall that the largest monomial, with respect to $\prec$, present in a polynomial $f \in K[X]$ is called the initial term of f, and is denoted by $in(f)$. Note that the initial term, with respect to $\prec$, of a minor of X is equal to the product of its elements on the skew diagonal.

Given an ideal $I \subset K[X]$, denote by $in(I)$ the ideal generated by the initial terms of the elements in I. Recall that a set $G \subset I$ is called a Gröbner basis of I (with respect to the monomial order $\prec$), if $in(I)$ is generated by the initial terms of the elements in G. Note that a Gröbner basis of I generates I as an ideal.

We recall the following (see [64]).

12.2.5. Theorem. *Let $M = [i_1, \ldots, i_r | j_1, \ldots, j_r]$ be a minor of X, and I the ideal of $K[X]$ cogenerated by M. For $1 \leq s \leq r+1$, let G_s be the set of all s-minors $[i'_1, \ldots, i'_s | j'_1, \ldots, j'_s]$ satisfying the conditions*

(1) $i'_s \leq i_r, i'_{s-1} \leq i_{r-1}, \ldots, i'_2 \leq i_{r-s+2}.$
(2) $j'_{s-1} \geq j_{s-1}, \ldots, j'_2 \geq j_2, j'_1 \geq j_1$
(3) *if $s \leq r$, then $i'_1 > i_{r-s+1}$ or $j'_s < j_s$.*

Then the set $G = \cup_{i=1}^{r+1} G_i$ is a Gröbner basis for the ideal I with respect to the monomial order $\prec$.

12.2.6. Theorem. *Let $1 \leq t \leq \min\{m, n\}$. Let $\mathcal{F}_t = \{M' \mid M'$ is a minor of X of size $t\}$. Then $\mathcal{F}_t$ is a Gröbner basis for $I_t(X)$.*

PROOF. We have D_t is generated by $\mathcal{F}_t$. Let M be as in Lemma 12.2.3, namely, the $(t-1)$-minor $[A|B]$ of X, where $A = \{m+2-t, m+3-t, \ldots, m\}$, $B = \{1, 2, \ldots, t-1\}$. Let us write $A = \{i_1, \ldots, i_r\}$, $B = \{j_1, \ldots, j_r\}$. We have $r = t-1$, $i_l = m+l-r$, $j_l = l$, $1 \leq l \leq r$.

Let $M' \in G$, G being as in Theorem 12.2.5, and let $M' = [i'_1, \ldots, i'_s | j'_1, \ldots, j'_s]$ be a minor of X of size s, $s \leq r+1 (= t)$. The inequalities regarding i's and i''s in condition (1) of Theorem 12.2.5 are redundant, since $i_l = m+l-r$, $1 \leq l \leq r$. Similarly, the inequalities regarding j's and j''s in condition (1) of Theorem 12.2.5 are again redundant since $j_l = l$, $1 \leq l \leq r$; also, condition (2) reduces to the condition that if $s \leq r(= t-1)$, then $i'_1 > i_{r-s+1}$ since $j_s = s$. Therefore, for the above choice of M, conditions (1) and (2) of Theorem 12.2.5 are equivalent to

$$\text{if } s \leq r\,(= t-1), \text{ then } i'_1 > i_{r-s+1}\,(= i_{t-s}).$$

Let $s \leq t-1$. The condition that $i'_1 > i_{t-s} (= m+1-s)$ implies $i'_s > m+1$, which is not possible. Hence $G_s = \emptyset$, if $s \leq t - 1$; thus we obtain $G = G_t$. Now for $s = t$, as discussed above, condition (1) of Theorem 12.2.5 is satisfied by any t-minor M' while condition (2) is vacuous. Hence we obtain that G_t is the set of all t-minors, i.e., $G = \mathcal{F}_t$. The result now follows from Lemma 12.2.3 and Theorem 12.2.5. □

Next we identify D_t with Y_ϕ, the opposite cell (cf.Chapter 4, §4.4) in $X_{P_d}(\phi)$ for a suitable ϕ in $I_{d,n}$.

12.2.7. Definition. The set Z_t and the elements $\tau, \phi \in I_{d,n}$. Let $G = SL(n)$. Let r, d be such that $r + d = n$. Let us identify $\mathcal{O}_d$, the opposite cell in G/P_d, as all matrices of the form

$$\mathcal{O}_d = \left\{ \begin{pmatrix} I_d \\ X \end{pmatrix} \right\}$$

where X is a generic $r \times d$ matrix. As in §12.1.2, we have a bijection between {Plücker coordinates $p_{\underline{i}}$, $\underline{i} \neq \{1, 2, \ldots, d\}$} and {minors of X; note that if $\underline{i} = \{1, 2, \ldots, d\}$}, then $p_{\underline{i}} = $ the constant function 1. Given a Plücker coordinate $p_{\underline{i}}$ (on $\mathcal{O}_d$), we denote the associated minor (of X) by $\Delta_{\underline{i}}$.

For example, take $r = 3 = d$. We have

$$\mathcal{O}_3 = \left\{ \begin{pmatrix} I_3 \\ X_{3 \times 3} \end{pmatrix} \right\}.$$

We have $p_{(1,2,4)} = [\{1\}|\{3\}]$, $p_{(2,4,6)} = [\{1,3\}|\{1,3\}]$. Set

(12.2.8) $Z_t = \{ \underline{i} \in I_{d,n} \mid \Delta_{\underline{i}} \text{ is a } t \text{ minor of } X \}$,

(12.2.9) $\tau = (1, 2, \ldots, d-t, d+1, d+2, \ldots, d+t)$,

(12.2.10) $\phi = (t, t+1, \ldots, d, n+2-t, n+3-t, \ldots, n)$.

Note that ϕ consists of the two blocks of consecutive integers.

12.2.11. Lemma. *Let notations be as above. Then $\tau \in Z_t$. Further, τ is the unique smallest element (under $\geq$ on $I_{d,n}$) in Z_t.*

PROOF. Clearly $\tau \in Z_t$. Let Δ be a t-minor of X, and p_i the associated Plücker coordinate. Let $\underline{i} = (i_1, \ldots, i_d)$. We have, for $1 \leq k \leq d-t$, $i_k \leq d$, and for $d-t+1 \leq k \leq d$, $i_k > d$. Clearly τ is the smallest such d-tuple. □

12.2.12. Remark. With τ as in Lemma 12.2.11, note that the associated minor Δ_τ of X is $[\{d+1-t, d+2-t, \ldots, d\} \mid \{d+1, d+2, \ldots, d+t\}]$, i.e., the rightmost top corner t-minor of X, the rows of X being indexed as $d+1, \ldots, n$. Furthermore, note that τ is the smallest d-tuple $(j_1, \ldots, j_d)$ such that $j_{d-t+1} \geq d+1$.

12.2.13. Lemma. *Let* $N_t = \{\underline{i} \in I_{d,n} |\ p_\tau|_{X_{P_d}(\underline{i})} = 0\}$. *Then* ϕ *is the unique largest element in* N_t.

PROOF. Let $\underline{i} \in N_t$, say, $\underline{i} = (i_1, \dots, i_d)$. We have, $p_\tau|_{X_{P_d}(\underline{i})} = 0$ if and only if $\underline{i} \not\geq \tau$, i.e., if and only if $i_{d-t+1} \not\geq d+1$ (cf. Remark 12.2.12), i.e., $i_{d-t+1} \leq d$. Now it is easily checked that ϕ is the largest d-tuple $(j_1, \dots, j_d)$ such that $j_{d-t+1} \leq d$. □

12.2.14. Remark. Let ϕ be as in Lemma 12.2.13. As observed in the proof of Lemma 12.2.13, we have ϕ is the largest d-tuple $(j_1, \dots, j_d)$ such that $j_{d-t+1} \leq d$.

12.2.15. Corollary. *Let* $\underline{i} \in Z_t$. *Then* $p_{\underline{i}}|_{X_{P_d}(\phi)} = 0$, ϕ *being as in Lemma 12.2.13.*

PROOF. We have (cf. Lemma 12.2.13) that $p_\tau|_{X_{P_d}(\phi)} = 0, \tau$ being as in Lemma 12.2.11, and hence $\phi \not\geq \tau$ which in turn implies (in view of Lemma 12.2.11), $\phi \not\geq \underline{i}$, $\underline{i} \in Z_t$. Hence we obtain $p_{\underline{i}}|_{X_{P_d}(\phi)} = 0$. □

12.2.16. Theorem. *Let* ϕ *be as in Lemma 12.2.13, and let* $Y_\phi = \mathcal{O}_d \cap X_{P_d}(\phi)$, *the opposite cell in* $X_{P_d}(\phi)$. *Then* $\sqrt{I_t(X)} = I_t(X)$, *and* $D_t \cong Y_\phi$.

PROOF. Let I_ϕ be the ideal defining Y_ϕ in $\mathcal{O}_d$. We have (cf. §12.1.4) that I_ϕ is generated by $N_t := \{p_{\underline{i}},\ \underline{i} \not\leq \phi\}$. Also, $I_t(X)$ is generated by $\{p_{\underline{i}},\ \underline{i} \in Z_t\}$.

Let $\underline{i} \in Z_t$, say, $\underline{i} = (i_1, \dots, i_d)$. We have (cf. Lemma 12.2.13) $\underline{i} \not\leq \phi$, and hence $p_{\underline{i}} \in N_t$.

Let now $p_{\underline{i}} \in N_t$, say, $\underline{i} = (i_1, \dots, i_d)$. This implies that $i_{d-t+1} \geq d+1$ (cf. Remark 12.2.14), and hence it corresponds to an s-minor in X, where $s \geq t$. From this it follows that $p_{\underline{i}} \in$ the ideal generated by $\{p_{\underline{i}},\ \underline{i} \in Z_t\}$, i.e., $p_{\underline{i}}$ belongs to $I_t(X)$. Thus we have shown $I_\phi = I_t(X)$ and the two assertions follow from this. □

12.2.17. Corollary. $\dim(D_t) = (t-1)(n-(t-1))$, *where note that* $n = r + d$.

PROOF. We have $\dim(D_t) = \dim X(\phi) = (t-1)(n-(t-1))$ (cf. §3.1.3). □

12.2.18. Corollary. D_t *is normal, Cohen–Macaulay, and has rational singularities.*

PROOF. This follows from the fact that Schubert varieties are normal, Cohen–Macaulay, and have rational singularities. □

12.2.19. The singular locus of D_t. Let $G = SL(n)$, and $P = P_d$, the maximal parabolic subgroup with the associated set of simple roots being $S\backslash\{\alpha_d\}$. Then as seen in §3.1.6, G/P gets identified with the Grassmannian variety $G_{d,n}$ = the set of d-dimensional subspaces of K^n. We will determine the singular locus of D_t by identifying D_t with Y_ϕ.

In this section, we follow the Young diagram representation of Schubert varieties in the Grassmannian (cf. §9.3). Recall, to $(a_1, \ldots, a_d) \in W^{P_d}$, we associate the partition $\mathbf{a} := (\mathbf{a}_1, \ldots, \mathbf{a}_d)$, where $\mathbf{a}_i = a_{d-i+1} - d - i + 1$. For a partition $\mathbf{a} = (\mathbf{a}_1, \ldots, \mathbf{a}_d)$, we shall denote by $X_{\mathbf{a}}$ the Schubert variety corresponding to $(a_1, \ldots, a_d)$. Then $\dim X_{\mathbf{a}} = |\mathbf{a}| = \mathbf{a}_1 + \cdots + \mathbf{a}_d$. It is clear that $\mathbf{a}_i \leq n - d$.

Let

$$\mathbf{a} = (p_1^{q_1}, \ldots, p_r^{q_r}) = (\underbrace{p_1, \ldots, p_1}_{q_1 \text{ times}}, \ldots, \underbrace{p_r, \ldots, p_r}_{q_r \text{ times}})$$

(we say that $\mathbf{a}$ consists of r rectangles: $p_1 \times q_1, \ldots, p_r \times q_r$). Let

$$\mathbf{a}'_i = (p_1^{q_1}, \ldots, p_{i-1}^{q_{i-1}}, p_i^{q_i-1}, (p_{i+1}-1)^{q_{i+1}+1}, p_{i+2}^{q_{i+2}}, \ldots, p_r^{q_r})$$

for each $1 \leq i \leq r - 1$. It was shown in Theorem 9.3.1 that $\mathrm{Sing}X_{\mathbf{a}}$ has $r - 1$ components $X_{\mathbf{a}'_1}, \ldots, X_{\mathbf{a}'_{r-1}}$. Furthermore, $X_{\mathbf{a}}$ is smooth if and only if $\mathbf{a}$ consists of one rectangle. We will use these facts to prove the following theorem.

12.2.20. Theorem. $\mathrm{Sing}D_t = D_{t-1}$.

PROOF. We have (cf. Theorem 12.2.16) that $D_t \cong Y_\phi$ where

$$\phi = (t, t+1, \ldots, d, n+2-t, n+3-t, \ldots, n)$$

and Y_ϕ is the opposite cell in $X_{P_d}(\phi)$ ($\subset G_{d,n}$). Note that ϕ consists of the two blocks $[t, d]$, $[n+2-t, n]$ of consecutive integers. Here, for $a, b \in \mathbb{N}$, $a < b$, $[a, b]$ denotes the set $\{a, a+1, \ldots, b\}$. The partition determined by ϕ is given by

$$\mathbf{a} = (\underbrace{n-d, \ldots, n-d}_{t-1 \text{ times}}, \underbrace{t-1, \ldots, t-1}_{d+1-t \text{ times}}).$$

Now by Theorem 9.3.1, $\mathrm{Sing}X_{\mathbf{a}} = X_{\mathbf{b}}$, where $\mathbf{b}$ is the partition

$$\mathbf{b} = (\underbrace{n-d, \ldots, n-d}_{t-2 \text{ times}}, \underbrace{t-2, \ldots, t-2}_{d+2-t \text{ times}}).$$

Now the d-tuple τ associated to $\mathbf{b}$ is given by $\tau = (t-1, t, \ldots, d, n+3-t, n+4-t, \ldots, n) = ([t-1, d], [n+3-t, n])$. Hence $\mathrm{Sing}Y_\phi = Y_\tau$. Hence we obtain $\mathrm{Sing}D_t = D_{t-1}$ in view of Th. 12.2.16. $\qquad\square$

Alternate proof. We can give an independent proof of the above result (not using Schubert varieties) as follows. Let $\mathcal{J}$ be the Jacobian matrix of

D_t (considered as a subvariety of $M_{r,d} \cong \mathbb{A}^{rd}$). Let $\mathcal{M}_t$ denote the set of all t-minors of the generic $r \times d$ matrix X. We shall index the rows of $\mathcal{J}$ by $\mathcal{M}_t$ and the columns by the set of all variables in X. Given $M \in \mathcal{M}_t$, and τ, a variable in X, the (M, τ)-th entry in $\mathcal{J}$ is nonzero if and only if τ is an entry in M, in which case it is equal to $\pm \det M'$, where M' is the matrix obtained from M by deleting the row and column containing τ.

Let $z \in D_t$ and let $\mathcal{J}_z$ be the Jacobian matrix $\mathcal{J}$ evaluated at z. First, let z be such that all $(t-1)$-minors of X vanish at z. Then clearly $\mathcal{J}_z = 0$, and z is a singular point of D_t.

Otherwise, Let z be such that some $(t-1)$-minor M of X is nonzero at z. We shall now show that z is a smooth point of D_t. Let B denote the set of all variables in X not appearing in any row or column of X given by the rows and columns of M. Let $\tau \in B$, and let M_τ be the t-minor of X obtained from M by adding the row and column of X through τ to M. Then the (M_τ, τ)-th entry in $\mathcal{J}_z$ is equal to $\pm(\det M)(z)$ and hence is nonzero; also for $\sigma \in B$, $\sigma \neq \tau$, the (M_τ, σ)-th entry in $\mathcal{J}_z$ is 0. From this, it follows that the minor in $\mathcal{J}_z$ with row indices given by $\{M_\tau, \ \tau \in B\}$ and column indices given by B is nonzero. Hence we obtain rank $\mathcal{J}_z \geq (r - (t-1))(d - (t-1))(= \#B) = rd - (t-1)(r + d - (t-1)) = \operatorname{codim}_z D_t$, where recall that $Z = M_{m,n}$. From this it follows that rank $\mathcal{J}_z = \operatorname{codim}_z D_t$, and hence z is a smooth point of D_t by the Jacobian criterion §4.5.

12.3. Ladder determinantal varieties

Let $X = (x_{ba})$, $1 \leq b \leq m$, $1 \leq a \leq n$, be a $m \times n$ matrix of indeterminates. Given $1 \leq b_1 < \cdots < b_h < m$, $1 < a_1 < \cdots < a_h \leq n$, we consider the subset of X, defined by

$$L = \{x_{ba} \mid \text{ there exists } 1 \leq i \leq h \text{ such that } b_i \leq b \leq m, 1 \leq a \leq a_i\}.$$

We call L an *one-sided ladder* in X, defined by the *outside corners* $\omega_i = x_{b_i a_i}$, $1 \leq i \leq h$. For simplicity of notation, we identify the variable x_{ba} with just (b, a). We think of L as a set of entries below the main diagonal of an $\ell \times \ell$, for some $\ell \geq \max\{m, n\}$.

For example, the following ladder

$$\begin{array}{cccc} * & * & & \\ * & * & & \\ * & * & * & \\ * & * & * & \end{array}$$

may be thought of as certain entries below the main diagonal of a 6×6 matrix; we may take $b_i = 3, 5$, $a_i = 2, 3$, $i = 1, 2$. Thus

$$L = \{x_{31}, x_{32}, x_{41}, x_{42}, x_{51}, x_{52}, x_{53}, x_{61}, x_{62}, x_{63}\}.$$

Let $2 \leq t \leq \min\{m - b_i + 1, a_i\}$, for all $1 \leq i \leq h$. For $1 \leq i \leq h$, let

$$L_i = \{x_{ba} \in L \mid b_i \leq b \leq m\}.$$

Let $K[L]$ denote the polynomial ring $K[x_{ba} \mid x_{ba} \in L]$, and let $\mathbb{A}(L) = \mathbb{A}^{|L|}$ be the associated affine space. Let $I_t(L)$ be the ideal in $K[L]$ generated by all the t-minors contained in L, and $D_t(L) \subset \mathbb{A}(L)$ the variety defined by the ideal $\sqrt{I_t(L)}$. We call $D_t(L)$ a *ladder determinantal variety* (associated to a one-sided ladder).

12.3.1. Remark. Ladder determinantal varieties were first introduced by Abhyankar (cf. [**3**]).

For $1 \leq i \leq h$, let

$$L(i) = \{x_{ba} \in L_i \mid 1 \leq a \leq a_i\}.$$

12.3.2. Remark. Given a t-minor Δ, say with row indices $\{r_1, \ldots, r_t\}$, and column indices $\{c_1, \ldots, c_t\}$, let i, $1 \leq i \leq h$ be such that $x_{r_1 c_t}$ belongs to $L(i)$, note that $L = \cup_i L(i)$. Then Δ is contained in $L(i)$, $L(i)$ being considered as a matrix. Thus the ideal $I_t(L)$ is generated by the t-minors of X contained in $L(i)$, $1 \leq i \leq h$.

From now on we shall suppose that $m = n$, with n large enough such that L is situated below the main diagonal, i.e., $b_i \geq a_i + 1$, $1 \leq i \leq h$. Let $G = SL(n)$, $Q = P_{a_1} \cap \cdots \cap P_{a_h}$. Let O^- be the opposite big cell in G/Q. Let H be the one-sided ladder defined by the outside corners $(a_i + 1, a_i)$, $1 \leq i \leq h$.

Let Z be the variety in $\mathbb{A}(H) \cong O^-$ defined by the vanishing of the t-minors in $L(i)$, $1 \leq i \leq h$. Note that $Z \cong D_t(L) \times \mathbb{A}(H \setminus L) \cong D_t(L) \times \mathbb{A}^r$, where $r = \dim SL(n)/Q - |L|$.

We shall now define an element $w \in W_Q^{\min}$, such that the variety Z identifies with the opposite cell in the Schubert variety $X(w)$ in G/Q. We define $w \in W_Q^{\min}$ by specifying $w^{(a_i)} \in W^{a_i}$ $1 \leq i \leq h$, where $\pi_i(X(w)) = X(w^{(a_i)})$ under the projection $\pi_i : G/Q \to G/P_{a_i}$.

Define $w^{(a_i)}$, $1 \leq i \leq h$, inductively, as the (unique) maximal element in W^{a_i} such that

1. $w^{(a_i)}(a_i - t + 1) = b_i - 1$.
2. If $i > 1$, then $w^{(a_{i-1})} \subset w^{(a_i)}$.

Note that $w^{(a_i)}$, $1 \leq i \leq h$, is well-defined in W^{a_i}, and w is well-defined as an element in $W_Q^{\min}$.

12.3.3. Remark. Note that $w^{(a_i)}$, $1 \leq i \leq h$, consists of several blocks of consecutive integers ending with $b_j - 1$ at the $(a_j - t + 1)$-th place for some j's belonging to $\{1, \ldots, i\}$, and a last block of length $(t - 1)$ ending with n at the a_j-th place. This is the best description that we have to describe the class of permutations corresponding to these ladder determinantal varieties.

Example. Let $n = 13$, $h = 3$, $t = 3$, $(b_1, a_1) = (6, 4)$, $(b_2, a_2) = (8, 5)$, $(b_3, a_3) = (11, 9)$. Then

$$w^{(4)} = (4, 5, \ 12, 13), \quad w^{(5)} = (4, 5, \ 7, \ 12, 13),$$

$$w^{(9)} = (4, 5, 6, 7, 8, 9, 10, \ 12, 13).$$

12.3.4. Theorem. *The variety Z $(= D_t(L) \times \mathbb{A}^r)$ identifies with the opposite cell in $X(w)$, i.e., $Z = X(w) \cap O^-$ (scheme theoretically). Further, $\sqrt{I_t(L)} = I_t(L)$.*

PROOF. Let $f = \det M$, where M is a $t \times t$ matrix contained in $L(i)$ for some $1 \leq i \leq h$, be a generator of $I(Z)$ (cf. Remark 12.3.2). By Lemma 12.1.7, f can be written in the form $f = \sum g_\phi p_\phi|_{O^-}$, with $\phi \in W^{a_i}$ such that $\{\phi(1), \ldots, \phi(a_i)\} \cap \{a_i + 1, \ldots, n\} = \{r_1, \ldots, r_t\}$, and $g_\phi \in K[H]$; here $r_1, \ldots, r_t$ are the row indices of M. In particular, we have $\phi(a_i - t + 1) = r_1$. Since M is contained in $L(i)$, we have $r_1 \geq b_i$, and hence $\phi(a_i - t + 1) \geq b_i$. We have $w^{(a_i)}(a_i - t + 1) = b_i - 1$, and hence $\phi(a_i - t + 1) > w^{(a_i)}(a_i - t + 1)$. This shows that $\phi \not\leq w^{(a_i)}$, and therefore $p_\phi \in I(X(w) \cap O^-)$. Thus $f \in I(X(w) \cap O^-)$.

Let now g be a generator of the ideal $I(X(w) \cap O^-)$, i.e., $g = p_\tau|_{O^-}$, with $\tau \in W^{a_i}$ for some i, $1 \leq i \leq h$, such that $\tau \not\leq w^{(a_i)}$. Since $w^{(a_i)}$ consists of several blocks of consecutive integers ending with $b_j - 1$ at the a_j-th place for some $j \in \{1, \ldots, i\}$, and a last block ending with n at the a_j-th place (cf. Remark 12.3.3), it follows that $\tau(a_j - t + 1) \geq b_j$ for some $j \in \{1, \ldots, i\}$. Using Lemma 12.1.8, we deduce that $p_\tau|_{O^-}$ belongs to the ideal of $K[H]$ generated by t-minors in L with row indices $\geq b_j$, and column indices $\leq a_j$. Thus $p_\tau|_{O^-}$ belongs to the ideal generated by t-minors contained in $L(j)$, which shows that $g \in I(Z)$. The two assertions now follow. $\square$

Since the Schubert varieties are irreducible, normal, Cohen–Macaulay, and have rational singularities (cf. [**81, 138, 139, 140**]), as a consequence of Theorem 12.3.4 we obtain

12.3.5. Theorem. *The variety $D_t(L)$ is irreducible, normal, Cohen–Macaulay, and has rational singularities.*

12.3.6. A Gröbner basis for $D_t(L)$. We recall the following well-known lemma.

12.3.7. Lemma. *Let $K[X]$ be the polynomial ring in the set of indeterminates X, I an ideal of $K[X]$, and G a Gröbner basis of I with respect to a certain monomial order. Let $L \subset X$ such that*

$$\text{if } f \in G \text{ and } in(f) \in K[L], \text{ then } f \in K[L].$$

Then the set $G \cap K[L]$ is a Gröbner basis of the ideal $I \cap K[L]$.

PROOF. Let $g \in I \cap K[L]$. Since G is a Gröbner basis of I, there exists $f \in G$ such that $in(g) = \langle in(f) \rangle$. Since $g \in K[L]$, we have $in(g) \in K[L]$, and hence $in(f) \in K[L]$. By hypothesis, $f \in K[L]$, and hence $f \in G \cap K[L]$. Therefore, the initial terms of the elements of $G \cap K[L]$ generate the ideal $in(I \cap K[L])$. $\square$

As a direct consequence, we obtain the following:

12.3.8. Proposition. *Let $L \subset X$ be a one-sided ladder and $\mathcal{G}_t(L) = \{all\ t\text{-}minors\ in\ L\}$. Then $\mathcal{G}_t(L)$ is a Gröbner basis of $I_t(L)$ with respect to the monomial order $\prec$.*

PROOF. By Theorem 12.2.6, $\mathcal{G}_t$ ({all t-minors in X}) is a Gröbner basis of $I_t(X)$. Let $\Delta \in \mathcal{G}_t$ be such that $in(\Delta) \in K[L]$. Now $in(\Delta)$ (with respect to the monomial order $\prec$) is simply the product of the entries on the anti-diagonal of Δ. Hence the fact that $in(\Delta) \in K[L]$ implies that $(\Delta) \in K[L]$. From this it follows by Lemma 12.3.7, $\mathcal{G}_t(L) = (\mathcal{G}_t \cap K[L])$ is a Gröbner basis of the ideal $I_t(X) \cap K[L]$. On the other hand we have that $\mathcal{G}_t(L)$ generates $I_t(L)$, and the result follows. $\square$

12.3.9. The dimension of $D_t(L)$.

12.3.10. Definition. The set $\mathcal{C}_t(L)$. Let $\mathcal{C}_t(X)$ be the set of variables in the submatrix obtained from X by deleting the first $t - 1$ columns and the last $t - 1$ rows. Set

(12.3.11) $\mathcal{C}_t(L) = \mathcal{C}_t(X) \cap L$.

For $i \leq h$, let $X = \{x_{ba} \in X \mid b_i \leq b \leq m\}$, $\tilde{X}_i = X_i \setminus X_{i+1}$, where, if $i = h$, then X_{i+1} is supposed to be the empty set. Let $\mathcal{C}_i(X) = \mathcal{C}_t(X) \cap \tilde{X}_i$, note that $\mathcal{C}_t(X) = \dot{\cup} \mathcal{C}_i(X)$.

For a one-sided ladder $L \subset X$, let $\mathcal{C}_i(L) = \mathcal{C}_i(X) \cap L$.

Note that in a solid minor in L, i.e., a minor with consecutive row indices and consecutive column indices, the smallest (for the order in §12.2.4) element belongs to $\mathcal{C}_t(L)$, and conversely, an element $\alpha \in \mathcal{C}_t(L)$ determines uniquely a solid minor in L having α as the smallest element. Hence the number of elements in $\mathcal{C}_t(L)$ is equal to the number of solid minors in the set L.

The following is a generalization of [**54**, Prop. 8].

12.3.12. Proposition. *Let $L \subset X$ an one-sided ladder. Then*

$$codim_{\mathbb{A}(L)} D_t(L) = |\mathcal{C}_t(L)|.$$

PROOF. By Proposition 12.3.8, the ideal $I_t(L)$ and the ideal $J_t(L)$ of its initial terms determine graded quotient rings of $K[L]$ having the same Hilbert series, and hence the codimension of the variety $D_t(L)$ is equal to the height of the monomial ideal $J_t(L)$. In general, the height of a monomial ideal J in a polynomial ring $K[x_1, \ldots, x_N]$ is equal to the minimal cardinality of a set $\mathcal{C} \subset \{x_1, \ldots, x_N\}$ of variables such that

(12.3.13)   each monomial in a set of monomial generators for J contains a variable from $\mathcal{C}$.

Let $J = J_t(L)$ and $\mathcal{C} = \mathcal{C}_t(L)$. Then it is easy to see that $\mathcal{C}$ satisfies (12.3.13), the set of monomial generators being the set of the initial terms of all the t-minors in L; note that for a t-minor $\Delta \in L$, $\mathrm{in}(\Delta)$ (with respect to the monomial order $\prec$) is simply the product of the entries on the anti-diagonal of Δ. Let us denote $\Delta_k = \{x_{ba} \in L \mid b + a = k + 1\}$, $k \geq 1$. Then $L = \dot{\bigcup}_{k \geq 1} \Delta_k$, and $\mathcal{C} = \dot{\bigcup}_{k \geq 1}(\mathcal{C} \cap \Delta_k)$.

Let now $\mathcal{C}' \subset \{x_{ba} \mid x_{ba} \in L\}$ be a set such that $|\mathcal{C}'| < |\mathcal{C}|$. Then there exists a k such that $|\mathcal{C}' \cap \Delta_k| < |\mathcal{C} \cap \Delta_k|$ (in particular $\mathcal{C} \cap \Delta_k \neq \emptyset$). Let $i \in \{1, \ldots, l\}$ be the largest such that $\Delta_k \cap \mathcal{C} \subset L_i$. Then

$$|\mathcal{C}' \cap \Delta_k| < |\mathcal{C} \cap \Delta_k| = |\Delta_k| - (t - 1).$$

Therefore there exist t distinct variables in $\Delta_k \setminus \mathcal{C}'$. Thus the initial term of the t-minor in L_i having these elements on the skew diagonal does not contain any variable in $\mathcal{C}'$, and hence $\mathcal{C}'$ does not satisfy (12.3.13).

Therefore $\mathcal{C}$ is a set of minimal cardinality among the sets satisfying $(*)$, and the required result follows. $\qquad\square$

12.3.14. The singular locus of $D_t(L)$. Let $X = (x_{ba})$, $1 \leq b < m$, $1 < a \leq n$ be a $m \times n$ matrix of indeterminates. Let $L \subset X$ be a one-sided ladder defined by the outside corners $\omega_i = x_{b_i a_i}$, $1 \leq i \leq h$, $1 \leq b_1 < \cdots < b_h \leq m$, $1 \leq a_1 < \cdots < a_h \leq n$. Let $V = D_t(L)$, $\mathcal{C} = \mathcal{C}_t(L)$.

For $1 \leq i \leq h$, let $V_i \subset \mathbb{A}(L)$ be the variety defined by the vanishing of the t-minors in $L(j)$, with $j \in \{1, \ldots, l\} \setminus \{i\}$, and the $(t-1)$-minors in $L(i)$.

12.3.15. Theorem. *With notation as above, we have*

$$\mathrm{Sing}\, D_t(L) = \cup_{i=1}^h V_i.$$

PROOF. For simplicity of notation, we identify the variable x_{ba} with the element (b, a).

First, we prove that $V_i \subset \mathrm{Sing}\, D_t(L)$, for all $1 \leq i \leq h$. Let $x \in V_i$ for some $1 \leq i \leq h$. Let $\mathcal{J}$ be the Jacobian matrix associated to the variety $D_t(L) \subset \mathbb{A}(L)$, evaluated at x. Then the rows of $\mathcal{J}$ are indexed by t-minors in $L(j)$, $1 \leq j \leq h$, and the columns are indexed by the elements $\alpha \in L$. The (M, α)-th entry in $\mathcal{J}$ is equal to $\pm(\det M')(x)$, where M' is the matrix obtained from M by deleting the row and the column containing α, if α appears in M, and 0 otherwise.

Fix i, $1 \leq i \leq h$. It is easily seen that $\omega_i \in \mathcal{C}_t(L)$. Now consider the one-sided ladder L' obtained from L by deleting the element ω_i, i.e., the one-sided ladder defined by the outside corners

$$\omega_1 = (b_1, a_1), \ldots, \omega_{i-1} = (b_{i-1}, a_{i-1}), \omega_{i-} = (b_i, a_i - 1),$$
$$\omega_{i+} = (b_i + 1, a_i), \omega_{i+1} = (b_{i+1}, a_{i+1}), \ldots, \omega_h = (b_h, a_h),$$

where ω_{i-} is present only if $a_i - 1 > a_{i-1}$, and ω_{i+} is present only if $b_i + 1 < b_{i+1}$.

Since $x \in V_i$, a row of $\mathcal{J}$ indexed by a t-minor involving $\omega_i = x_{b_i a_i}$ is 0; note that such a t-minor is completely contained in $L(i)$. Also, the column of $\mathcal{J}$ indexed by ω_i is 0; note again that a t-minor involving ω_i is completely contained in $L(i)$. Let $\mathcal{J}'$ be the matrix obtained from $\mathcal{J}$ by deleting the column indexed by ω_i and the rows indexed by t-minors containing ω_i. Then

$$\operatorname{rank} \mathcal{J} = \operatorname{rank} \mathcal{J}',$$

since $\mathcal{J}'$ is obtained from $\mathcal{J}$ by deleting zero rows and columns. Let $x = (x_\alpha)_{\alpha \in L}$, $x' = (x_\alpha)_{\alpha \in L'}$. Then $x' \in D_t(L')$, and $\mathcal{J}'$ is the Jacobian matrix associated to the variety $D_t(L') \subset \mathbb{A}(L')$, evaluated at x'. Thus

$$\operatorname{rank} \mathcal{J}' \leq \operatorname{codim}_{\mathbb{A}(L')} D_{\mathbf{s},\mathbf{t}}(L').$$

Now, using Proposition 12.3.12 we obtain

$$\operatorname{codim}_{\mathbb{A}(L')} D_t(L') = |\mathcal{C}_t(L')| = |\mathcal{C}_t(L) \setminus \{\omega_i\}| < |\mathcal{C}_t(L)| = \operatorname{codim}_{\mathbb{A}(L)} D_t(L).$$

Hence $\operatorname{rank} \mathcal{J}' < \operatorname{codim}_{\mathbb{A}(L)} D_t(L)$, which implies $\operatorname{rank} \mathcal{J} < \operatorname{codim}_{\mathbb{A}(L)} D_t(L)$, i.e., $x \in \operatorname{Sing} D_t(L)$.

Now we prove that $\operatorname{Sing} D_t(L) \subset \cup_{i=1}^l V_i$. Let $\mathcal{C} = \mathcal{C}_t(L)$; note that $\mathcal{C} = \cup_{i=1}^h \mathcal{C}_i$, as defined in §12.3.11.

We introduce a total order on the set of minors of L of size r, with $r \geq 1$ fixed, as follows: $[i_1, \ldots, i_r | j_1, \ldots, j_r] < [i'_1, \ldots, i'_r | j'_1, \ldots, j'_r]$ if there exists $1 \leq k \leq r$ such that

$$\text{either } i_1 = i'_1, \ldots, i_{k-1} = i'_{k-1}, i_k < i'_k,$$
$$\text{or } i_1 = i'_1, \ldots, i_r = i'_r, j_1 = j'_1, \ldots, j_{k-1} = j'_{k-1}, j_k < j'_k.$$

This is simply the lexicographic order on $\{i_1, \ldots, i_r, j_1, \ldots, j_r\}$.

Let $x \in D_t(L) \setminus \cup_{i=1}^h V_i$. For each $1 \leq i \leq h$, let M_i be the largest $(t-1)$-minor in $L(i)$ such that $(\det M_i)(x) \neq 0$. Let $\mathcal{T}_h$ be the set of elements in $L(h)$ not in the rows or the columns given by the rows and the columns of M_h. Clearly, $|\mathcal{T}_h| = |\mathcal{C}_h|$. By (decreasing) induction on i, suppose that, for some i, $1 < i \leq h$, the sets $\mathcal{T}_i, \ldots, \mathcal{T}_h$ have been constructed, such that

$(1)_i$ $\mathcal{T}_j \subset L(j)$, $i \leq j \leq h$,

$(2)_i$ the sets $\mathcal{T}_i, \ldots, \mathcal{T}_l$ are pairwise disjoint,

$(3)_i$ $|\mathcal{T}_j| = |\mathcal{C}_j|$, $i \leq j \leq h$,

$(4)_i$ $\mathcal{T}_j$ contains no elements appearing in the rows or in the columns of L given by the rows and the columns of M_j, $i \leq j \leq h$,

$(5)_i$ there exist $t - 1$ rows in $L(i)$ not containing any element from $\mathcal{T}_i \cup \cdots \cup \mathcal{T}_h$.

We define the set $\mathcal{T}_{i-1}$ as follows. Let r be the number of the rows of M_{i-1} contained in $\tilde{L}(i-1) := L(i-1) \setminus L(i)$, note that $r \leq t - 1$. Then $\mathcal{T}_{i-1}$ is obtained from $\tilde{L}(i-1)$ by deleting the r rows given by the rows of M_{i-1}, and then adding r rows from the $t - 1$ rows of $L(i)$ in $(5)_i$ which are not rows of M_{i-1}, intersected with $L(i-1)$ (this is possible, since there are $t - 1 - r$ rows of M_{i-1} in $L(i)$, and hence at least $(t-1) - (t-1-r) = r$ rows from the $t - 1$ rows of $L(i)$ in $(5)_i$ are not rows of M_{i-1}), followed by the deletion of the $t - 1$ columns in $\tilde{L}(i-1)$ obtained by intersecting the columns of M_{i-1} with $\tilde{L}(i-1)$. Again, the properties $(1)_{i-1}$-$(4)_{i-1}$ are obvious; the r rows of M_{i-1} which were deleted from $\tilde{L}(i-1)$, and the $(t-1) - r$ rows from the $t - 1$ rows in $(5)_i$ which were not used while defining $\mathcal{T}_{i-1}$, intersected with $L(i-1)$, give $t - 1$ rows of $L(i-1)$ not containing any elements in $\mathcal{T}_{i-1} \cup \mathcal{T}_i \cup \cdots \cup \mathcal{T}_h$, so that we have $(5)_{i-1}$.

Thus, using induction, we obtain the disjoint sets $\mathcal{T}_j \subset L(j)$, $1 \leq j \leq h$, such that $|\mathcal{T}_j| = |\mathcal{C}_j|$, and $\mathcal{T}_j$ contains no elements in the rows or columns of L given by the rows and columns of M_j.

For $\tau \in \mathcal{T}_i \subset \mathcal{T}$, $1 \leq i \leq h$, let M^τ be the t-minor obtained from M_i by adding the row and the column containing τ. Obviously, $M^\tau \neq M^{\tau'}$ for τ, $\tau' \in \mathcal{T}$, with $\tau \neq \tau'$.

We now take a total order on $\mathcal{T}$, namely $(b, a) > (b', a')$ if either $b > b'$, or $b = b'$ and $a > a'$.

Let us fix $\tau \in \mathcal{T}$, say $\tau \in \mathcal{T}_i$ for some i, $1 \leq i \leq l$. Then the (M^τ, τ)-th entry in $\mathcal{J}$ is equal to $\pm(\det M_i)(x)$, so it is nonzero. Let now $\sigma \in \mathcal{T}$, $\sigma < \tau$. If σ is not an entry of M^τ, then the (M^τ, σ)-th entry of $\mathcal{J}$ is equal to 0. Assume now that σ is the (r, s)-th entry of M^τ. Then the (M^τ, σ)-th entry of $\mathcal{J}$ is equal to $\pm(\det M')(x)$, where M' is the $(t-1) \times (t-1)$ matrix obtained from M^τ by deleting the r-th row and the s-th column. Let $\tau = (b, a)$, $\sigma = (b', a')$. If $b' < b$, then the indices of the first $r - 1$ rows of M' and M_i are the same, while the index of the r-th row of M' is $> b'$, which is the index of the r-th row of M_i. Thus, $M' > M_i$, and by the maximality of M_i, we obtain $(\det M')(x) = 0$. If $b' = b$, then $a' < a$. The indices of all the rows and those of the first $s - 1$ columns are the same, while the index of the s-th column in M' is $> a'$, which is the index of the s-th column of M_i. Thus $M' > M_i$, and the maximality of M_i implies that $(\det M')(x) = 0$. Thus, for $\sigma < \tau$, the (M^τ, σ)-th entry in $\mathcal{J}$ is 0.

Let $\mathcal{J}'$ be the submatrix of $\mathcal{J}$ given by the rows indexed by M^τ's and the columns indexed by τ's, with $\tau \in \mathcal{T}$. We suppose that both rows and columns of $\mathcal{J}'$ are indexed by the elements in $\mathcal{T}$, and we arrange them increasingly, with respect to the total order on $\mathcal{T}$ defined above. Then $\mathcal{J}'$ is upper triangular, and all the diagonal entries are nonzero. Thus $\det \mathcal{J}' \neq 0$, and this implies that

$$\operatorname{rank} \mathcal{J}' = |\mathcal{T}| = |\mathcal{C}| = \operatorname{codim}_{\mathbb{A}(L)} D_t(L).$$

Consequently $\operatorname{rank} \mathcal{J} = \operatorname{codim}_{\mathbb{A}(L)} D_t(L)$, i.e., $x \notin \operatorname{Sing} D_t(L)$. □

12.3.16. The irreducible components of $\operatorname{Sing} D_t(L)$ **and** $\operatorname{Sing} X(w)$.
We preserve the notation of §12.3.14.

Let us fix $j \in \{1, \ldots, l\}$, and let $Z_j = V_j \times \mathbb{A}(H \setminus L)$. We shall now define $\theta_j \in W_Q^{\min}$ such that the variety Z_j identifies with the opposite cell in the Schubert variety $X(\theta_j)$ in G/Q (recall $Q = P_{a_1} \cap \cdots \cap P_{a_h}$.)

Note that $w^{(a_j)}(a_j - t + 1) = b_j - 1$, and $b_j - 1$ is the end of a block of consecutive integers in $w^{(a_j)}$. Also, the beginning of this block is ≥ 2; if the block started with 1, we would have $a_j - t + 1 = b_j - 1$, which would then imply $2 - t = b_j - a_j > 0$, i.e., $t < 2$ which is not possible, since $t \geq 2$. Let $u_j + 1$ be the beginning of this block, where $u_j \geq 1$. The fact that $u_j \notin w^{(a_j)}$ implies that $u_j \notin w^{(a_i)}$, $i \leq j$. For $i > j$, if $b_j - 1$ is the end of a block in $w^{(a_i)}$, then the beginning of the block is $u_j + 1$, since $w^{(a_i)}(a_j - t + 1) = b_j - 1 \; (= w^{(a_j)}(a_j - t + 1))$ (cf. Remark 12.3.3) and $w^{(a_i)} \supset w^{(a_j)}$. For each i, $1 \leq i \leq h$, such that $u_j \notin w^{(a_i)}$, let v_i be the smallest entry in $w^{(a_i)}$ which is bigger than $b_j - 1$.

12.3.17. Remark. Note that for $i \leq j$, $v_i = n - t + 2$, and for $i > j$ such that $u_j \notin w^{(a_i)}$, $v_i = w^{(a_i)}(a_j - t + 2)$.

Define $\theta_j^{(a_i)}$, $1 \leq i \leq h$, as follows.

1. If $b_j - 1 \notin w^{(a_i)}$ (equivalently $i < j$), let $\theta_j^{(a_i)} = w^{(a_i)} \setminus \{v_i\} \cup \{b_j - 1\}$.
2. If $b_j - 1 \in w^{(a_i)}$ (equivalently $i \geq j$) and $u_j \notin w^{(a_i)}$, then $\theta_j^{(a_i)} = w^{(a_i)} \setminus \{v_i\} \cup \{u_j\}$.
3. If $b_j - 1 \in w^{(a_i)}$ and $u_j \in w^{(a_i)}$, then $\theta_j^{(a_i)} = w^{(a_i)}$.

Note that θ_j is well-defined as an element in $W_Q^{\min}$, and $\theta_j \leq w$.

12.3.18. Remark. Note that for $i < j$, $\theta_j^{(a_i)}$ differs from $w^{(a_i)}$ only at the $a_i - t + 1$-th place ($w^{(a_i)}(a_i - t + 1) = n - t + 2$, $\theta_j^{(a_i)}(a_i - t + 1) = b_j - 1$). Also, for $i \geq j$ such that $u_j \notin w^{(a_i)}$, $\theta_j^{(a_i)}(a_j - t + 2) = b_j - 1$, while $w^{(a_i)}(a_j - t + 2) = v_i$; further, there is a block in $\theta_j^{(a_i)}$ ending with $b_j - 1$ at the $(a_j - t + 2)$-th place. In particular, we have $\theta_j^{(a_j)}(a_j - t + 2) = b_j - 1$.

Example. Let $n = 13$, $h = 3$, $t = 3$, $(b_1, a_1) = (6, 4)$, $(b_2, a_2) = (8, 5)$, $(b_3, a_3) = (11, 9)$. Then we have seen

$$w^{(4)} = (4, 5, \ 12, 13), \ w^{(5)} = (4, 5, \ 7, \ 12, 13),$$
$$w^{(9)} = (4, 5, 6, 7, 8, 9, 10, \ 12, 13).$$

We have

$$\theta_1^{(4)} = (3, 4, 5, \ 13), \ \theta_1^{(5)} = (3, 4, 5, \ 12, 13), \ \theta_1^{(9)} = (3, 4, 5, \ 7, 8, 9, 10, \ 12, 13),$$

$$\theta_2^{(4)} = (4, 5, \ 7, \ 13), \ \theta_2^{(5)} = (4, 5, 6, 7, \ 13), \ \theta_2^{(9)} = (4, 5, 6, 7, 8, 9, 10, \ 12, 13),$$

$$\theta_3^{(4)} = (4, 5, \ 10, 13), \ \theta_3^{(5)} = (4, 5, \ 7, \ 10, 13), \ \theta_3^{(9)} = (3, 4, 5, 6, 7, 8, 9, 10, \ 13).$$

12.3.19. Theorem. *The subvariety $Z_j \subset Z$ identifies with the opposite cell in $X(\theta_j)$, i.e., $Z_j = X(\theta_j) \cap O^-$ (scheme theoretically).*

PROOF. Let $f = \det M$, M being either a t-minor contained in $L(i)$, $i \in \{1, \ldots, h\} \setminus \{j\}$, or a $(t-1)$-minor contained in $L(j)$ be a generator of $I(Z_j)$. In the former case we have $f \in I(Z)$, and Theorem 12.3.4 implies that $f \in I(X(w) \cap O^-) \subset I(X(\theta_j) \cap O^-)$. In the latter case, M is contained in H_j (of the one-sided ladder H with outside corners (a_{i+1}, a_i), $1 \leq i \leq h$). By Lemma 12.1.7, f can be written in the form $f = \sum g_\phi p_\phi|_{O^-}$, with $\phi \in W^{a_j}$ such that $\{\phi(1), \ldots, \phi(a_j)\} \cap \{a_j + 1, \ldots, n\} = \{r_1, \ldots, r_{t-1}\}$, and $g_\phi \in K[H]$; here $r_1, \ldots, r_{t-1}$ are the row indices of M. In particular we have $\phi(a_j - t + 2) = r_1$. Since M is contained in $L(j)$, we deduce that $r_1 \geq b_j$, and hence $\phi(a_j - t + 2) \geq b_j$. We have $\theta_j^{(a_j)}(a_j - t + 2) = b_j - 1$, and hence $\phi(a_j - t + 2) > \theta_j^{(a_j)}(a_j - t + 2)$. This shows that $\phi \not\leq \theta_j^{(a_j)}$, and therefore $p_\phi \in I(X(\theta_j) \cap O^-)$. Thus $f \in I(X(\theta_j) \cap O^-)$.

Let now $g = p_\tau|_{O^-}$, with $\tau \in W^{a_i}$ for some i, $1 \leq i \leq h$, such that $\tau \not\leq \theta_j^{(a_i)}$, be a generator of the ideal $I(X(\theta_j) \cap O^-)$. Since $I(X(w) \cap O^-) = I(Z)$, $I(Z) \subset I(Z_j)$, we may suppose $p_\tau \notin I(X(w) \cap O^-)$, and thus may suppose $\tau \leq w^{(a_i)}$, $\tau \not\leq \theta_j^{(a_i)}$.

Let $i \leq j$. The facts that $\tau \leq w^{(a_i)}$, $\tau \not\leq \theta_j^{(a_i)}$, and $i \leq j$ imply that $\tau(a_i - t + 2) \geq b_j$. Hence using Lemma 12.1.8, we deduce that $p_\tau|_{O^-}$ belongs to the ideal of $K[H]$ generated by $(t-1)$-minors with row indices $\geq b_j$, and column indices $\leq a_i$, note that such minors are contained in $L(j)$. Thus $p_\tau|_{O^-}$ belongs to the ideal generated by $(t-1)$-minors contained in $L(j)$, which implies that $g \in I(Z_j)$.

Let $i > j$. We have $w^{(a_i)} \neq \theta_j^{(a_i)}$, since $\tau \leq w^{(a_i)}$, $\tau \not\leq \theta_j^{(a_i)}$, and hence $u_j \notin w^{(a_i)}$ by definition of θ_j. Now $\theta_j^{(a_i)}$ consists of several blocks of consecutive integers ending with $b_m - 1$ at the $(a_m - t + 1)$-th place, for some $m \in \{1, \ldots, i\} \setminus \{j\}$, a block ending with $b_j - 1$ at the $(a_j - t + 2)$-th place (cf. Remark 12.3.18), and a last block ending with n at the a_i-th place. As above, the facts that $\tau \leq w^{(a_i)}$, $\tau \not\leq \theta_j^{(a_i)}$ imply that $\tau(a_j - t + 2) \geq b_j$.

Using Lemma 12.1.8, we deduce that $p_\tau|_{O^-}$ belongs to the ideal of $K[H]$ generated by $(t-1)$-minors with row indices $\geq b_j$, and column indices $\leq a_j$. Thus $p_\tau|_{O^-}$ belongs to the ideal generated by $(t-1)$-minors contained in $L(j)$, which implies that $g \in I(Z_j)$. $\square$

12.3.20. Theorem. *The irreducible components of* $\mathrm{Sing}\, D_{\mathbf{s},\mathbf{t}}(L)$ *are precisely the* V_j*'s,* $1 \leq j \leq h$.

PROOF. In view of Theorem 12.3.19, we obtain that V_j, $1 \leq j \leq h$, is irreducible, and the required result follows from Theorem 12.3.15. $\square$

Let $X(w^{\mathrm{max}})$, resp. $X(\theta_j^{\mathrm{max}})$, $1 \leq j \leq h$, be the pull-back in $SL(n)/B$ of $X(w)$, resp. $X(\theta_j)$, $1 \leq j \leq h$, under the canonical projection $\pi : SL(n)/B \to SL(n)/Q$. Then using Theorems 12.3.15, 12.3.4 and 12.3.19, we obtain

12.3.21. Theorem. *The irreducible components of* $\mathrm{Sing}\, X(w^{\mathrm{max}})$ *are precisely* $X(\theta_j^{\mathrm{max}})$, $1 \leq j \leq h$.

12.3.22. The validity of the conjecture 8.2.12 for $X(w^{\mathrm{max}})$. Let $G = SL(n)$. In this section we show that conjecture of [**107**] holds for $X(w^{\mathrm{max}})$.

For $\tau \in W$, let P_τ, resp. Q_τ, be the maximal element of the set of parabolic subgroups which leave $\overline{B\tau B}$ (in G) stable under multiplication on the left, resp. right.

By Theorem 8.2.6, we have

$$S_{P_\tau} = \{\alpha \in S \mid \tau^{-1}(\alpha) \in R^-\},$$
$$S_{Q_\tau} = \{\alpha \in S \mid \tau(\alpha) \in R^-\}.$$

Given parabolic subgroups P, Q, as in Definition 8.2.2, we say that $\overline{B\tau B}$ is P-Q stable if $P \subset P_\tau$ and $Q \subset Q_\tau$.

12.3.23. Lemma. *Let* $G = SL(n)$. *Let* $\tau \in \mathcal{S}_n$, *say* $\tau = (a_1, \ldots, a_n)$. *Let* $\alpha_i = \epsilon_i - \epsilon_{i+1}$. *Then*
(1) $\tau(\alpha_i) \in R^-$ *if and only if* $a_i > a_{i+1}$.
(2) $\tau^{-1}(\alpha_i) \in R^-$ *if and only if* $i+1$ *occurs before* i *in* τ.

PROOF. We have $\tau(\alpha_i) = \epsilon_{a_i} - \epsilon_{a_{i+1}}$ and $\tau^{-1}(\alpha_i) = \epsilon_j - \epsilon_k$, where $a_j = i$ and $a_k = i+1$. The results follow from this. $\square$

Let $\eta = (a_1 \ldots a_n) \in \mathcal{S}_n$. Let $\mathrm{Sing}X(\eta) \neq \emptyset$. Let (a, b, c, d) be four distinct entries in $\{1, \ldots, n\}$ such that $a < b < c < d$. A pattern in η of the form d, b, c, a (or 4231), where $d = a_i$, $b = a_j$, $c = a_k$, $a = a_m$, $i < j < k < m$, will be referred to as a *Type I bad pattern in* η. A pattern in η of the form (c, d, a, b) (or 3412), where $c = a_i$, $d = a_j$, $a = a_k$, $b = a_m$, $i < j < k < m$, will be referred to as a *Type II bad pattern in* η. Let (d, b, c, a) (resp. (c', d', a', b')) be a bad pattern of Type I (resp. Type II), where $a < b < c < d$ (resp. $a' < b' < c' < d'$). Let θ, θ' be both $\leq w$.

Further, let b, a, d, c (resp. a', c', b', d') appear in that order in θ (resp. θ'). By abuse of language, we shall refer to (b, a, d, c) (resp. (a', c', b', d')) as a bad pattern in θ (resp. θ') corresponding to the bad pattern (d, b, c, a) (resp. (c', d', a', b')) in η.

Let $\tau \in W_Q^{\min}$. We have $\pi^{-1}(X_Q(\tau)) = X_B(\tau^{\max})$, where $\tau^{\max}$, as a permutation, is given by $\tau^{(a_1)}$ arranged in descending order, followed by $\tau^{(a_2)} \setminus \tau^{(a_1)}$ arranged in descending order, etc.. We shall refer to the set $\tau^{(a_i)} \setminus \tau^{(a_{i-1})}$, $1 \le i \le h+1$, arranged in descending order, as the i-th block in $\tau^{\max}$; here, $\tau^{(a_0)} = \emptyset$, and $\tau^{(a_{h+1})}$ is the set $\{1, \ldots, n\} \setminus \tau^{(a_h)}$ arranged in descending order.

Example. Let $n = 13$, $h = 3$, $t = 3$, $(b_1, a_1) = (6, 4)$, $(b_2, a_2) = (8, 5)$, $(b_3, a_3) = (11, 9)$. Then

$$w^{\max} = (13, 12, 5, 4, \ 7, \ 10, 9, 8, 6, \ 11, 3, 2, 1),$$
$$\theta_1^{\max} = (13, 5, 4, 3, \ 12, \ 10, 9, 8, 7, \ 11, 6, 2, 1),$$
$$\theta_2^{\max} = (13, 7, 5, 4, \ 6, \ 12, 10, 9, 8, \ 11, 3, 2, 1),$$
$$\theta_3^{\max} = (13, 10, 5, 4, \ 7, \ 9, 8, 6, 3, \ 12, 11, 2, 1).$$

For the rest of this section, w and Q will be as in §12.3.

12.3.24. Remark. Set $b_{h+1} - 1 = n - t + 1$. All of the entries in the i-th block in $w^{\max}$ are $\le b_i - 1$, $2 \le i \le h+1$. In particular, for $1 \le j \le h$, b_j occurs after $b_j - 1$ in $w^{\max}$ in view of Lemma 5.2.8.

12.3.25. Lemma. *We have*

1. $Q_{w^{\max}} = Q$.
2. *Let* $I_{w^{\max}} = \{\epsilon_i - \epsilon_{i+1} \mid i = b_j - 1, 1 \le j \le h\}$. *Then* $S_{P_{w^{\max}}} = S \setminus I_{w^{\max}}$.

The assertions are clear from the description of $w^{\max}$ in view of Lemma 12.3.23 and Remark 12.3.24.

12.3.26. Lemma. *Let* $P = P_{w^{\max}}$, $Q = Q_{w^{\max}}$. *Then* $\overline{B\theta_j^{\max}B}$ *is P-Q stable.*

PROOF. The Q-stability of $\overline{B\theta_j^{\max}B}$ on the right is obvious. Regarding the P-stability of $\overline{B\theta_j^{\max}B}$ on the left, we shall now show that if $x \ne b_l$, for any l, $1 \le l \le h$, then $x - 1$ occurs after x in $\theta_j^{\max}$; note that $S_P = \{\epsilon_i - \epsilon_{i+1} \mid i \ne b_j - 1, 1 \le j \le h\}$ (cf. Lemma 12.3.25). Let u_j appear in the k-th block in $w^{\max}$, for some $k > j$. Fix $x \in \{1, \ldots, n\}$, $x \ne b_l$, for any l, $1 \le l \le h$. If $x \ne v_i$, $j \le i < k$ (v_i being as in the definition of θ_j), then clearly $x - 1$ occurs after x in $\theta_j^{\max}$ also. Otherwise, if $x = v_i$, for some i, $j \le i < k$. The fact that $v_i \ne b_l$, for any l, $1 \le l \le h$ implies that $v_i - 1 \ne b_l - 1$, for any l, $1 \le l \le h$; in particular, $v_i - 1 \ne b_j - 1$. Hence we conclude, by the definition of v_i, that $v_i - 1$ occurs in the m-th block in $w^{\max}$, for some $i < m \le k$. This implies that for $i \le l < m$, $v_l = v_i$, and if

$m < k$, then $v_m \leq v_{i-1}$. From this it follows that in $\theta_j^{\max}$, v_i occurs in the m-th block while v_{i-1} occurs in the q-th block where $q \geq m$. From this it follows that $v_i - 1$ appears after v_i in $\theta_j^{\max}$ also.   □

12.3.27. Lemma. *Any bad pattern in $w^{\max}$ is of Type I.*

PROOF. Let $w^{\max} = (a_1 \ldots a_n)$. Assume that (c, d, a, b) is a bad pattern of Type II in $w^{\max}$, where $a < b < c < d$. Clearly, c and d, resp. a and b, cannot both appear in the same block , in view of the description of $w^{\max}$. Let then c, d, a, b appear in the r-th, i-th, j-th, k-th blocks, respectively, where $r < i \leq j < k$. This implies that $a < b < c < d \leq b_i - 1$ (cf. Remark 12.3.24). But now, a and b are both $< b_i - 1$, and they both appear after $b_i - 1$; further, a appears before b in $w^{\max}$, which is not possible by the construction of $w^{\max}$, note that $a < b$. The required result follows from this.   □

12.3.28. Remark. Of course, there are several bad patterns in $w^{\max}$ of Type I. For example, fix some j, $1 \leq j \leq h$. Observe that b_j appears after $b_j - 1$ (cf. Remark 12.3.24), and u_j appears after b_j in $w^{\max}$ (notation being as in §12.3.16). Take d, to be any entry in $\{n - t + 2, \ldots, n\}$, $b = b_j - 1$, $c = b_j$, $a = u_j$. Then d, b, c, a occur in the 1-st, j-th, k-th, m-th blocks respectively, where $m \geq k > j$. This provides an example of a Type I bad pattern in $w^{\max}$.

12.3.29. Lemma. *Let (d, b, c, a) be a Type I bad pattern in $w^{\max}$, where $a < b < c < d$. Assume that b belongs to the i-th block, for some i, note that $i \leq h$, since $b < c$. Then*

1. $c < n - t + 2$
2. $b \leq b_i - 1$
3. $d \geq n - t + 2$.

PROOF. Let (d, b, c, a) occur in the r-th, i-th, j-th, k-th blocks respectively in $w^{\max}$, where $r \leq i < j \leq k$. The hypothesis that $b < c$ implies that $j > 1$. Hence we obtain $c \leq b_j - 1$ (cf. Remark 12.3.24), and (1) follows. Now, if $i \geq 2$, then the assertion (2) follows from Remark 12.3.24. If $i = 1$, then the assertion (2) follows from the fact that $b < c < n - t + 2$.

Next, we claim $d > b_i - 1$. Assume that $d \leq b_i - 1$. Then the assumption implies $c < b_i - 1$ (since $c < d$). Now both c and b are $< b_i - 1$, and b belongs to the i-th block in $w^{\max}$. This implies that c should occur before b, which is not possible. Hence our assumption is wrong, and the claim follows.

Note that the claim and Remark 12.3.24 imply that $d \geq n - t + 2$, and d appears in the first block.   □

12.3.30. Lemma. *Fix j, $1 \leq j \leq h$. Then $\theta_j^{\max}$ is the unique maximal element of the set $\{\tau \in W \mid \tau \leq w^{\max}, \tau^{(a_j)}(a_j - t + 2) \leq b_j - 1\}$.*

The proof is clear from the definition of $\theta_j^{\max}$.

12.3.31. Proposition. *The maximal elements in $F_{w^{\max}}$ are precisely $\theta_j^{\max}$,* $1 \leq i \leq h$; *here $F_{w^{\max}}$ is as in §8.2.12.*

PROOF. We first observe that $\theta_j^{\max} \in F_{w^{\max}}$; since corresponding to the bad pattern $d = n - t + 2$, $b = b_j - 1$, $c = b_j$, $a = u_j$ (cf. Remark 12.3.28), we have the bad pattern (b, a, d, c); note that b, a, d, c occur in that order in $\theta_j^{\max}$. Let us denote $\theta_j^{\max}$ by τ'. Let w', resp. τ, be the element of S_n obtained from τ', resp. w, by replacing b, a, d, c, resp. d, b, c, a, by d, b, c, a, resp. b, a, d, c. Then clearly $\tau \leq \tau'$, and $w' \leq w$. Further, $\overline{B\theta_j^{\max}B}$ is P-Q stable (cf. Lemma 12.3.26). Thus $\theta_j^{\max} \in F_{w^{\max}}$.

Let now $\tau' \in F_{w^{\max}}$. In particular, we have $\tau' \in W_Q^{\max}$.

We have a bad occurrence in τ' which has to be of the form (b, a, d, c), $a < b < c < d$, corresponding to the occurrence (d, b, c, a) in $w^{\max}$ (cf. Lemma 12.3.27). Let b, a, d, c occur in the p-th, q-th, r-th, s-th blocks respectively in τ', where $p \leq q < r \leq s$ (note that $\tau' \in W_Q^{\max}$).

We have

$$w'^{(a_q)}(a_q - t + 1) \leq w^{(a_q)}(a_q - t + 1) = b_q - 1$$

(here w' is as in §8.2.12). Further, $\tau'^{(a_q)}$ is obtained from $w'^{(a_q)}$ by replacing d by a, where $a(< b) < n - t + 2 \leq d$ (cf. Lemma 12.3.29). Hence we obtain $a \leq b_q - 1$ (since $\tau'^{(a_q)} \leq w'^{(a_q)}$), and

$$\tau'^{(a_q)}(a_q - t + 2) \leq w'^{(a_q)}(a_q - t + 1) \leq b_q - 1.$$

This implies $\tau' \leq \theta_q^{\max}$ (cf. Lemma 12.3.30). $\qquad\square$

12.3.32. Theorem. *The conjecture 8.2.12 holds for $X(w^{\max})$.*

PROOF. In view of Theorem 12.3.21, $X(\theta_j^{\max})$, $1 \leq j \leq h$ are precisely the irreducible components of $X(w^{\max})$. On the other hand, we have (cf. Proposition 12.3.31) that the maximal elements in $F_{w^{\max}}$ are precisely $\theta_j^{\max}$, $1 \leq j \leq h$. Hence the irreducible components of $\operatorname{Sing} X(w^{\max})$ are precisely $\{X(\theta) \mid \theta \text{ a maximal element of } F_{w^{\max}}\}$. Thus the conjecture holds for $X(w^{\max})$. $\qquad\square$

12.3.33. More general ladder determinantal varieties. Let $X = (x_{ba})$, $1 \leq b \leq m$, $1 \leq a \leq n$ be an $m \times n$ matrix of indeterminates.

Given $1 \leq b_1 < \cdots < b_h < m$, $1 < a_1 < \cdots < a_h \leq n$, we consider the subset of X defined by

$$L = \{x_{ba} \mid \text{ there exists } 1 \leq i \leq h \text{ such that } b_i \leq b \leq m, 1 \leq a \leq a_i\}.$$

Again, we call L an *one-sided ladder* in X, defined by the *outside corners* $\omega_i = x_{b_i a_i}$, $1 \leq i \leq h$. For simplicity of notation, we identify the variable x_{ba} with just (b, a).

Let $\mathbf{s} = (s_1, s_2 \ldots, s_l) \in \mathbb{Z}_+^l$, $\mathbf{t} = (t_1, t_2 \ldots, t_l) \in \mathbb{Z}_+^l$ such that

$$b_1 = s_1 < s_2 < \cdots < s_l \leq m,$$

(L1) $t_1 \geq t_2 \geq \cdots \geq t_l$, $1 \leq t_i \leq \min\{m - s_i + 1, a_{i*}\}$ for $1 \leq i \leq l$, and

$$s_i - s_{i-1} > t_{i-1} - t_i \text{ for } 1 < i \leq l,$$

where i^* is the largest integer such that $b_{i*} \leq s_i$ For $1 \leq i \leq l$, let

$$L_i = \{x_{ba} \in L \mid s_i \leq b \leq m\}.$$

12.3.34. Definition. Generalized Ladder determinantal varieties, $D_{\mathbf{s},\mathbf{t}}$. Let $K[L]$ denote the polynomial ring $K[x_{ba} \mid x_{ba} \in L]$, and let $\mathbb{A}(L) = \mathbb{A}^{|L|}$ be the associated affine space. Let $I_{\mathbf{s},\mathbf{t}}(L)$ be the ideal in $K[L]$ generated by all the t_i-minors contained in L_i, $1 \leq i \leq l$, and $D_{\mathbf{s},\mathbf{t}}(L) \subset \mathbb{A}(L)$ the variety defined by the ideal $I_{\mathbf{s},\mathbf{t}}(L)$. We call $D_{\mathbf{s},\mathbf{t}}(L)$ a *ladder determinantal variety* (associated to a one-sided ladder).

Let $\Omega = \{\omega_1, \ldots, \omega_h\}$. For each $1 < j \leq l$, let

$$\Omega_j = \{\omega_i \mid 1 \leq i \leq h \text{ such that } s_{j-1} < b_i < s_j \text{ and } s_j - b_i \leq t_{j-1} - t_j\}.$$

Let

$$\Omega' = (\Omega \setminus \bigcup_{j=2}^{l} \Omega_j) \bigcup_{\Omega_j \neq \emptyset} \{(s_j, a_{j*})\}.$$

Let L' be the one-sided ladder in X defined by the set of outside corners Ω'. Then it is easily seen that $D_{\mathbf{s},\mathbf{t}}(L) \cong D_{\mathbf{s},\mathbf{t}}(L') \times \mathbb{A}^d$, where $d = |L| - |L'|$.

Let $\omega_k' = (b_k', a_k') \in \Omega'$, for some k, $1 \leq k \leq h'$, where $h' = |\Omega'|$. If $b_k' \notin \{s_1, \ldots, s_l\}$, then $b_k' = b_i$ for some i, $1 \leq i \leq h$, and we define $s_{j-} = b_i$, $t_{j-} = t_{j-1}$, $s_{j+} = s_j$, $t_{j+} = t_j$, where j is the unique integer such that $s_j < b_i < s_{j+1}$. Let $\mathbf{s}'$, resp. $\mathbf{t}'$, be the sequence obtained from $\mathbf{s}$, resp. $\mathbf{t}$, by replacing s_j, resp. t_j with s_{j-} and s_{j+}, resp. t_{j-} and t_{j+}, for all k such that $b_k' \notin \{s_1, \ldots, s_l\}$, j being the unique integer such that $s_{j-1} < b_i < s_j$, and i being given by $b_k' = b_i$. Let $l' = |\mathbf{s}'|$. Then $\mathbf{s}'$ and $\mathbf{t}'$ satisfy (L1), and in addition we have $\{b_1', \ldots, b_{h'}'\} \subset \{s_1', \ldots, s_{l'}'\}$. It is easily seen that $D_{\mathbf{s},\mathbf{t}}(L') = D_{\mathbf{s}',\mathbf{t}'}(L')$, and hence

$$D_{\mathbf{s},\mathbf{t}}(L) \cong D_{\mathbf{s}',\mathbf{t}'}(L') \times \mathbb{A}^d.$$

Therefore it is enough to study $D_{\mathbf{s},\mathbf{t}}(L)$ with $\mathbf{s}, \mathbf{t} \in \mathbb{Z}_+^l$ such that

(L2) $\{s_1, \ldots, s_l\} \supset \{b_1, \ldots, b_h\}.$

Without loss of generality, we can also assume that

(L3) $t_l \geq 2$, and $t_{i-1} > t_i$ if $s_i \notin \{b_1, \ldots, b_h\}, 1 < i \leq l.$

For $1 \leq i \leq l$, let

$$L(i) = \{x_{ba} \mid s_i \leq b \leq m, 1 \leq a \leq a_{i*}\}.$$

Note that the ideal $I_{\mathbf{s},\mathbf{t}}(L)$ is generated by the t_i-minors of X contained in $L(i)$, $1 \leq i \leq l$.

The ladder determinantal varieties (associated to one-sided ladders) get related to Schubert varieties (cf. [56]). We describe below the main results of [56].

12.3.35. Definition. The variety Z. Let $G = SL(n)$, $Q = P_{a_1} \cap \cdots \cap P_{a_h}$. Let O^- be the opposite big cell in G/Q. Let H be the one-sided ladder defined by the outside corners $(a_i + 1, a_i)$, $1 \leq i \leq h$. Let $\mathbf{s}, \mathbf{t} \in \mathbb{Z}_+^l$ satisfy (L1), (L2), (L3) above. For each $1 \leq i \leq l$, let

$$(12.3.36) \qquad L(i) = \{x_{ba} \mid s_j \leq b \leq n, 1 \leq a \leq a_{i*}\}.$$

Let Z be the variety in $\mathbb{A}(H) \cong O^-$ defined by the vanishing of the t_i-minors in $L(i)$, $1 \leq i \leq l$. Note that $Z \cong D_{\mathbf{s},\mathbf{t}}(L) \times \mathbb{A}(H \setminus L) \cong D_{\mathbf{s},\mathbf{t}}(L) \times \mathbb{A}^r$, where $r = \dim SL(n)/Q - |L|$.

We shall now define an element $w \in W_Q^{\min}$, such that the variety Z identifies with the opposite cell in the Schubert variety $X(w)$ in G/Q. We define $w \in W_Q^{\min}$ by specifying $w^{(a_i)} \in W^{a_i}$ $1 \leq i \leq h$, where $\pi_i(X(w)) = X(w^{(a_i)})$ under the projection $\pi_i : G/Q \to G/P_{a_i}$.

Define $w^{(a_i)}$, $1 \leq i \leq h$, inductively, as the (unique) maximal element in W^{a_i} such that

(1) $w^{(a_i)}(a_i - t_j + 1) = s_j - 1$ for all $j \in \{1, \ldots, l\}$ such that $s_j \geq b_i$, and $t_j \neq t_{j-1}$ if $j > 1$.

(2) if $i > 1$, then $w^{(a_{i-1})} \subset w^{(a_i)}$.

Note that $w^{(a_i)}$, $1 \leq i \leq h$, is well-defined in W^{a_i}, and w is well-defined as an element in $W_Q^{\min}$.

12.3.37. Theorem. [56] *The variety Z $(= D_{\mathbf{s},\mathbf{t}}(L) \times \mathbb{A}^r)$ identifies with the opposite cell in $X(w)$, i.e., $Z = X(w) \cap O^-$ (scheme theoretically).*

As a consequence of the above theorem, we obtain (cf. [56]):

12.3.38. Theorem. *The variety $D_{\mathbf{s},\mathbf{t}}(L)$ is irreducible, normal, Cohen–Macaulay, and has rational singularities.*

12.3.39. Definition. The varieties V_i, $1 \leq i \leq l$. Let V_i, $1 \leq i \leq l$ be the subvariety of $D_{\mathbf{s},\mathbf{t}}(L)$ defined by the vanishing of all $(t_i - 1)$-minors in $L(i)$, where $L(i)$ is as in (12.3.36).

In [56] the singular locus of $D_{\mathbf{s},\mathbf{t}}(L)$ has also been determined, as described below.

12.3.40. Theorem. [56] $\operatorname{Sing} D_{\mathbf{s},\mathbf{t}}(L) = \cup_{i=1}^l V_i$.

12.3.41. Definition. The varieties $X(\theta_j)$, $1 \leq j \leq l$. Let us fix $j \in \{1, \ldots, l\}$, and let $Z_j = V_j \times \mathbb{A}(H \setminus L)$. We shall now define $\theta_j \in W_Q^{\min}$

such that the variety Z_j identifies with the opposite cell in the Schubert variety $X(\theta_j)$ in G/Q.

Note that $w^{(a_r)}(a_r - t_j + 1) = s_j - 1$, and $s_j - 1$ is the end of a block of consecutive integers in $w^{(a_r)}$, where $r = j^*$ is the largest integer such that $b_r \leq s_j$. Also, the beginning of this block is ≥ 2; if the block started with 1, we would have $a_r - t_j + 1 = s_j - 1 \geq b_r - 1 \geq a_r$, which is not possible, since $t_j \geq 2$. Let $u_j + 1$ be the beginning of this block, where $u_j \geq 1$. It is easily seen that if $s_j - 1$ is the end of a block in $w^{(a_i)}$, $1 \leq i \leq h$, then the beginning of the block is $u_j + 1$. For each i, $1 \leq i \leq h$, such that $u_j \notin w^{(a_i)}$, let v_i be the smallest entry in $w^{(a_i)}$ which is bigger than $s_j - 1$. Note that $v_i = w^{(a_i)}(a_k - t_j + 2)$, where $k \in \{1, \ldots, i\}$ is the largest such that $b_k \leq s_j$.

Define $\theta_j^{(a_i)}$, $1 \leq i \leq h$, as follows:

- If $s_j - 1 \notin w^{(a_i)}$ (which is equivalent to $j > 1$, $t_{j-1} = t_j$ and $i < r$), let $\theta_j^{(a_i)} = w^{(a_i)} \setminus \{v_i\} \cup \{s_j - 1\}$.
- If $s_j - 1 \in w^{(a_i)}$ and $u_j \notin w^{(a_i)}$, then $\theta_j^{(a_i)} = w^{(a_i)} \setminus \{v_i\} \cup \{u_j\}$.
- If $s_j - 1$ and $u_j \in w^{(a_i)}$, then $\theta_j^{(a_i)} = w^{(a_i)}$ (note that in this case $i > r$).

Note that θ_j is well-defined as an element in $W_Q^{\min}$, and $\theta_j \leq w$.

12.3.42. Remark. An equivalent description of θ_j is the following. Let $t_{i_k} < t_j \leq t_{i_{k-1}}$.

1. If $j \notin \{i_1, \ldots, i_m\}$ (i.e., $j > 1$ and $t_{j-1} = t_j$), then
   (a) For $i < r$, $\theta_j^{(a_i)} = w_j^{(a_i)} \setminus \{e_{i_k}\} \cup \{s_j - 1\}$.
   (b) For $i = r$, $\theta_j^{(a_r)} = w_j^{(a_r)} \setminus \{e_{i_k}\} \cup \{u_j\}$, where u_j is the largest entry in $\{1, \ldots, s_j - 1\} \setminus w^{(a_r)}$.
   (c) For $i > r$ and $u_j \in w^{(a_i)}$, $\theta_j^{(a_i)} = w_j^{(a_i)}$.
   (d) For $i > r$ and $u_j \notin w^{(a_i)}$, $\theta_j^{(a_i)} = w_j^{(a_i)} \setminus \{v_i\} \cup \{u_j\}$, where v_i is the smallest entry in $w^{(a_i)} \setminus \theta_j^{(a_{i-1})}$.
2. If $j \in \{i_1, \ldots, i_m\}$, (i.e., $t_{j-1} > t_j$ if $j > 1$), then
   (a) For $i \leq r$, $\theta_j^{(a_i)} = w_j^{(a_i)} \setminus \{e_{i_k}\} \cup \{u_j\}$, where u_j is the largest entry in $\{1, \ldots, s_j - 1\} \setminus w^{(a_r)}$.
   (b) For $i > r$ and $u_j \in w^{(a_i)}$, $\theta_j^{(a_i)} = w_j^{(a_i)}$.
   (c) For $i > r$ and $u_j \notin w^{(a_i)}$, $\theta_j^{(a_i)} = w_j^{(a_i)} \setminus \{v_i\} \cup \{u_j\}$, where v_i is the smallest entry in $w^{(a_i)} \setminus \theta_j^{(a_{i-1})}$.

12.3.43. Theorem. [56] *The subvariety $Z_j \subset Z$ identifies with the opposite cell in $X(\theta_j)$, i.e., $Z_j = X(\theta_j) \cap O^-$ (scheme theoretically).*

As a consequence of the above theorem, we obtain (cf. [56]):

12.3.44. Theorem. *The irreducible components of* $\mathrm{Sing}\,D_{\mathbf{s},\mathbf{t}}(L)$ *are precisely the* V_j*'s,* $1 \leq j \leq l$.

Let $X(w^{\max})$, resp. $X(\theta_j^{\max})$, $1 \leq j \leq l$ be the pull-back in $SL(n)/B$ of $X(w)$, resp. $X(\theta_j)$, $1 \leq j \leq l$ under the canonical projection $\pi :$ $SL(n)/B \to SL(n)/Q$ (here B is a Borel subgroup of $SL(n)$ such that $B \subset Q$). Then using Theorems 12.3.37, 12.3.43 and 12.3.44 above, we obtain (cf. [**56**]):

12.3.45. Theorem. *The irreducible components of* $\mathrm{Sing}\,X(w^{\max})$ *are precisely* $X(\theta_j^{\max})$, $1 \leq j \leq l$.

In [**56**], it is also shown that the conjecture of [**107**] on the irreducible components of $\mathrm{Sing}\,X(\theta)$, $\theta \in W$ holds for $X(w^{\max})$.

12.3.46. Remark. A similar identification as in Theorem 12.3.37 for the case $t_1 = \cdots = t_l$ has also been obtained by Mulay (cf. [**129**]).

12.3.47. Remark. The w's as above arising from ladder determinantal varieties are covexillary in the sense of Lascoux and Schützenberger (cf. [**119**]), namely, the associated permutations avoid the pattern $c, d.a, b$ where $a < b < c < d$.

12.3.48. Remark. In [**56**], the theory of Schubert varieties and the theory of ladder determinantal varieties complement each other. To be more precise, geometric properties such as normality, Cohen–Macaulayness, etc., for ladder determinantal varieties are concluded by relating these varieties to Schubert varieties. The components of singular loci of Schubert varieties are determined by first determining them for ladder determinantal varieties, and then using the above-mentioned relationship between ladder determinantal varieties and Schubert varieties.

12.4. Quiver varieties

Fix an $(h+1)$-tuple of nonnegative integers $\mathbf{n} = (n_1, \ldots, n_{h+1})$ and a list of vector spaces $V_1, \ldots, V_{h+1}$ over an arbitrary field K with respective dimensions $n_1, \ldots, n_{h+1}$. Let Z be the affine space of all h-tuples of linear maps $(f_1, \ldots, f_h)$:

$$V_1 \overset{f_1}{\to} V_2 \overset{f_2}{\to} \cdots \overset{f_{h-1}}{\to} V_h \overset{f_h}{\to} V_{h+1} \ .$$

If we endow each V_i with a basis, we get $V_i \cong K^{n_i}$ and

$$Z \cong M(n_2 \times n_1) \times \cdots \times M(n_{h+1} \times n_h),$$

where $M(l \times m)$ denotes the affine space of matrices over K with l rows and m columns. The group

$$G_{\mathbf{n}} = GL(n_1) \times \cdots \times GL(n_{h+1})$$

acts on $(f_1, f_2, \ldots, f_h) \in Z$ by

$$(g_1, g_2, \ldots, g_{h+1}) \cdot (f_1, f_2, \ldots, f_h) = (g_2 f_1 g_1^{-1}, g_3 f_2 g_2^{-1}, \ldots, g_{h+1} f_h g_h^{-1}).$$

Now, let $\mathbf{r} = (r_{ij})_{1 \le i \le j \le h+1}$ be an array of nonnegative integers with $r_{ii} = n_i$, and define $r_{ij} = 0$ for any indices other than $1 \le i \le j \le h + 1$. Define the sets

$$Z^\circ(\mathbf{r}) = \{(f_1, \ldots, f_h) \in Z \mid \forall i < j, \ \operatorname{rank} (f_{j-1} \cdots f_i : V_i \to V_j) = r_{ij}\},$$

$$Z(\mathbf{r}) = \overline{Z^\circ(\mathbf{r})}.$$

(These sets might be empty for a bad choice of $\mathbf{r}$.)

12.4.1. Proposition. [2] *The $G_{\mathbf{n}}$-orbits of Z are exactly the sets $Z^\circ(\mathbf{r})$ for $\mathbf{r} = (r_{ij})$ with*

$$r_{ij} - r_{i,j+1} - r_{i-1,j} + r_{i-1,j+1} \ge 0, \quad \forall \ 1 \le i < j \le h + 1.$$

Furthermore, $Z(\mathbf{r}) = \{(f_1, \ldots, f_h) \in Z \mid \forall i < j, \ \operatorname{rank} (f_{j-1} \cdots f_i : V_i \to V_j) \le r_{ij}\}$.

12.4.2. Quiver varieties and Schubert varieties. Given $\mathbf{n} = (n_1, \ldots, n_{h+1})$ and the corresponding list of vector spaces $V_1, \ldots, V_{h+1}$, for $1 \le i \le h + 1$, let

$$a_i = n_1 + n_2 + \cdots + n_i, \qquad a_0 = 0, \qquad \text{and} \qquad n = n_1 + \cdots + n_{h+1}.$$

For positive integers $i \le j$, we shall frequently use the notation

$$[i, j] = \{i, i+1, \ldots, j\}, \ [i] = [1, i], \ [0] = \{\}, \text{ the empty set.}$$

Let $K^n \cong V_1 \oplus \cdots \oplus V_{h+1}$ have basis $e_1, \ldots, e_n$ compatible with the V_i. Consider its general linear group $GL(n)$, the subgroup B of upper-triangular matrices, and the parabolic subgroup Q given by

$$Q = \{(a_{ij}) \in GL(n) \mid a_{ij} = 0 \text{ whenever } j \le a_k < i \text{ for some } k\}.$$

Consider the Schubert varieties in G/Q. Given $\mathbf{r} = (r_{ij})_{1 \le i \le j \le h+1}$ define the a_i-tuple $\tau_{\mathbf{r}}^{(a_i)}$ as

$$\tau_{\mathbf{r}}^{(a_i)} = \{ \underbrace{1 \ldots a_{i-1}}_{a_{i-1}} \underbrace{\ldots \ldots a_i}_{r_{ii} - r_{i,i+1}} \underbrace{\ldots \ldots a_{i+1}}_{r_{i,i+1} - r_{i,i+2}} \underbrace{\ldots \ldots a_{i+2}}_{r_{i,i+2} - r_{i,i+3}} \cdots \underbrace{\ldots \ldots n}_{r_{i,h+1}} \},$$

where we use the visual notation

$$\underbrace{\cdots \cdots a}_{b} = [a - b + 1, a].$$

Recall that $a_j = a_{j-1} + n_j$ and $0 \le r_{ij} - r_{i,j+1} \le n_j$, so that each $\tau_{\mathbf{r}}^{(a_i)}$ is an increasing list of integers. Also $r_{ij} - r_{i,j+1} \le r_{i+1,j} - r_{i+1,j+1}$, so that

$\tau_{\mathbf{r}}^{(a_i)} \subset \tau_{\mathbf{r}}^{(a_{i+1})}$. Thus $\tau_{\mathbf{r}}$ belong to W^Q, the set of minimal representatives in W of W/W_Q.

We define $X_Q(\mathbf{r}) = X_Q(\tau_{\mathbf{r}})$. Let $\mathcal{O}^-$ denote the opposite big cell in G/Q. Let

(12.4.3) $$Y_Q(\mathbf{r}) := X_Q(\mathbf{r}) \cap \mathcal{O}^-,$$

the opposite cell in $X_Q(\mathbf{r})$. Let $z = (A_1, A_2, \dots, A_h) \in Z$. Define $f : Z \to \mathcal{O}^-$ by

$$f(z) = \begin{pmatrix} I_1 & 0 & 0 & 0 & \cdots \\ A_1 & I_2 & 0 & 0 & \cdots \\ A_2 A_1 & A_2 & I_3 & 0 & \cdots \\ A_3 A_2 A_1 & A_3 A_2 & A_3 & I_4 & \cdots \\ \vdots & \vdots & \vdots & \vdots & \vdots \end{pmatrix} \pmod{Q} .$$

Denote a generic element of $Z = M(n_2 \times n_1) \times \cdots \times M(n_{h+1} \times n_h)$ by $(A_1, \dots, A_h)$, so that the coordinate ring of Z is the polynomial ring in the entries of all the matrices A_i.

Let $\mathcal{J}(\mathbf{r}) \subset K[Z]$ be the ideal generated by the determinantal conditions implied by the definition of $Z(\mathbf{r})$:

$$\mathcal{J}(\mathbf{r}) = \left\langle \det(A_{j-1} A_{j-2} \cdots A_i)_{\lambda \times \mu} \; \middle| \; \begin{matrix} j > i, \ \lambda \subset [n_j], \ \mu \subset [n_i] \\ \#\lambda = \#\mu = r_{ij} + 1 \end{matrix} \right\rangle .$$

12.4.4. Theorem. [103]

1. $\mathcal{J}(\mathbf{r})$ is a prime ideal and is the vanishing ideal of $Z(\mathbf{r}) \subset Z$.
2. The restriction $f \mid_{Z(\mathbf{r})}$ defines an isomorphism of $Z(\mathbf{r})$ onto $Y_Q(\mathbf{r})$.

12.5. Variety of complexes

In this section we consider a special case of the quiver varieties called the variety of complexes. Let V be a variety of complexes, namely, the subvariety of Z consisting of $\{(A_1, A_2, \dots, A_h) \mid A_i A_{i-1} = 0, \ 2 \le i \le h\}$. We have from the previous section

$$V = \bigcup_{\mathbf{r}} Z(\mathbf{r}),$$

where the union is taken over all $\mathbf{r} = (r_{ij})$ such that $r_{ij} = 0$ for all $j \ne i, i+1$. Let us denote $k_i = r_{i,i+1}$, $\mathbf{k} = (k_1, k_2, \dots, k_h)$, and $V(\mathbf{k}) = Z(\mathbf{r})$. Let $\mathbf{n} = (n_1, n_2, \dots, n_{h+1})$, and

$$A_{\mathbf{n}} = \{ \mathbf{k} = (k_1, k_2, \dots, k_h) \mid k_i \le \min\{n_i, n_{i+1}\}, \quad 1 \le i \le h$$
$$k_{i-1} + k_i \le n_i, \ 2 \le i \le h \}.$$

Then

$$V = \bigcup_{\mathbf{k} \in A_{\mathbf{n}}} V(\mathbf{k}).$$

Let us denote $\tau_{\mathbf{k}} = \tau_{\mathbf{r}}$ (cf. §12.4.2).

12.5.1. A partial order on $\{(k_1, k_2, \ldots, k_h)\}$. The partial order on the set $\{Z(\mathbf{k}),\ \mathbf{k} \in A_{\mathbf{n}}\}$ of $G_{\mathbf{n}}$-orbit closures in V given by inclusion induces a partial order $\geq$ on $A_{\mathbf{n}}$, namely, for $\mathbf{k} = (k_1, k_2, \ldots, k_h)$, $\mathbf{k}' = (k_1', k_2', \ldots, k_h')$ in $A_{\mathbf{n}}$,

$$\mathbf{k} \geq \mathbf{k}' \iff k_t \geq k_t',\ 1 \leq t \leq h.$$

12.5.2. Theorem. [103, 133]
  1. $V(\mathbf{k}) \cong Y_Q(\tau_{\mathbf{k}})$ *as in* (12.4.3).
  2. *The irreducible components of V are given by*
     $\{V(\mathbf{k}) \mid \mathbf{k}$ *is maximal for the above partial order*$\}$.

12.5.3. Proposition. *(cf.* [103, 133]*) Let $\mathbf{k} \in A_{\mathbf{n}}$. Then* $\dim V(\mathbf{k}) = \sum_{1 \leq i \leq h+1} (n_i - k_i)(k_{i-1} + k_i)$, *where $k_0 = k_{h+1} = 0$.*

12.5.4. Singular locus of $V(k_1, k_2)$. In this section, we take $h = 2$, and (k_1, k_2) is a maximal element in $A_{\mathbf{n}}$, so that $V(k_1, k_2)$ is an irreducible component of V.

12.5.5. Theorem. [101] $V(k_1, k_2)$ *being an irreducible component of V, we have* $\mathrm{Sing} V(k_1, k_2) = V(k_1 - 1, k_2 - 1)$.

12.5.6. Validity of the conjecture of 8.2.12 for Schubert varieties associated to varieties of complexes. We preserve the notation of §12.5. Let $V(k_1, k_2)$ be an irreducible component of V, the variety of complexes, namely, the subvariety of $Z(= M_{n_2 \times n_1} \times M_{n_3 \times n_2})$ consisting of $\{(A_1, A_2) \mid A_2 A_1 = 0\}$. Note that (k_1, k_2) is a maximal element of $A_{\mathbf{n}}$ for the partial order on $A_{\mathbf{n}}$ defined in §12.5.

Let $V(k_1, k_2)$ be an irreducible component of V. Let $n = a_3 (= n_1 + n_2 + n_3)$, $G = SL(n)$, and $Q = P_{a_1} \cap P_{a_2}$. Let $w, \theta \in W^Q$ be defined by

$$w^{(a_1)} = (\underbrace{\ldots a_1}_{a_1 - k_1}\ \underbrace{\ldots a_2}_{k_1}),\ w^{(a_2)} = (\underbrace{1 \ldots a_1}_{a_1}\ \underbrace{\ldots a_2}_{n_2 - k_2}\ \underbrace{\ldots a_3}_{k_2}),$$

$$\theta^{(a_1)} = (\underbrace{\ldots a_1}_{a_1 - k_1 + 1}\ \underbrace{\ldots a_2}_{k_1 - 1}),\ \theta^{(a_2)} = (\underbrace{1 \ldots a_1}_{a_1}\ \underbrace{\ldots a_2}_{n_2 - k_2 + 1}\ \underbrace{\ldots a_3}_{k_2 - 1}).$$

Let w^{max}, respectively θ^{max}, denote the element in W^Q representing the coset wW_Q, respectively θW_Q. Note that $X(w^{\mathrm{max}})$, respectively $X(\theta^{\mathrm{max}})$ is the inverse image of $X_Q(w)$, respectively $X_Q(\theta)$ under the canonical projection $G \to G/Q$.

12.5.7. Theorem. $\mathrm{Sing} X(w^{\mathrm{max}}) = X(\theta^{\mathrm{max}})$

PROOF. By Theorems 12.5.2, 12.5.5, we have that $V(k_1, k_2) \cong Y_Q(w)$, $\operatorname{Sing}V(k_1, k_2) \cong Y_Q(\theta)$, and the result follows from this. $\square$

12.5.8. Proposition. [101] *If $X(w^{\max})$ is not smooth, all bad pattern in $w^{\max}$ has to be of Type II, i.e., a 3412 pattern.*

12.5.9. Corollary. [101] *The conjecture §8.2.12 holds for $X(w^{\max})$.*

12.5.10. Remark. Theorem 12.5.7 has been generalized to the case of an arbitrary $\mathbf{n}$ in [55] as described below.

Let $\mathbf{n} = (n_1, \ldots, n_{h+1})$. Fix $(k_1, k_2, \ldots, k_h)$ so that $k_i \leq \min \{n_i, n_{i+1}\}$, $1 \leq i \leq h$, $k_{i-1} + k_i \leq n_i$, $2 \leq i \leq h$, where $k_0 = k_{h+1} = 0$. Let $V_{\mathbf{k}} = V(k_1, k_2, \ldots, k_h) := \{(f_1, \ldots, f_h) \mid \operatorname{rank}(f_i) \leq k_i\}$, $f_i \circ \phi_{i-1} = 0$, $2 \leq i \leq h\}$, $f_1, \ldots, f_h$ being as in §12.4. For $1 \leq i \leq h$, let

$$V_i = V(k_1, \ldots, k_{i-1}, k_i - 1, k_{i+1}, \ldots, k_h),$$

and for $1 \leq j \leq h - 1$, let

$$V_{j,j+1} = V(k_1, \ldots, k_{j-1}, k_j - 1, k_{j+1} - 1, k_{j+2}, \ldots, k_h).$$

12.5.11. Theorem. 1. *The irreducible components of $\operatorname{Sing}V_{\mathbf{k}}$ are V_i, $i \in \Omega$, and $V_{j,j+1}$, $j \notin \Omega$, where Ω is the set of all $1 \leq i \leq h$ such that $k_{i-1} + k_i < n_i$ and $k_i + k_{i+1} < n_{i+1}$.*

2. *Let $V_{\mathbf{k}}$ be an irreducible component of the variety of complexes V (cf. §12.5). Then the irreducible components of $\operatorname{Sing}V_{\mathbf{k}}$ are $V_{1,2,}, \ldots, V_{h-1,h+1}$.*

12.5.12. Remark. The w's as above arising from the variety of complexes are non-covexillary, i.e., the associated permutations involve the patterns c, d, a, b, where $a < b < c < d$ (see [119] for the definition of covexillary permutations).

Example. Let $\mathbf{n} = (2, 3, 2)$. We have $n = 7$, $Q = P_2 \cap P_5$. Further, $V(1, 2)$, $V(2, 1)$ are the two irreducible components of V. Let w_1, w_2 be the elements of W^Q such that $V(1, 2) \cong Y_Q(w_1)$, $V(2, 1) \cong Y_Q(w_2)$. We have,

$$w_1^{(2)} = (2, 5), \quad w_1^{(5)} = (1, 2\,5\,6\,7),$$
$$w_2^{(2)} = (4, 5), \quad w_2^{(5)} = (1, 2\,4, 5\,7).$$

Let θ_1, θ_2 be the elements of W^Q such that $V(0, 1) \cong Y_Q(\theta_1)$, $V(1, 0) \cong Y_Q(\theta_2)$. We have

$$\theta_1^{(2)} = (1, 2), \quad \theta_1^{(5)} = (1, 2\,4, 5\,7),$$
$$\theta_2^{(2)} = (2, 5), \quad \theta_2^{(5)} = (1, 2\,3, 4, 5).$$

Further, we have

$$w_1^{\max} = (5276143), \quad \theta_1^{\max} = (2175463),$$
$$w_2^{\max} = (5472163), \quad \theta_2^{\max} = (5243176).$$

Note that w_1 (resp. w_2) is non-covexillary with $5, 6, 1, 4$ (resp. $4, 7, 2, 3$) as a Type II occurrence. Note also that the occurrence $1, 5, 4, 6$ in θ_1 (resp. $2, 4, 3, 7$ in θ_2) corresponds to the occurrence $5, 6, 1, 4$ in w_1 (resp. $4, 7, 2, 3$ in w_1).

12.5.13. Remark. The varieties of complexes are the only quiver varieties for which singular loci are known.

CHAPTER 13

Addendum

13.1. Dynkin Diagrams

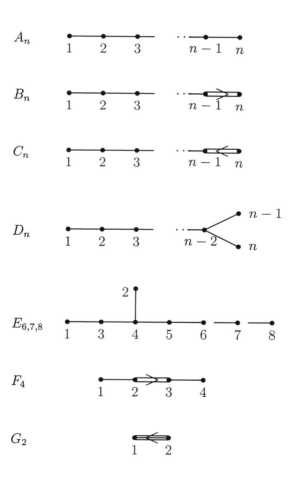

13.2. Summary of Smoothness Criteria

Let X be a variety and $P \in X$. We can assume X is an affine, say $X \subset \mathbb{A}^d$ variety since smoothness is a local condition. Then the following are equivalent:

1. X is smooth at P.
2. $\mathcal{O}_{X,P}$ is a regular local ring.
3. $\mathrm{gr}(\mathcal{O}_{X,P}, \mathfrak{m}_P) = \sum_{n \geq 0} \mathfrak{m}_P^n / \mathfrak{m}_P^{n+1}$ is a polynomial algebra generated by degree one elements.
4. The dimensions of X and of the tangent space to X at P are equal.
5. The rank of the Jacobian matrix $J_P = \mathrm{codim}_{\mathbb{A}^d} X$.
6. $\mathrm{mult}_P X = 1$.

Let $X(w)$ be a Schubert variety in G/B with a T-fixed point e_v. Each of the four lists below give equivalent conditions pertaining to singularities of Schubert varieties. Criteria for special cases such as Grassmannians have not been included here.

Smooth Schubert varieties.

1. $X(w)$ is smooth at every point
2. $X(w)$ is smooth at e_{id}.
3. Type A: w avoids 4231 and 3412.
4. Types A,B,C, D: For each subsequence $i_1 < i_2 < i_3 < i_4$ such that $w' = \mathrm{fl}(w_{i_1} w_{i_2} w_{i_3} w_{i_4}) \in W_4$, $X(w')$ is a smooth Schubert variety. In other words, the one-line notation for w avoids the patterns in Section 13.
5. Assuming G contains no G_2 factor, the reduced tangent cone at every point in $X(w)$ is linear.

Smooth points.

1. $X(w)$ is smooth at e_v.
2. The following identity holds in the nil-Hecke ring:

$$c_{w,v} = (-1)^{l(w) - l(v)} \prod_{\beta \in S(w,v)} \beta^{-1}$$

where $S(w, v) = \{\alpha \in R^+ \mid s_\alpha v \leq w\}$.
3. Assuming G contains no G_2 factor, the reduced tangent cones of $X(w)$ at all $x \in [v, w]$ are linear.

Rationally smooth Schubert varieties.

1. $X(w)$ is rationally smooth at every point.
2. $X(w)$ is rationally smooth at e_{id}.
3. The Poincaré polynomial $p_w(q)$ of $X(w)$ is symmetric.

4. The Bruhat graph $\Gamma(\mathrm{id}, w)$ is regular, i.e., every vertex has the same number of edges.
5. Types A, B, C, D: For each subsequence $i_1 < i_2 < i_3 < i_4$ such that $w' = \mathrm{fl}(w_{i_1} w_{i_2} w_{i_3} w_{i_4}) \in W_4$, $X(w')$ is a rationally smooth Schubert variety. In other words, the one-line notation for w avoids the patterns in Section 13.

Rationally smooth points.

1. $X(w)$ is rationally smooth at e_v.
2. The Kazhdan–Lusztig polynomial $P_{x,w} = 1$ for all $v \leq x \leq w$.
3. $P_{v,w} = 1$.
4. For all $v \leq x \leq w$,

$$\sum_{x \leq y \leq w} R_{x,y} = q^{l(w)-l(x)}.$$

5. $\#\{r \in \mathcal{R} \mid x < rx \leq w\} = l(w) - l(x),\ \forall x \in [v, w]$.
6. The Bruhat graph $\Gamma(v, w)$ is regular, i.e., every vertex has $l(w) - l(v)$ edges.
7. For $v \leq x \leq w$, $\#\{r \in \mathcal{R} \mid ry \leq w\} = l(w)$, i.e., e_x lies on exactly $l(w)$ T - invariant curves on $X(w)$.
8. For each $v \leq x \leq w$ there exists a positive integer d (depending on w, x) such that

$$c_{w,x} = d(-1)^{l(w)-l(x)} \prod_{\beta \in S(w,x)} \beta^{-1}.$$

where $S(w, v) = \{\alpha \in R^+ \mid s_\alpha v \leq w\}$.

13.3. Table of Minimal Bad Patterns

13.3.1. Patterns for rational smoothness (equivalently smoothness) in type A.

$$(13.3.2) \qquad 3412 \quad 4231$$

13.3.3. Patterns for rational smoothness in types B and C.

Signed notation:

$(13.3.4)$

$\overline{1}23$	$1\overline{2}3$	$12\overline{3}$	$1\overline{3}2$	$\overline{2}\overline{1}3$	$\overline{2}13$	$2\overline{1}3$
$2\overline{3}1$	$\overline{3}12$	$\overline{3}\overline{2}1$	$\overline{3}21$	$\overline{3}\overline{2}1$	$\overline{3}\overline{2}1$	$\overline{3}\overline{2}1$
$\overline{2}431$	$2\overline{4}31$	$\overline{3}\overline{4}\overline{1}\overline{2}$	$\overline{3}4\overline{1}2$	$\overline{3}412$	$34\overline{1}2$	3412
$4\overline{1}3\overline{2}$	$4\overline{1}32$	$\overline{4}231$	$423\overline{1}$	4231		

Extended notation:

$(13.3.5)$

653421	624351	623451	645231	642531
635241	563412	536142	462513	456123
451623	426153	356124	351624	62518473
62418573	65872143	35127846	34172856	34127856
52318674	42863175	42836175	42381675	42318675

13.3.6. Remark.
Note, there are 26 patterns in the signed notation but only 25 patterns in the unsigned notation above. The signed pattern $\overline{3}4\overline{1}2$ is 35172846 in unsigned notation; 35172846 contains the pattern 351624 so it is redundant.

13.3.7. Additional pattern for smoothness in types B and C.

$$(13.3.8) \qquad \begin{array}{lll} \text{Type B} & 3412 & \equiv \ \overline{2}\overline{1} \\ \text{Type C} & 4231 & \equiv \ 1\overline{2} \end{array}$$

13.3.9. Patterns for (rational) smoothness in type D.

Signed notation:

$(13.3.10)$

$12\overline{3}$	$\overline{1}2\overline{3}$	$\overline{1}\overline{3}2$	$1\overline{3}\overline{2}$	$\overline{2}\overline{1}3$	$\overline{3}\overline{2}1$	
$\overline{1}4\overline{3}2$	$\overline{2}1\overline{3}4$	$2\overline{1}\overline{3}4$	$21\overline{3}4$	$\overline{2}31\overline{4}$	$2\overline{3}1\overline{4}$	
$2\overline{4}31$	$\overline{2}\overline{4}3\overline{1}$	$\overline{2}4\overline{3}\overline{1}$	$24\overline{3}\overline{1}$	$\overline{2}43\overline{1}$	$\overline{2}431$	
$\overline{2}431$	$31\overline{2}\overline{4}$	$31\overline{2}4$	$3\overline{2}1\overline{4}$	$3\overline{2}4\overline{1}$	$\overline{3}41\overline{2}$	
$\overline{3}4\overline{1}2$	$\overline{3}412$	$34\overline{1}2$	$3\overline{4}\overline{1}2$	3412	$\overline{3}4\overline{1}\overline{2}$	
$34\overline{1}\overline{2}$	$\overline{3}\overline{4}21$	$34\overline{2}\overline{1}$	$3\overline{4}21$	$3\overline{4}21$	$\overline{4}132$	
4132	$\overline{4}\overline{1}3\overline{2}$	$4\overline{1}3\overline{2}$	$\overline{4}13\overline{2}$	$4\overline{1}32$	$\overline{4}1\overline{3}2$	$4\overline{2}1\overline{3}$
$4\overline{2}3\overline{1}$	$\overline{4}\overline{2}31$	$4\overline{2}31$	$423\overline{1}$	$\overline{4}23\overline{1}$	4231	$4\overline{2}\overline{3}1$
$4\overline{3}1\overline{2}$	$4\overline{3}12$	$\overline{4}\overline{3}12$	$4\overline{3}12$	$4\overline{3}\overline{2}1$		

Extended notation:

(13.3.11)

623451	624351	563412	645231	456123	
37154826	87463521	87536421	87436521	84763251	84736251
42836175	52863174	57163824	57136824	52836174	47163825
42863175	86527431	86427531	84627351	48627315	64872153
65827143	34172856	35127846	35172846	34127856	65872143
64827153	46872135	56127834	46172835	46827135	37581426
62418573	62581473	62518473	62481573	37518426	37481526
74618352	57618324	47681325	42381675	52318674	52381674
42318675	47618325	65718243	35718246	34781256	64718253
46718235	78654312	76854132	68754213		

13.4. Singular loci of A_5, B_4, C_4, D_4

The tables below can be read as follows:
Perm:(2 4 5 6 1 3) Sing:((2 4 1 5 3 6) (2 1 5 4 3 6) (2 1 4 6 3 5))
means $\text{Sing}(X(245613))$ consists of the union of three Schubert varieties,
namely $X(241536)$, $X(215436)$, and $X(214635)$.

13.4.1. Singular locus for $\mathcal{S}_6$.

Perm:(1 2 5 6 3 4) Sing:((1 2 3 5 4 6))
Perm:(1 2 6 4 5 3) Sing:((1 2 4 3 6 5))
Perm:(1 3 5 6 2 4) Sing:((1 3 2 5 4 6))
Perm:(1 3 6 4 5 2) Sing:((1 3 4 2 6 5))
Perm:(1 4 5 2 3 6) Sing:((1 2 4 3 5 6))
Perm:(1 4 5 2 6 3) Sing:((1 2 4 3 6 5))
Perm:(1 4 5 6 2 3) Sing:((1 4 2 5 3 6)(1 2 5 4 3 6)(1 2 4 6 3 5))
Perm:(1 4 6 2 3 5) Sing:((1 2 4 3 6 5))
Perm:(1 4 6 2 5 3) Sing:((1 2 4 3 6 5))
Perm:(1 4 6 3 5 2) Sing:((1 4 3 2 6 5))
Perm:(1 4 6 5 2 3) Sing:((1 2 6 4 3 5))
Perm:(1 5 2 6 3 4) Sing:((1 3 2 5 4 6))
Perm:(1 5 3 4 2 6) Sing:((1 3 2 5 4 6))
Perm:(1 5 3 4 6 2) Sing:((1 3 2 5 6 4))
Perm:(1 5 3 6 2 4) Sing:((1 3 2 5 4 6))
Perm:(1 5 3 6 4 2) Sing:((1 3 2 6 5 4))
Perm:(1 5 4 6 2 3) Sing:((1 5 2 4 3 6))
Perm:(1 5 6 2 3 4) Sing:((1 3 5 2 4 6)(1 2 5 4 3 6)(1 2 5 3 6 4))
Perm:(1 5 6 2 4 3) Sing:((1 2 5 4 6 3))
Perm:(1 5 6 3 2 4) Sing:((1 3 5 4 2 6))
Perm:(1 5 6 3 4 2) Sing:((1 5 3 2 6 4)(1 3 6 2 5 4)(1 3 5 4 6 2))
Perm:(1 5 6 4 2 3) Sing:((1 2 5 4 3 6))
Perm:(1 6 2 4 5 3) Sing:((1 4 2 3 6 5))
Perm:(1 6 3 4 2 5) Sing:((1 3 2 6 4 5))
Perm:(1 6 3 4 5 2) Sing:((1 4 3 2 6 5)(1 3 2 6 5 4))
Perm:(1 6 3 5 2 4) Sing:((1 3 2 6 5 4))
Perm:(1 6 3 5 4 2) Sing:((1 3 2 6 5 4))
Perm:(1 6 4 2 5 3) Sing:((1 4 3 2 6 5))
Perm:(1 6 4 3 5 2) Sing:((1 4 3 2 6 5))
Perm:(1 6 4 5 2 3) Sing:((1 6 2 4 3 5)(1 4 3 6 2 5)(1 4 2 6 5 3))
Perm:(1 6 4 5 3 2) Sing:((1 4 3 6 5 2))
Perm:(1 6 5 3 4 2) Sing:((1 6 3 2 5 4))
Perm:(2 1 5 6 3 4) Sing:((2 1 3 5 4 6))
Perm:(2 1 6 4 5 3) Sing:((2 1 4 3 6 5))
Perm:(2 3 5 6 1 4) Sing:((2 3 1 5 4 6))
Perm:(2 3 6 4 5 1) Sing:((2 3 4 1 6 5))
Perm:(2 4 5 1 3 6) Sing:((2 1 4 3 5 6))
Perm:(2 4 5 1 6 3) Sing:((2 1 4 3 6 5))

Perm:(2 4 5 6 1 3) Sing:((2 4 1 5 3 6)(2 1 5 4 3 6)(2 1 4 6 3 5))
Perm:(2 4 6 1 3 5) Sing:((2 1 4 3 6 5))
Perm:(2 4 6 1 5 3) Sing:((2 1 4 3 6 5))
Perm:(2 4 6 3 5 1) Sing:((2 4 3 1 6 5))
Perm:(2 4 6 5 1 3) Sing:((2 1 6 4 3 5))
Perm:(2 5 1 6 3 4) Sing:((2 3 1 5 4 6))
Perm:(2 5 3 4 1 6) Sing:((2 3 1 5 4 6))
Perm:(2 5 3 4 6 1) Sing:((2 3 1 5 6 4))
Perm:(2 5 3 6 1 4) Sing:((2 3 1 5 4 6))
Perm:(2 5 3 6 4 1) Sing:((2 3 1 6 5 4))
Perm:(2 5 4 6 1 3) Sing:((2 5 1 4 3 6))
Perm:(2 5 6 1 3 4) Sing:((2 3 5 1 4 6)(2 1 5 4 3 6)(2 1 5 3 6 4))
Perm:(2 5 6 1 4 3) Sing:((2 1 5 4 6 3))
Perm:(2 5 6 3 1 4) Sing:((2 3 5 4 1 6))
Perm:(2 5 6 3 4 1) Sing:((2 5 3 1 6 4)(2 3 6 1 5 4)(2 3 5 4 6 1))
Perm:(2 5 6 4 1 3) Sing:((2 1 5 4 3 6))
Perm:(2 6 1 4 5 3) Sing:((2 4 1 3 6 5))
Perm:(2 6 3 4 1 5) Sing:((2 3 1 6 4 5))
Perm:(2 6 3 4 5 1) Sing:((2 4 3 1 6 5)(2 3 1 6 5 4))
Perm:(2 6 3 5 1 4) Sing:((2 3 1 6 5 4))
Perm:(2 6 3 5 4 1) Sing:((2 3 1 6 5 4))
Perm:(2 6 4 1 5 3) Sing:((2 4 3 1 6 5))
Perm:(2 6 4 3 5 1) Sing:((2 4 3 1 6 5))
Perm:(2 6 4 5 1 3) Sing:((2 6 1 4 3 5)(2 4 3 6 1 5)(2 4 1 6 5 3))
Perm:(2 6 4 5 3 1) Sing:((2 4 3 6 5 1))
Perm:(2 6 5 3 4 1) Sing:((2 6 3 1 5 4))
Perm:(3 1 5 6 2 4) Sing:((3 1 2 5 4 6))
Perm:(3 1 6 4 5 2) Sing:((3 1 4 2 6 5))
Perm:(3 2 5 6 1 4) Sing:((3 2 1 5 4 6))
Perm:(3 2 6 4 5 1) Sing:((3 2 4 1 6 5))
Perm:(3 4 1 2 5 6) Sing:((1 3 2 4 5 6))
Perm:(3 4 1 2 6 5) Sing:((1 3 2 4 6 5))
Perm:(3 4 1 5 2 6) Sing:((1 3 2 5 4 6))
Perm:(3 4 1 5 6 2) Sing:((1 3 2 5 6 4))
Perm:(3 4 1 6 2 5) Sing:((1 3 2 6 4 5))
Perm:(3 4 1 6 5 2) Sing:((1 3 2 6 5 4))
Perm:(3 4 5 1 2 6) Sing:((3 1 4 2 5 6)(1 4 3 2 5 6)(1 3 5 2 4 6))
Perm:(3 4 5 1 6 2) Sing:((3 1 4 2 6 5)(1 4 3 2 6 5)(1 3 5 2 6 4))
Perm:(3 4 5 6 1 2) Sing:((3 4 1 5 2 6)(3 1 5 4 2 6)(1 4 5 3 2 6)
                         (3 1 4 6 2 5)(1 4 3 6 2 5)(1 3 5 6 2 4))
Perm:(3 4 6 1 2 5) Sing:((3 1 4 2 6 5)(1 4 3 2 6 5)(1 3 6 2 4 5))
Perm:(3 4 6 1 5 2) Sing:((3 1 4 2 6 5)(1 4 3 2 6 5)(1 3 6 2 5 4))
Perm:(3 4 6 2 5 1) Sing:((3 4 2 1 6 5))
Perm:(3 4 6 5 1 2) Sing:((3 1 6 4 2 5)(1 4 6 3 2 5)(1 3 6 5 2 4))
Perm:(3 5 1 2 4 6) Sing:((1 3 2 5 4 6))
Perm:(3 5 1 2 6 4) Sing:((1 3 2 5 6 4))

Perm:(3 5 1 4 2 6)   Sing:((1 3 2 5 4 6))
Perm:(3 5 1 4 6 2)   Sing:((1 3 2 5 6 4))
Perm:(3 5 1 6 2 4)   Sing:((3 2 1 5 4 6)(1 3 2 6 5 4))
Perm:(3 5 1 6 4 2)   Sing:((1 3 2 6 5 4))
Perm:(3 5 2 4 1 6)   Sing:((3 2 1 5 4 6))
Perm:(3 5 2 4 6 1)   Sing:((3 2 1 5 6 4))
Perm:(3 5 2 6 1 4)   Sing:((3 2 1 5 4 6))
Perm:(3 5 2 6 4 1)   Sing:((3 2 1 6 5 4))
Perm:(3 5 4 1 2 6)   Sing:((1 5 3 2 4 6))
Perm:(3 5 4 1 6 2)   Sing:((1 5 3 2 6 4))
Perm:(3 5 4 6 1 2)   Sing:((3 5 1 4 2 6)(1 5 4 3 2 6)(1 5 3 6 2 4))
Perm:(3 5 6 1 2 4)   Sing:((3 2 5 1 4 6)(3 1 5 4 2 6)(3 1 5 2 6 4)
                            (1 5 3 2 6 4)(1 3 6 2 5 4))
Perm:(3 5 6 1 4 2)   Sing:((1 5 3 2 6 4)(1 3 6 2 5 4)(3 1 5 4 6 2))
Perm:(3 5 6 2 1 4)   Sing:((3 2 5 4 1 6))
Perm:(3 5 6 2 4 1)   Sing:((3 5 2 1 6 4)(3 2 6 1 5 4)(3 2 5 4 6 1))
Perm:(3 5 6 4 1 2)   Sing:((3 1 5 4 2 6)(1 5 6 3 2 4))
Perm:(3 6 1 2 4 5)   Sing:((1 3 2 6 4 5))
Perm:(3 6 1 2 5 4)   Sing:((1 3 2 6 5 4))
Perm:(3 6 1 4 2 5)   Sing:((1 3 2 6 4 5))
Perm:(3 6 1 4 5 2)   Sing:((3 4 1 2 6 5)(1 3 2 6 5 4))
Perm:(3 6 1 5 2 4)   Sing:((1 3 2 6 5 4))
Perm:(3 6 1 5 4 2)   Sing:((1 3 2 6 5 4))
Perm:(3 6 2 4 1 5)   Sing:((3 2 1 6 4 5))
Perm:(3 6 2 4 5 1)   Sing:((3 4 2 1 6 5)(3 2 1 6 5 4))
Perm:(3 6 2 5 1 4)   Sing:((3 2 1 6 5 4))
Perm:(3 6 2 5 4 1)   Sing:((3 2 1 6 5 4))
Perm:(3 6 4 1 2 5)   Sing:((1 6 3 2 4 5))
Perm:(3 6 4 1 5 2)   Sing:((3 4 2 1 6 5)(1 6 3 2 5 4))
Perm:(3 6 4 2 5 1)   Sing:((3 4 2 1 6 5))
Perm:(3 6 4 5 1 2)   Sing:((3 6 1 4 2 5)(1 6 4 3 2 5)(3 4 2 6 1 5)
                            (1 6 3 5 2 4)(3 4 1 6 5 2))
Perm:(3 6 4 5 2 1)   Sing:((3 4 2 6 5 1))
Perm:(3 6 5 1 2 4)   Sing:((1 6 3 2 5 4))
Perm:(3 6 5 1 4 2)   Sing:((1 6 3 2 5 4))
Perm:(3 6 5 2 4 1)   Sing:((3 6 2 1 5 4))
Perm:(3 6 5 4 1 2)   Sing:((1 6 5 3 2 4))
Perm:(4 1 5 2 3 6)   Sing:((2 1 4 3 5 6))
Perm:(4 1 5 2 6 3)   Sing:((2 1 4 3 6 5))
Perm:(4 1 5 6 2 3)   Sing:((4 1 2 5 3 6)(2 1 5 4 3 6)(2 1 4 6 3 5))
Perm:(4 1 6 2 3 5)   Sing:((2 1 4 3 6 5))
Perm:(4 1 6 2 5 3)   Sing:((2 1 4 3 6 5))
Perm:(4 1 6 3 5 2)   Sing:((4 1 3 2 6 5))
Perm:(4 1 6 5 2 3)   Sing:((2 1 6 4 3 5))
Perm:(4 2 3 1 5 6)   Sing:((2 1 4 3 5 6))
Perm:(4 2 3 1 6 5)   Sing:((2 1 4 3 6 5))

```
Perm:(4 2 3 5 1 6)   Sing:((2 1 4 5 3 6))
Perm:(4 2 3 5 6 1)   Sing:((2 1 4 5 6 3))
Perm:(4 2 3 6 1 5)   Sing:((2 1 4 6 3 5))
Perm:(4 2 3 6 5 1)   Sing:((2 1 4 6 5 3))
Perm:(4 2 5 1 3 6)   Sing:((2 1 4 3 5 6))
Perm:(4 2 5 1 6 3)   Sing:((2 1 4 3 6 5))
Perm:(4 2 5 3 1 6)   Sing:((2 1 5 4 3 6))
Perm:(4 2 5 3 6 1)   Sing:((2 1 5 4 6 3))
Perm:(4 2 5 6 1 3)   Sing:((4 2 1 5 3 6)(2 1 5 4 3 6)(2 1 4 6 3 5))
Perm:(4 2 5 6 3 1)   Sing:((2 1 5 6 4 3))
Perm:(4 2 6 1 3 5)   Sing:((2 1 4 3 6 5))
Perm:(4 2 6 1 5 3)   Sing:((2 1 4 3 6 5))
Perm:(4 2 6 3 1 5)   Sing:((2 1 6 4 3 5))
Perm:(4 2 6 3 5 1)   Sing:((4 2 3 1 6 5)(2 1 6 4 5 3))
Perm:(4 2 6 5 1 3)   Sing:((2 1 6 4 3 5))
Perm:(4 2 6 5 3 1)   Sing:((2 1 6 5 4 3))
Perm:(4 3 5 1 2 6)   Sing:((4 1 3 2 5 6))
Perm:(4 3 5 1 6 2)   Sing:((4 1 3 2 6 5))
Perm:(4 3 5 6 1 2)   Sing:((4 3 1 5 2 6)(4 1 5 3 2 6)(4 1 3 6 2 5))
Perm:(4 3 6 1 2 5)   Sing:((4 1 3 2 6 5))
Perm:(4 3 6 1 5 2)   Sing:((4 1 3 2 6 5))
Perm:(4 3 6 2 5 1)   Sing:((4 3 2 1 6 5))
Perm:(4 3 6 5 1 2)   Sing:((4 1 6 3 2 5))
Perm:(4 5 1 2 3 6)   Sing:((2 4 1 3 5 6)(1 4 3 2 5 6)(1 4 2 5 3 6))
Perm:(4 5 1 2 6 3)   Sing:((2 4 1 3 6 5)(1 4 3 2 6 5)(1 4 2 5 6 3))
Perm:(4 5 1 3 2 6)   Sing:((1 4 3 5 2 6))
Perm:(4 5 1 3 6 2)   Sing:((1 4 3 5 6 2))
Perm:(4 5 1 6 2 3)   Sing:((4 2 1 5 3 6)(2 5 1 4 3 6)(2 4 1 6 3 5)
                          (1 4 3 6 2 5)(1 4 2 6 5 3))
Perm:(4 5 1 6 3 2)   Sing:((1 4 3 6 5 2))
Perm:(4 5 2 1 3 6)   Sing:((2 4 3 1 5 6))
Perm:(4 5 2 1 6 3)   Sing:((2 4 3 1 6 5))
Perm:(4 5 2 3 1 6)   Sing:((4 2 1 5 3 6)(2 5 1 4 3 6)(2 4 3 5 1 6))
Perm:(4 5 2 3 6 1)   Sing:((4 2 1 5 6 3)(2 5 1 4 6 3)(2 4 3 5 6 1))
Perm:(4 5 2 6 1 3)   Sing:((4 2 1 5 3 6)(2 5 1 4 3 6)(2 4 3 6 1 5))
Perm:(4 5 2 6 3 1)   Sing:((4 2 1 6 5 3)(2 5 1 6 4 3)(2 4 3 6 5 1))
Perm:(4 5 3 1 2 6)   Sing:((1 4 3 2 5 6))
Perm:(4 5 3 1 6 2)   Sing:((1 4 3 2 6 5))
Perm:(4 5 3 6 1 2)   Sing:((4 5 1 3 2 6)(1 4 3 6 2 5))
Perm:(4 5 6 1 2 3)   Sing:((4 2 5 1 3 6)(2 5 4 1 3 6)(4 1 5 3 2 6)
                          (1 5 4 3 2 6)(2 4 6 1 3 5)(1 4 6 3 2 5)
                          (4 1 5 2 6 3)(1 5 4 2 6 3)(1 4 6 2 5 3))
Perm:(4 5 6 1 3 2)   Sing:((4 1 5 3 6 2)(1 5 4 3 6 2)(1 4 6 3 5 2))
Perm:(4 5 6 2 1 3)   Sing:((4 2 5 3 1 6)(2 5 4 3 1 6)(2 4 6 3 1 5))
Perm:(4 5 6 2 3 1)   Sing:((4 5 2 1 6 3)(4 2 6 1 5 3)(2 5 6 1 4 3)
                          (4 2 5 3 6 1)(2 5 4 3 6 1)(2 4 6 3 5 1))
```

Perm:(4 5 6 3 1 2) Sing:((4 1 5 3 2 6)(1 5 4 3 2 6)(1 4 6 3 2 5))
Perm:(4 6 1 2 3 5) Sing:((2 4 1 3 6 5)(1 4 3 2 6 5)(1 4 2 6 3 5))
Perm:(4 6 1 2 5 3) Sing:((2 4 1 3 6 5)(1 4 3 2 6 5)(1 4 2 6 5 3))
Perm:(4 6 1 3 2 5) Sing:((1 4 3 6 2 5))
Perm:(4 6 1 3 5 2) Sing:((4 3 1 2 6 5)(1 4 3 6 5 2))
Perm:(4 6 1 5 2 3) Sing:((2 6 1 4 3 5)(1 4 3 6 2 5)(1 4 2 6 5 3))
Perm:(4 6 1 5 3 2) Sing:((1 4 3 6 5 2))
Perm:(4 6 2 1 3 5) Sing:((2 4 3 1 6 5))
Perm:(4 6 2 1 5 3) Sing:((2 4 3 1 6 5))
Perm:(4 6 2 3 1 5) Sing:((4 2 1 6 3 5)(2 6 1 4 3 5)(2 4 3 6 1 5))
Perm:(4 6 2 3 5 1) Sing:((4 3 2 1 6 5)(4 2 1 6 5 3)(2 6 1 4 5 3)
                          (2 4 3 6 5 1))
Perm:(4 6 2 5 1 3) Sing:((2 6 1 4 3 5)(2 4 3 6 1 5)(4 2 1 6 5 3))
Perm:(4 6 2 5 3 1) Sing:((4 2 1 6 5 3)(2 6 1 5 4 3)(2 4 3 6 5 1))
Perm:(4 6 3 1 2 5) Sing:((1 4 3 2 6 5))
Perm:(4 6 3 1 5 2) Sing:((4 3 2 1 6 5))
Perm:(4 6 3 2 5 1) Sing:((4 3 2 1 6 5))
Perm:(4 6 3 5 1 2) Sing:((4 6 1 3 2 5)(4 3 2 6 1 5)(4 3 1 6 5 2))
Perm:(4 6 3 5 2 1) Sing:((4 3 2 6 5 1))
Perm:(4 6 5 1 2 3) Sing:((2 6 4 1 3 5)(1 6 4 3 2 5)(1 6 4 2 5 3))
Perm:(4 6 5 1 3 2) Sing:((1 6 4 3 5 2))
Perm:(4 6 5 2 1 3) Sing:((2 6 4 3 1 5))
Perm:(4 6 5 2 3 1) Sing:((4 6 2 1 5 3)(2 6 5 1 4 3)(2 6 4 3 5 1))
Perm:(4 6 5 3 1 2) Sing:((1 6 4 3 2 5))
Perm:(5 1 2 6 3 4) Sing:((3 1 2 5 4 6))
Perm:(5 1 3 4 2 6) Sing:((3 1 2 5 4 6))
Perm:(5 1 3 4 6 2) Sing:((3 1 2 5 6 4))
Perm:(5 1 3 6 2 4) Sing:((3 1 2 5 4 6))
Perm:(5 1 3 6 4 2) Sing:((3 1 2 6 5 4))
Perm:(5 1 4 6 2 3) Sing:((5 1 2 4 3 6))
Perm:(5 1 6 2 3 4) Sing:((3 1 5 2 4 6)(2 1 5 4 3 6)(2 1 5 3 6 4))
Perm:(5 1 6 2 4 3) Sing:((2 1 5 4 6 3))
Perm:(5 1 6 3 2 4) Sing:((3 1 5 4 2 6))
Perm:(5 1 6 3 4 2) Sing:((5 1 3 2 6 4)(3 1 6 2 5 4)(3 1 5 4 6 2))
Perm:(5 1 6 4 2 3) Sing:((2 1 5 4 3 6))
Perm:(5 2 1 6 3 4) Sing:((3 2 1 5 4 6))
Perm:(5 2 3 1 4 6) Sing:((2 1 5 3 4 6))
Perm:(5 2 3 1 6 4) Sing:((2 1 5 3 6 4))
Perm:(5 2 3 4 1 6) Sing:((3 2 1 5 4 6)(2 1 5 4 3 6))
Perm:(5 2 3 4 6 1) Sing:((3 2 1 5 6 4)(2 1 5 4 6 3))
Perm:(5 2 3 6 1 4) Sing:((3 2 1 5 4 6)(2 1 5 6 3 4))
Perm:(5 2 3 6 4 1) Sing:((3 2 1 6 5 4)(2 1 5 6 4 3))
Perm:(5 2 4 1 3 6) Sing:((2 1 5 4 3 6))
Perm:(5 2 4 1 6 3) Sing:((2 1 5 4 6 3))
Perm:(5 2 4 3 1 6) Sing:((2 1 5 4 3 6))
Perm:(5 2 4 3 6 1) Sing:((2 1 5 4 6 3))

Perm:(5 2 4 6 1 3) Sing:((5 2 1 4 3 6)(2 1 5 6 4 3))
Perm:(5 2 4 6 3 1) Sing:((2 1 5 6 4 3))
Perm:(5 2 6 1 3 4) Sing:((3 2 5 1 4 6)(2 1 5 4 3 6)(2 1 5 3 6 4))
Perm:(5 2 6 1 4 3) Sing:((2 1 5 4 6 3))
Perm:(5 2 6 3 1 4) Sing:((3 2 5 4 1 6)(2 1 6 5 3 4))
Perm:(5 2 6 3 4 1) Sing:((5 2 3 1 6 4)(3 2 6 1 5 4)(2 1 6 5 4 3)
                          (3 2 5 4 6 1))
Perm:(5 2 6 4 1 3) Sing:((2 1 6 5 4 3))
Perm:(5 2 6 4 3 1) Sing:((2 1 6 5 4 3))
Perm:(5 3 1 4 2 6) Sing:((3 2 1 5 4 6))
Perm:(5 3 1 4 6 2) Sing:((3 2 1 5 6 4))
Perm:(5 3 1 6 2 4) Sing:((3 2 1 5 4 6))
Perm:(5 3 1 6 4 2) Sing:((3 2 1 6 5 4))
Perm:(5 3 2 4 1 6) Sing:((3 2 1 5 4 6))
Perm:(5 3 2 4 6 1) Sing:((3 2 1 5 6 4))
Perm:(5 3 2 6 1 4) Sing:((3 2 1 5 4 6))
Perm:(5 3 2 6 4 1) Sing:((3 2 1 6 5 4))
Perm:(5 3 4 1 2 6) Sing:((5 1 3 2 4 6)(3 2 5 1 4 6)(3 1 5 4 2 6))
Perm:(5 3 4 1 6 2) Sing:((5 1 3 2 6 4)(3 2 5 1 6 4)(3 1 5 4 6 2))
Perm:(5 3 4 2 1 6) Sing:((3 2 5 4 1 6))
Perm:(5 3 4 2 6 1) Sing:((3 2 5 4 6 1))
Perm:(5 3 4 6 1 2) Sing:((5 3 1 4 2 6)(5 1 4 3 2 6)(5 1 3 6 2 4)
                          (3 2 5 6 1 4)(3 1 5 6 4 2))
Perm:(5 3 4 6 2 1) Sing:((3 2 5 6 4 1))
Perm:(5 3 6 1 2 4) Sing:((3 2 5 1 4 6)(3 1 5 4 2 6)(5 1 3 2 6 4))
Perm:(5 3 6 1 4 2) Sing:((5 1 3 2 6 4)(3 2 6 1 5 4)(3 1 5 4 6 2))
Perm:(5 3 6 2 1 4) Sing:((3 2 5 4 1 6))
Perm:(5 3 6 2 4 1) Sing:((5 3 2 1 6 4)(3 2 6 1 5 4)(3 2 5 4 6 1))
Perm:(5 3 6 4 1 2) Sing:((5 1 6 3 2 4)(3 2 6 5 1 4)(3 1 6 5 4 2))
Perm:(5 3 6 4 2 1) Sing:((3 2 6 5 4 1))
Perm:(5 4 1 6 2 3) Sing:((5 2 1 4 3 6))
Perm:(5 4 2 3 1 6) Sing:((5 2 1 4 3 6))
Perm:(5 4 2 3 6 1) Sing:((5 2 1 4 6 3))
Perm:(5 4 2 6 1 3) Sing:((5 2 1 4 3 6))
Perm:(5 4 2 6 3 1) Sing:((5 2 1 6 4 3))
Perm:(5 4 3 6 1 2) Sing:((5 4 1 3 2 6))
Perm:(5 4 6 1 2 3) Sing:((5 2 4 1 3 6)(5 1 4 3 2 6)(5 1 4 2 6 3))
Perm:(5 4 6 1 3 2) Sing:((5 1 4 3 6 2))
Perm:(5 4 6 2 1 3) Sing:((5 2 4 3 1 6))
Perm:(5 4 6 2 3 1) Sing:((5 4 2 1 6 3)(5 2 6 1 4 3)(5 2 4 3 6 1))
Perm:(5 4 6 3 1 2) Sing:((5 1 4 3 2 6))
Perm:(5 6 1 2 3 4) Sing:((3 5 1 2 4 6)(2 5 1 4 3 6)(1 5 4 2 3 6)
                          (2 5 1 3 6 4)(1 5 3 2 6 4)(1 5 2 6 3 4))
Perm:(5 6 1 2 4 3) Sing:((2 5 1 4 6 3)(1 5 4 2 6 3)(1 5 2 6 4 3))
Perm:(5 6 1 3 2 4) Sing:((3 5 1 4 2 6)(1 5 4 3 2 6)(1 5 3 6 2 4))

Perm:(5 6 1 3 4 2)   Sing:((5 3 1 2 6 4)(3 6 1 2 5 4)(3 5 1 4 6 2)
                              (1 5 4 3 6 2)(1 5 3 6 4 2))
Perm:(5 6 1 4 2 3)   Sing:((2 5 1 4 3 6)(1 5 4 6 2 3))
Perm:(5 6 1 4 3 2)   Sing:((1 5 4 6 3 2))
Perm:(5 6 2 1 3 4)   Sing:((3 5 2 1 4 6)(2 5 4 1 3 6)(2 5 3 1 6 4))
Perm:(5 6 2 1 4 3)   Sing:((2 5 4 1 6 3))
Perm:(5 6 2 3 1 4)   Sing:((3 5 2 4 1 6)(2 5 4 3 1 6)(5 2 1 6 3 4)
                              (2 6 1 5 3 4)(2 5 3 6 1 4))
Perm:(5 6 2 3 4 1)   Sing:((5 3 2 1 6 4)(3 6 2 1 5 4)(5 2 1 6 4 3)
                              (2 6 1 5 4 3)(3 5 2 4 6 1)(2 5 4 3 6 1)
                              (2 5 3 6 4 1))
Perm:(5 6 2 4 1 3)   Sing:((5 2 1 6 4 3)(2 6 1 5 4 3)(2 5 4 6 1 3))
Perm:(5 6 2 4 3 1)   Sing:((5 2 1 6 4 3)(2 6 1 5 4 3)(2 5 4 6 3 1))
Perm:(5 6 3 1 2 4)   Sing:((3 5 4 1 2 6)(1 5 3 2 6 4))
Perm:(5 6 3 1 4 2)   Sing:((5 3 2 1 6 4)(3 6 2 1 5 4)(3 5 4 1 6 2))
Perm:(5 6 3 2 1 4)   Sing:((3 5 4 2 1 6))
Perm:(5 6 3 2 4 1)   Sing:((5 3 2 1 6 4)(3 6 2 1 5 4)(3 5 4 2 6 1))
Perm:(5 6 3 4 1 2)   Sing:((5 6 1 3 2 4)(5 3 2 6 1 4)(3 6 2 5 1 4)
                              (5 3 1 6 4 2)(3 6 1 5 4 2)(3 5 4 6 1 2))
Perm:(5 6 3 4 2 1)   Sing:((5 3 2 6 4 1)(3 6 2 5 4 1)(3 5 4 6 2 1))
Perm:(5 6 4 1 2 3)   Sing:((2 5 4 1 3 6)(1 5 4 3 2 6)(1 5 4 2 6 3))
Perm:(5 6 4 1 3 2)   Sing:((1 5 4 3 6 2))
Perm:(5 6 4 2 1 3)   Sing:((2 5 4 3 1 6))
Perm:(5 6 4 2 3 1)   Sing:((5 6 2 1 4 3)(2 5 4 3 6 1))
Perm:(5 6 4 3 1 2)   Sing:((1 5 4 3 2 6))
Perm:(6 1 2 4 5 3)   Sing:((4 1 2 3 6 5))
Perm:(6 1 3 4 2 5)   Sing:((3 1 2 6 4 5))
Perm:(6 1 3 4 5 2)   Sing:((4 1 3 2 6 5)(3 1 2 6 5 4))
Perm:(6 1 3 5 2 4)   Sing:((3 1 2 6 5 4))
Perm:(6 1 3 5 4 2)   Sing:((3 1 2 6 5 4))
Perm:(6 1 4 2 5 3)   Sing:((4 1 3 2 6 5))
Perm:(6 1 4 3 5 2)   Sing:((4 1 3 2 6 5))
Perm:(6 1 4 5 2 3)   Sing:((6 1 2 4 3 5)(4 1 3 6 2 5)(4 1 2 6 5 3))
Perm:(6 1 4 5 3 2)   Sing:((4 1 3 6 5 2))
Perm:(6 1 5 3 4 2)   Sing:((6 1 3 2 5 4))
Perm:(6 2 1 4 5 3)   Sing:((4 2 1 3 6 5))
Perm:(6 2 3 1 4 5)   Sing:((2 1 6 3 4 5))
Perm:(6 2 3 1 5 4)   Sing:((2 1 6 3 5 4))
Perm:(6 2 3 4 1 5)   Sing:((3 2 1 6 4 5)(2 1 6 4 3 5))
Perm:(6 2 3 4 5 1)   Sing:((4 2 3 1 6 5)(3 2 1 6 5 4)(2 1 6 4 5 3))
Perm:(6 2 3 5 1 4)   Sing:((3 2 1 6 5 4)(2 1 6 5 3 4))
Perm:(6 2 3 5 4 1)   Sing:((3 2 1 6 5 4)(2 1 6 5 4 3))
Perm:(6 2 4 1 3 5)   Sing:((2 1 6 4 3 5))
Perm:(6 2 4 1 5 3)   Sing:((4 2 3 1 6 5)(2 1 6 4 5 3))
Perm:(6 2 4 3 1 5)   Sing:((2 1 6 4 3 5))
Perm:(6 2 4 3 5 1)   Sing:((4 2 3 1 6 5)(2 1 6 4 5 3))

Perm:(6 2 4 5 1 3) Sing:((6 2 1 4 3 5)(4 2 3 6 1 5)(4 2 1 6 5 3)
                          (2 1 6 5 4 3))
Perm:(6 2 4 5 3 1) Sing:((2 1 6 5 4 3)(4 2 3 6 5 1))
Perm:(6 2 5 1 3 4) Sing:((2 1 6 5 3 4))
Perm:(6 2 5 1 4 3) Sing:((2 1 6 5 4 3))
Perm:(6 2 5 3 1 4) Sing:((2 1 6 5 3 4))
Perm:(6 2 5 3 4 1) Sing:((6 2 3 1 5 4)(2 1 6 5 4 3))
Perm:(6 2 5 4 1 3) Sing:((2 1 6 5 4 3))
Perm:(6 2 5 4 3 1) Sing:((2 1 6 5 4 3))
Perm:(6 3 1 4 2 5) Sing:((3 2 1 6 4 5))
Perm:(6 3 1 4 5 2) Sing:((4 3 1 2 6 5)(3 2 1 6 5 4))
Perm:(6 3 1 5 2 4) Sing:((3 2 1 6 5 4))
Perm:(6 3 1 5 4 2) Sing:((3 2 1 6 5 4))
Perm:(6 3 2 4 1 5) Sing:((3 2 1 6 4 5))
Perm:(6 3 2 4 5 1) Sing:((4 3 2 1 6 5)(3 2 1 6 5 4))
Perm:(6 3 2 5 1 4) Sing:((3 2 1 6 5 4))
Perm:(6 3 2 5 4 1) Sing:((3 2 1 6 5 4))
Perm:(6 3 4 1 2 5) Sing:((6 1 3 2 4 5)(3 2 6 1 4 5)(3 1 6 4 2 5))
Perm:(6 3 4 1 5 2) Sing:((4 3 2 1 6 5)(6 1 3 2 5 4)(3 2 6 1 5 4)
                          (3 1 6 4 5 2))
Perm:(6 3 4 2 1 5) Sing:((3 2 6 4 1 5))
Perm:(6 3 4 2 5 1) Sing:((4 3 2 1 6 5)(3 2 6 4 5 1))
Perm:(6 3 4 5 1 2) Sing:((6 3 1 4 2 5)(6 1 4 3 2 5)(4 3 2 6 1 5)
                          (6 1 3 5 2 4)(3 2 6 5 1 4)(4 3 1 6 5 2)
                          (3 1 6 5 4 2))
Perm:(6 3 4 5 2 1) Sing:((4 3 2 6 5 1)(3 2 6 5 4 1))
Perm:(6 3 5 1 2 4) Sing:((6 1 3 2 5 4)(3 2 6 1 5 4)(3 1 6 5 2 4))
Perm:(6 3 5 1 4 2) Sing:((6 1 3 2 5 4)(3 2 6 1 5 4)(3 1 6 5 4 2))
Perm:(6 3 5 2 1 4) Sing:((3 2 6 5 1 4))
Perm:(6 3 5 2 4 1) Sing:((6 3 2 1 5 4)(3 2 6 5 4 1))
Perm:(6 3 5 4 1 2) Sing:((6 1 5 3 2 4)(3 2 6 5 1 4)(3 1 6 5 4 2))
Perm:(6 3 5 4 2 1) Sing:((3 2 6 5 4 1))
Perm:(6 4 1 2 5 3) Sing:((4 3 1 2 6 5))
Perm:(6 4 1 3 5 2) Sing:((4 3 1 2 6 5))
Perm:(6 4 1 5 2 3) Sing:((6 2 1 4 3 5)(4 3 1 6 2 5)(4 2 1 6 5 3))
Perm:(6 4 1 5 3 2) Sing:((4 3 1 6 5 2))
Perm:(6 4 2 1 5 3) Sing:((4 3 2 1 6 5))
Perm:(6 4 2 3 1 5) Sing:((6 2 1 4 3 5))
Perm:(6 4 2 3 5 1) Sing:((4 3 2 1 6 5)(6 2 1 4 5 3))
Perm:(6 4 2 5 1 3) Sing:((6 2 1 4 3 5)(4 3 2 6 1 5)(4 2 1 6 5 3))
Perm:(6 4 2 5 3 1) Sing:((6 2 1 5 4 3)(4 3 2 6 5 1))
Perm:(6 4 3 1 5 2) Sing:((4 3 2 1 6 5))
Perm:(6 4 3 2 5 1) Sing:((4 3 2 1 6 5))
Perm:(6 4 3 5 1 2) Sing:((6 4 1 3 2 5)(4 3 2 6 1 5)(4 3 1 6 5 2))
Perm:(6 4 3 5 2 1) Sing:((4 3 2 6 5 1))

Perm:(6 4 5 1 2 3)   Sing:((6 2 4 1 3 5)(6 1 4 3 2 5)(4 3 6 1 2 5)
                            (6 1 4 2 5 3)(4 2 6 1 5 3)(4 1 6 5 2 3))
Perm:(6 4 5 1 3 2)   Sing:((6 1 4 3 5 2)(4 3 6 1 5 2)(4 1 6 5 3 2))
Perm:(6 4 5 2 1 3)   Sing:((6 2 4 3 1 5)(4 3 6 2 1 5)(4 2 6 5 1 3))
Perm:(6 4 5 2 3 1)   Sing:((6 4 2 1 5 3)(6 2 5 1 4 3)(6 2 4 3 5 1)
                            (4 3 6 2 5 1)(4 2 6 5 3 1))
Perm:(6 4 5 3 1 2)   Sing:((6 1 4 3 2 5)(4 3 6 5 1 2))
Perm:(6 4 5 3 2 1)   Sing:((4 3 6 5 2 1))
Perm:(6 5 1 3 4 2)   Sing:((6 3 1 2 5 4))
Perm:(6 5 2 3 1 4)   Sing:((6 2 1 5 3 4))
Perm:(6 5 2 3 4 1)   Sing:((6 3 2 1 5 4)(6 2 1 5 4 3))
Perm:(6 5 2 4 1 3)   Sing:((6 2 1 5 4 3))
Perm:(6 5 2 4 3 1)   Sing:((6 2 1 5 4 3))
Perm:(6 5 3 1 4 2)   Sing:((6 3 2 1 5 4))
Perm:(6 5 3 2 4 1)   Sing:((6 3 2 1 5 4))
Perm:(6 5 3 4 1 2)   Sing:((6 5 1 3 2 4)(6 3 2 5 1 4)(6 3 1 5 4 2))
Perm:(6 5 3 4 2 1)   Sing:((6 3 2 5 4 1))
Perm:(6 5 4 2 3 1)   Sing:((6 5 2 1 4 3))

13.4.2. Singular locus for B_4.

B_Perm:(1 8 3 4 5 6 7 2 9)   Sing:((1 3 2 6 5 4 8 7 9))
B_Perm:(1 8 3 6 5 4 7 2 9)   Sing:((1 3 2 6 5 4 8 7 9))
B_Perm:(1 8 7 4 5 6 3 2 9)   Sing:((1 4 3 2 5 8 7 6 9))
B_Perm:(9 2 3 4 5 6 7 8 1)   Sing:((2 1 7 4 5 6 3 9 8)(3 2 1 6 5 4 9 8 7))
B_Perm:(9 2 3 6 5 4 7 8 1)   Sing:((2 1 7 6 5 4 3 9 8)(3 2 1 6 5 4 9 8 7))
B_Perm:(9 2 7 4 5 6 3 8 1)   Sing:((2 1 7 6 5 4 3 9 8)(4 2 3 1 5 9 7 8 6))
B_Perm:(9 2 7 6 5 4 3 8 1)   Sing:((2 1 7 6 5 4 3 9 8))
B_Perm:(9 8 3 4 5 6 7 2 1)   Sing:((9 3 2 6 5 4 8 7 1))
B_Perm:(9 8 3 6 5 4 7 2 1)   Sing:((9 3 2 6 5 4 8 7 1))
B_Perm:(9 8 7 4 5 6 3 2 1)   Sing:((9 4 3 2 5 8 7 6 1))
B_Perm:(2 9 3 4 5 6 7 1 8)   Sing:((2 3 1 6 5 4 9 7 8))
B_Perm:(2 9 3 6 5 4 7 1 8)   Sing:((2 3 1 6 5 4 9 7 8))
B_Perm:(2 9 7 4 5 6 3 1 8)   Sing:((2 4 3 1 5 9 7 6 8))
B_Perm:(8 1 3 4 5 6 7 9 2)   Sing:((3 1 2 6 5 4 8 9 7))
B_Perm:(8 1 3 6 5 4 7 9 2)   Sing:((3 1 2 6 5 4 8 9 7))
B_Perm:(8 1 7 4 5 6 3 9 2)   Sing:((4 1 3 2 5 8 7 9 6))
B_Perm:(8 9 3 4 5 6 7 1 2)   Sing:((1 8 7 4 5 6 3 2 9)(8 3 1 6 5 4 9 7 2)
                                  (3 9 2 6 5 4 8 1 7)(3 8 6 1 5 9 4 2 7)
                                  (3 8 4 9 5 1 6 2 7)(6 8 1 3 5 7 9 2 4)
                                  (6 3 2 9 5 1 8 7 4)(4 8 1 7 5 3 9 2 6)
                                  (4 8 3 1 5 9 7 2 6))
B_Perm:(8 9 3 6 5 4 7 1 2)   Sing:((1 8 7 6 5 4 3 2 9)(8 3 1 6 5 4 9 7 2)
                                  (3 9 2 6 5 4 8 1 7)(3 8 6 9 5 1 4 2 7)
                                  (6 8 1 7 5 3 9 2 4)(6 3 2 9 5 1 8 7 4))
B_Perm:(8 9 7 4 5 6 3 1 2)   Sing:((1 8 7 6 5 4 3 2 9)(8 4 3 1 5 9 7 6 2)
                                  (4 9 3 2 5 8 7 1 6)(4 8 7 1 5 9 3 2 6))

B_Perm:(8 9 7 6 5 4 3 1 2)   Sing:((1 8 7 6 5 4 3 2 9))
B_Perm:(1 7 8 4 5 6 2 3 9)   Sing:((1 2 7 6 5 4 3 8 9)(1 4 7 2 5 8 3 6 9))
B_Perm:(1 7 8 6 5 4 2 3 9)   Sing:((1 2 7 6 5 4 3 8 9))
B_Perm:(9 3 2 4 5 6 8 7 1)   Sing:((3 2 1 6 5 4 9 8 7))
B_Perm:(9 3 2 6 5 4 8 7 1)   Sing:((3 2 1 6 5 4 9 8 7))
B_Perm:(9 3 8 4 5 6 2 7 1)   Sing:((3 2 9 6 5 4 1 8 7)(4 3 2 1 5 9 8 7 6))
B_Perm:(9 3 8 6 5 4 2 7 1)   Sing:((3 2 9 6 5 4 1 8 7))
B_Perm:(9 7 2 4 5 6 8 3 1)   Sing:((7 2 1 6 5 4 9 8 3)(4 3 2 1 5 9 8 7 6))
B_Perm:(9 7 2 6 5 4 8 3 1)   Sing:((7 2 1 6 5 4 9 8 3))
B_Perm:(9 7 8 4 5 6 2 3 1)   Sing:((9 2 7 6 5 4 3 8 1)(9 4 7 2 5 8 3 6 1)
                                   (7 6 9 2 5 8 1 4 3)(7 4 9 8 5 2 1 6 3))
B_Perm:(9 7 8 6 5 4 2 3 1)   Sing:((9 2 7 6 5 4 3 8 1)(7 6 9 8 5 2 1 4 3))
B_Perm:(3 9 2 4 5 6 8 1 7)   Sing:((3 2 1 6 5 4 9 8 7))
B_Perm:(3 9 2 6 5 4 8 1 7)   Sing:((3 2 1 6 5 4 9 8 7))
B_Perm:(3 9 8 4 5 6 2 1 7)   Sing:((3 4 2 1 5 9 8 6 7))
B_Perm:(7 1 8 4 5 6 2 9 3)   Sing:((2 1 7 6 5 4 3 9 8)(4 1 7 2 5 8 3 9 6))
B_Perm:(7 1 8 6 5 4 2 9 3)   Sing:((2 1 7 6 5 4 3 9 8))
B_Perm:(7 9 2 4 5 6 8 1 3)   Sing:((2 7 6 1 5 9 4 3 8)(2 7 4 9 5 1 6 3 8)
                                   (7 2 1 6 5 4 9 8 3)(4 7 2 1 5 9 8 3 6))
B_Perm:(7 9 2 6 5 4 8 1 3)   Sing:((2 7 6 9 5 1 4 3 8)(7 2 1 6 5 4 9 8 3))
B_Perm:(7 9 8 4 5 6 2 1 3)   Sing:((2 9 7 6 5 4 3 1 8)(7 4 2 1 5 9 8 6 3)
                                   (4 9 7 2 5 8 3 1 6))
B_Perm:(7 9 8 6 5 4 2 1 3)   Sing:((2 9 7 6 5 4 3 1 8))
B_Perm:(2 7 9 4 5 6 1 3 8)   Sing:((2 1 7 6 5 4 3 9 8)(2 4 7 1 5 9 3 6 8))
B_Perm:(2 7 9 6 5 4 1 3 8)   Sing:((2 1 7 6 5 4 3 9 8))
B_Perm:(8 3 1 4 5 6 9 7 2)   Sing:((3 2 1 6 5 4 9 8 7))
B_Perm:(8 3 1 6 5 4 9 7 2)   Sing:((3 2 1 6 5 4 9 8 7))
B_Perm:(8 3 9 4 5 6 1 7 2)   Sing:((3 2 9 6 5 4 1 8 7)(6 1 8 3 5 7 2 9 4)
                                   (4 1 8 7 5 3 2 9 6)(4 3 8 1 5 9 2 7 6))
B_Perm:(8 3 9 6 5 4 1 7 2)   Sing:((3 2 9 6 5 4 1 8 7)(6 1 8 7 5 3 2 9 4))
B_Perm:(8 7 1 4 5 6 9 3 2)   Sing:((4 3 1 2 5 8 9 7 6))
B_Perm:(8 7 9 4 5 6 1 3 2)   Sing:((8 1 7 6 5 4 3 9 2)(8 4 7 1 5 9 3 6 2)
                                   (4 3 9 2 5 8 1 7 6))
B_Perm:(8 7 9 6 5 4 1 3 2)   Sing:((8 1 7 6 5 4 3 9 2))
B_Perm:(3 8 1 4 5 6 9 2 7)   Sing:((1 3 2 6 5 4 8 7 9))
B_Perm:(3 8 1 6 5 4 9 2 7)   Sing:((1 3 2 6 5 4 8 7 9))
B_Perm:(3 8 9 4 5 6 1 2 7)   Sing:((1 6 8 3 5 7 2 4 9)(1 4 8 7 5 3 2 6 9)
                                   (3 1 8 6 5 4 2 9 7)(3 4 8 1 5 9 2 6 7))
B_Perm:(3 8 9 6 5 4 1 2 7)   Sing:((1 6 8 7 5 3 2 4 9)(3 1 8 6 5 4 2 9 7))
B_Perm:(7 2 9 4 5 6 1 8 3)   Sing:((2 1 7 6 5 4 3 9 8)(4 2 7 1 5 9 3 8 6))
B_Perm:(7 2 9 6 5 4 1 8 3)   Sing:((2 1 7 6 5 4 3 9 8))
B_Perm:(7 8 1 4 5 6 9 2 3)   Sing:((1 7 6 2 5 8 4 3 9)(1 7 4 8 5 2 6 3 9)
                                   (2 7 1 6 5 4 9 3 8)(4 7 1 2 5 8 9 3 6))
B_Perm:(7 8 1 6 5 4 9 2 3)   Sing:((1 7 6 8 5 2 4 3 9)(2 7 1 6 5 4 9 3 8))

B_Perm:(7 8 9 4 5 6 1 2 3)   Sing:((1 8 7 6 5 4 3 2 9)(2 7 9 6 5 4 1 3 8)
                             (7 1 8 6 5 4 2 9 3)(7 4 8 1 5 9 2 6 3)
                             (4 8 7 1 5 9 3 2 6)(4 7 9 2 5 8 1 3 6))
B_Perm:(7 8 9 6 5 4 1 2 3)   Sing:((1 8 7 6 5 4 3 2 9)(2 7 9 6 5 4 1 3 8)
                             (7 1 8 6 5 4 2 9 3))
B_Perm:(1 2 6 7 5 3 4 8 9)   Sing:((1 2 3 6 5 4 7 8 9))
B_Perm:(1 8 4 7 5 3 6 2 9)   Sing:((1 4 3 8 5 2 7 6 9))
B_Perm:(1 8 6 3 5 7 4 2 9)   Sing:((1 6 3 2 5 8 7 4 9))
B_Perm:(1 8 6 7 5 3 4 2 9)   Sing:((1 8 3 6 5 4 7 2 9))
B_Perm:(9 2 4 3 5 7 6 8 1)   Sing:((2 1 7 4 5 6 3 9 8))
B_Perm:(9 2 4 7 5 3 6 8 1)   Sing:((2 1 7 6 5 4 3 9 8)(4 2 3 9 5 1 7 8 6))
B_Perm:(9 2 6 3 5 7 4 8 1)   Sing:((2 1 7 6 5 4 3 9 8)(6 2 3 1 5 9 7 8 4))
B_Perm:(9 2 6 7 5 3 4 8 1)   Sing:((9 2 3 6 5 4 7 8 1))
B_Perm:(9 8 4 3 5 7 6 2 1)   Sing:((4 3 2 1 5 9 8 7 6))
B_Perm:(9 8 4 7 5 3 6 2 1)   Sing:((9 4 3 8 5 2 7 6 1))
B_Perm:(9 8 6 3 5 7 4 2 1)   Sing:((9 6 3 2 5 8 7 4 1))
B_Perm:(9 8 6 7 5 3 4 2 1)   Sing:((9 8 3 6 5 4 7 2 1))
B_Perm:(2 1 6 7 5 3 4 9 8)   Sing:((2 1 3 6 5 4 7 9 8))
B_Perm:(2 9 4 7 5 3 6 1 8)   Sing:((2 4 3 9 5 1 7 6 8))
B_Perm:(2 9 6 3 5 7 4 1 8)   Sing:((2 6 3 1 5 9 7 4 8))
B_Perm:(2 9 6 7 5 3 4 1 8)   Sing:((2 9 3 6 5 4 7 1 8))
B_Perm:(8 1 4 7 5 3 6 9 2)   Sing:((4 1 3 8 5 2 7 9 6))
B_Perm:(8 1 6 3 5 7 4 9 2)   Sing:((6 1 3 2 5 8 7 9 4))
B_Perm:(8 1 6 7 5 3 4 9 2)   Sing:((8 1 3 6 5 4 7 9 2))
B_Perm:(8 9 4 3 5 7 6 1 2)   Sing:((4 8 7 1 5 9 3 2 6))
B_Perm:(8 9 4 7 5 3 6 1 2)   Sing:((8 4 3 9 5 1 7 6 2)(4 9 3 8 5 2 7 1 6)
                             (4 8 7 9 5 1 3 2 6))
B_Perm:(8 9 6 3 5 7 4 1 2)   Sing:((8 6 3 1 5 9 7 4 2)(6 9 3 2 5 8 7 1 4)
                             (6 8 7 1 5 9 3 2 4))
B_Perm:(8 9 6 7 5 3 4 1 2)   Sing:((8 9 3 6 5 4 7 1 2)(6 8 7 9 5 1 3 2 4))
B_Perm:(1 6 2 7 5 3 8 4 9)   Sing:((1 3 2 6 5 4 8 7 9))
B_Perm:(1 6 8 3 5 7 2 4 9)   Sing:((1 3 6 2 5 8 4 7 9))
B_Perm:(1 6 8 7 5 3 2 4 9)   Sing:((1 3 8 6 5 4 2 7 9))
B_Perm:(9 4 2 3 5 7 8 6 1)   Sing:((7 2 1 4 5 6 9 8 3))
B_Perm:(9 4 2 7 5 3 8 6 1)   Sing:((7 2 1 6 5 4 9 8 3)(4 3 2 9 5 1 8 7 6))
B_Perm:(9 4 8 3 5 7 2 6 1)   Sing:((4 3 9 8 5 2 1 7 6))
B_Perm:(9 4 8 7 5 3 2 6 1)   Sing:((4 3 9 8 5 2 1 7 6))
B_Perm:(9 6 2 3 5 7 8 4 1)   Sing:((7 2 1 6 5 4 9 8 3)(6 3 2 1 5 9 8 7 4))
B_Perm:(9 6 2 7 5 3 8 4 1)   Sing:((9 3 2 6 5 4 8 7 1))
B_Perm:(9 6 8 3 5 7 2 4 1)   Sing:((9 3 6 2 5 8 4 7 1)(6 3 9 8 5 2 1 7 4))
B_Perm:(9 6 8 7 5 3 2 4 1)   Sing:((9 3 8 6 5 4 2 7 1))
B_Perm:(4 9 2 3 5 7 8 1 6)   Sing:((2 7 1 4 5 6 9 3 8)(2 4 3 9 5 1 7 6 8)
                             (4 2 1 7 5 3 9 8 6))
B_Perm:(4 9 2 7 5 3 8 1 6)   Sing:((2 7 1 6 5 4 9 3 8)(2 4 3 9 5 1 7 6 8)
                             (4 2 1 7 5 3 9 8 6))
B_Perm:(4 9 8 3 5 7 2 1 6)   Sing:((4 3 2 1 5 9 8 7 6))

B_Perm:(6 1 2 7 5 3 8 9 4) Sing:((3 1 2 6 5 4 8 9 7))
B_Perm:(6 1 8 3 5 7 2 9 4) Sing:((3 1 6 2 5 8 4 9 7))
B_Perm:(6 1 8 7 5 3 2 9 4) Sing:((3 1 8 6 5 4 2 9 7))
B_Perm:(6 9 2 3 5 7 8 1 4) Sing:((2 7 1 6 5 4 9 3 8)(2 6 3 9 5 1 7 4 8)
                                  (3 6 2 1 5 9 8 4 7)(6 2 1 7 5 3 9 8 4))
B_Perm:(6 9 2 7 5 3 8 1 4) Sing:((3 9 2 6 5 4 8 1 7)(6 2 1 7 5 3 9 8 4))
B_Perm:(6 9 8 3 5 7 2 1 4) Sing:((3 9 6 2 5 8 4 1 7)(6 3 2 1 5 9 8 7 4))
B_Perm:(6 9 8 7 5 3 2 1 4) Sing:((3 9 8 6 5 4 2 1 7))
B_Perm:(2 6 1 7 5 3 9 4 8) Sing:((2 3 1 6 5 4 9 7 8))
B_Perm:(2 6 9 3 5 7 1 4 8) Sing:((2 3 6 1 5 9 4 7 8))
B_Perm:(2 6 9 7 5 3 1 4 8) Sing:((2 3 9 6 5 4 1 7 8))
B_Perm:(8 4 1 7 5 3 9 6 2) Sing:((4 3 1 8 5 2 9 7 6))
B_Perm:(8 4 9 3 5 7 1 6 2) Sing:((4 1 8 7 5 3 2 9 6)(4 3 8 1 5 9 2 7 6))
B_Perm:(8 4 9 7 5 3 1 6 2) Sing:((4 3 9 8 5 2 1 7 6))
B_Perm:(8 6 1 3 5 7 9 4 2) Sing:((6 3 1 2 5 8 9 7 4))
B_Perm:(8 6 1 7 5 3 9 4 2) Sing:((8 3 1 6 5 4 9 7 2))
B_Perm:(8 6 9 3 5 7 1 4 2) Sing:((8 3 6 1 5 9 4 7 2)(6 1 8 7 5 3 2 9 4)
                                  (6 3 9 2 5 8 1 7 4))
B_Perm:(8 6 9 7 5 3 1 4 2) Sing:((8 3 9 6 5 4 1 7 2))
B_Perm:(4 2 9 3 5 7 1 8 6) Sing:((2 1 7 4 5 6 3 9 8))
B_Perm:(4 2 9 7 5 3 1 8 6) Sing:((2 1 7 6 5 4 3 9 8))
B_Perm:(4 8 1 3 5 7 9 2 6) Sing:((1 4 3 8 5 2 7 6 9))
B_Perm:(4 8 1 7 5 3 9 2 6) Sing:((1 4 3 8 5 2 7 6 9))
B_Perm:(4 8 9 3 5 7 1 2 6) Sing:((4 1 8 7 5 3 2 9 6)(4 3 8 1 5 9 2 7 6))
B_Perm:(4 8 9 7 5 3 1 2 6) Sing:((4 1 8 7 5 3 2 9 6))
B_Perm:(6 2 1 7 5 3 9 8 4) Sing:((3 2 1 6 5 4 9 8 7))
B_Perm:(6 2 9 3 5 7 1 8 4) Sing:((2 1 7 6 5 4 3 9 8)(3 2 6 1 5 9 4 8 7))
B_Perm:(6 2 9 7 5 3 1 8 4) Sing:((3 2 9 6 5 4 1 8 7))
B_Perm:(6 8 1 3 5 7 9 2 4) Sing:((1 6 3 8 5 2 7 4 9)(3 6 1 2 5 8 9 4 7))
B_Perm:(6 8 1 7 5 3 9 2 4) Sing:((3 8 1 6 5 4 9 2 7))
B_Perm:(6 8 9 3 5 7 1 2 4) Sing:((3 8 6 1 5 9 4 2 7)(3 6 9 2 5 8 1 4 7)
                                  (6 1 8 7 5 3 2 9 4)(6 3 8 1 5 9 2 7 4))
B_Perm:(6 8 9 7 5 3 1 2 4) Sing:((3 8 9 6 5 4 1 2 7)(6 1 8 7 5 3 2 9 4))
B_Perm:(1 3 6 8 5 2 4 7 9) Sing:((1 3 2 6 5 4 8 7 9))
B_Perm:(1 7 4 8 5 2 6 3 9) Sing:((1 4 2 7 5 3 8 6 9))
B_Perm:(1 7 6 8 5 2 4 3 9) Sing:((1 7 2 6 5 4 8 3 9))
B_Perm:(9 3 4 2 5 8 6 7 1) Sing:((3 2 9 4 5 6 1 8 7))
B_Perm:(9 3 4 8 5 2 6 7 1) Sing:((3 2 9 6 5 4 1 8 7)(4 3 2 9 5 1 8 7 6))
B_Perm:(9 3 6 2 5 8 4 7 1) Sing:((3 2 9 6 5 4 1 8 7)(6 3 2 1 5 9 8 7 4))
B_Perm:(9 3 6 8 5 2 4 7 1) Sing:((9 3 2 6 5 4 8 7 1))
B_Perm:(9 7 4 2 5 8 6 3 1) Sing:((7 6 2 1 5 9 8 4 3))
B_Perm:(9 7 4 8 5 2 6 3 1) Sing:((9 4 2 7 5 3 8 6 1)(7 6 2 9 5 1 8 4 3))
B_Perm:(9 7 6 2 5 8 4 3 1) Sing:((7 6 2 1 5 9 8 4 3))
B_Perm:(9 7 6 8 5 2 4 3 1) Sing:((9 7 2 6 5 4 8 3 1))
B_Perm:(3 1 6 8 5 2 4 9 7) Sing:((3 1 2 6 5 4 8 9 7))
B_Perm:(3 9 4 8 5 2 6 1 7) Sing:((3 4 2 9 5 1 8 6 7))

B_Perm:(3 9 6 2 5 8 4 1 7)   Sing:((3 6 2 1 5 9 8 4 7))
B_Perm:(3 9 6 8 5 2 4 1 7)   Sing:((3 9 2 6 5 4 8 1 7))
B_Perm:(7 1 4 8 5 2 6 9 3)   Sing:((4 1 2 7 5 3 8 9 6))
B_Perm:(7 1 6 8 5 2 4 9 3)   Sing:((7 1 2 6 5 4 8 9 3))
B_Perm:(7 9 4 2 5 8 6 1 3)   Sing:((2 7 6 1 5 9 4 3 8)(4 7 2 1 5 9 8 3 6))
B_Perm:(7 9 4 8 5 2 6 1 3)   Sing:((2 7 6 9 5 1 4 3 8)(7 4 2 9 5 1 8 6 3)
                                    (4 9 2 7 5 3 8 1 6))
B_Perm:(7 9 6 2 5 8 4 1 3)   Sing:((7 6 2 1 5 9 8 4 3))
B_Perm:(7 9 6 8 5 2 4 1 3)   Sing:((7 9 2 6 5 4 8 1 3))
B_Perm:(1 4 7 2 5 8 3 6 9)   Sing:((1 2 4 3 5 7 6 8 9))
B_Perm:(1 4 7 8 5 2 3 6 9)   Sing:((1 4 2 7 5 3 8 6 9))
B_Perm:(1 6 3 8 5 2 7 4 9)   Sing:((1 3 2 6 5 4 8 7 9))
B_Perm:(1 6 7 2 5 8 3 4 9)   Sing:((1 3 6 2 5 8 4 7 9))
B_Perm:(1 6 7 8 5 2 3 4 9)   Sing:((1 2 7 6 5 4 3 8 9)(1 3 6 8 5 2 4 7 9)
                                    (1 6 2 7 5 3 8 4 9))
B_Perm:(9 4 3 8 5 2 7 6 1)   Sing:((4 3 2 9 5 1 8 7 6))
B_Perm:(9 4 7 2 5 8 3 6 1)   Sing:((9 2 4 3 5 7 6 8 1)(7 2 6 1 5 9 4 8 3)
                                    (7 4 2 1 5 9 8 6 3)(4 2 9 7 5 3 1 8 6)
                                    (4 3 9 2 5 8 1 7 6))
B_Perm:(9 4 7 8 5 2 3 6 1)   Sing:((9 4 2 7 5 3 8 6 1)(7 2 6 9 5 1 4 8 3)
                                    (4 3 9 8 5 2 1 7 6))
B_Perm:(9 6 3 2 5 8 7 4 1)   Sing:((6 3 2 1 5 9 8 7 4))
B_Perm:(9 6 3 8 5 2 7 4 1)   Sing:((9 3 2 6 5 4 8 7 1))
B_Perm:(9 6 7 2 5 8 3 4 1)   Sing:((9 3 6 2 5 8 4 7 1)(7 6 2 1 5 9 8 4 3)
                                    (6 2 9 7 5 3 1 8 4))
B_Perm:(9 6 7 8 5 2 3 4 1)   Sing:((9 2 7 6 5 4 3 8 1)(9 3 6 8 5 2 4 7 1)
                                    (9 6 2 7 5 3 8 4 1))
B_Perm:(4 1 7 2 5 8 3 9 6)   Sing:((2 1 4 3 5 7 6 9 8))
B_Perm:(4 1 7 8 5 2 3 9 6)   Sing:((4 1 2 7 5 3 8 9 6))
B_Perm:(4 9 3 8 5 2 7 1 6)   Sing:((4 3 2 9 5 1 8 7 6))
B_Perm:(4 9 7 2 5 8 3 1 6)   Sing:((2 9 4 3 5 7 6 1 8)(2 7 6 1 5 9 4 3 8)
                                    (4 7 2 1 5 9 8 3 6))
B_Perm:(4 9 7 8 5 2 3 1 6)   Sing:((2 7 6 9 5 1 4 3 8)(4 9 2 7 5 3 8 1 6))
B_Perm:(6 1 3 8 5 2 7 9 4)   Sing:((3 1 2 6 5 4 8 9 7))
B_Perm:(6 1 7 2 5 8 3 9 4)   Sing:((3 1 6 2 5 8 4 9 7))
B_Perm:(6 1 7 8 5 2 3 9 4)   Sing:((2 1 7 6 5 4 3 9 8)(3 1 6 8 5 2 4 9 7)
                                    (6 1 2 7 5 3 8 9 4))
B_Perm:(6 9 3 2 5 8 7 1 4)   Sing:((3 6 2 1 5 9 8 4 7))
B_Perm:(6 9 3 8 5 2 7 1 4)   Sing:((3 9 2 6 5 4 8 1 7)(6 3 2 9 5 1 8 7 4))
B_Perm:(6 9 7 2 5 8 3 1 4)   Sing:((3 9 6 2 5 8 4 1 7)(6 7 2 1 5 9 8 3 4))
B_Perm:(6 9 7 8 5 2 3 1 4)   Sing:((2 9 7 6 5 4 3 1 8)(3 9 6 8 5 2 4 1 7)
                                    (6 9 2 7 5 3 8 1 4))
B_Perm:(3 4 1 2 5 8 9 6 7)   Sing:((1 3 2 4 5 6 8 7 9))
B_Perm:(3 4 1 8 5 2 9 6 7)   Sing:((1 3 2 6 5 4 8 7 9))
B_Perm:(3 6 1 2 5 8 9 4 7)   Sing:((1 3 2 6 5 4 8 7 9))
B_Perm:(3 6 1 8 5 2 9 4 7)   Sing:((3 2 1 6 5 4 9 8 7))

B_Perm:(3 6 9 2 5 8 1 4 7) Sing:((3 2 6 1 5 9 4 8 7))
B_Perm:(3 6 9 8 5 2 1 4 7) Sing:((3 2 9 6 5 4 1 8 7))
B_Perm:(7 4 1 8 5 2 9 6 3) Sing:((4 2 1 7 5 3 9 8 6))
B_Perm:(7 4 9 2 5 8 1 6 3) Sing:((2 1 7 6 5 4 3 9 8)(4 2 7 1 5 9 3 8 6))
B_Perm:(7 4 9 8 5 2 1 6 3) Sing:((4 2 9 7 5 3 1 8 6))
B_Perm:(7 6 1 8 5 2 9 4 3) Sing:((7 2 1 6 5 4 9 8 3))
B_Perm:(7 6 9 2 5 8 1 4 3) Sing:((7 2 6 1 5 9 4 8 3))
B_Perm:(7 6 9 8 5 2 1 4 3) Sing:((7 2 9 6 5 4 1 8 3))
B_Perm:(4 7 1 2 5 8 9 3 6) Sing:((1 4 2 7 5 3 8 6 9)(1 4 3 2 5 8 7 6 9)
                                      (2 4 1 3 5 7 9 6 8))
B_Perm:(4 7 1 8 5 2 9 3 6) Sing:((1 4 3 8 5 2 7 6 9)(4 2 1 7 5 3 9 8 6))
B_Perm:(4 7 9 2 5 8 1 3 6) Sing:((2 1 7 6 5 4 3 9 8)(2 4 9 3 5 7 1 6 8)
                                      (4 2 7 1 5 9 3 8 6))
B_Perm:(4 7 9 8 5 2 1 3 6) Sing:((4 2 9 7 5 3 1 8 6))
B_Perm:(6 3 1 8 5 2 9 7 4) Sing:((3 2 1 6 5 4 9 8 7))
B_Perm:(6 3 9 2 5 8 1 7 4) Sing:((3 2 6 1 5 9 4 8 7))
B_Perm:(6 3 9 8 5 2 1 7 4) Sing:((3 2 9 6 5 4 1 8 7))
B_Perm:(6 7 1 2 5 8 9 3 4) Sing:((1 6 2 7 5 3 8 4 9)(3 6 1 2 5 8 9 4 7))
B_Perm:(6 7 1 8 5 2 9 3 4) Sing:((2 7 1 6 5 4 9 3 8)(3 6 1 8 5 2 9 4 7)
                                      (6 2 1 7 5 3 9 8 4))
B_Perm:(6 7 9 2 5 8 1 3 4) Sing:((2 7 6 1 5 9 4 3 8)(3 6 9 2 5 8 1 4 7)
                                      (6 2 7 1 5 9 3 8 4))
B_Perm:(6 7 9 8 5 2 1 3 4) Sing:((2 7 9 6 5 4 1 3 8)(3 6 9 8 5 2 1 4 7)
                                      (6 2 9 7 5 3 1 8 4))
B_Perm:(2 3 6 9 5 1 4 7 8) Sing:((2 3 1 6 5 4 9 7 8))
B_Perm:(2 7 4 9 5 1 6 3 8) Sing:((2 4 1 7 5 3 9 6 8))
B_Perm:(2 7 6 9 5 1 4 3 8) Sing:((2 7 1 6 5 4 9 3 8))
B_Perm:(8 3 4 1 5 9 6 7 2) Sing:((3 1 8 4 5 6 2 9 7)(3 2 6 1 5 9 4 8 7)
                                      (6 1 3 2 5 8 7 9 4))
B_Perm:(8 3 4 9 5 1 6 7 2) Sing:((3 1 8 6 5 4 2 9 7)(3 2 6 9 5 1 4 8 7)
                                      (6 1 3 8 5 2 7 9 4)(4 3 1 8 5 2 9 7 6))
B_Perm:(8 3 6 1 5 9 4 7 2) Sing:((3 1 8 6 5 4 2 9 7)(3 2 6 1 5 9 4 8 7)
                                      (6 1 3 2 5 8 7 9 4))
B_Perm:(8 3 6 9 5 1 4 7 2) Sing:((8 3 1 6 5 4 9 7 2)(3 2 6 9 5 1 4 8 7))
B_Perm:(8 7 4 1 5 9 6 3 2) Sing:((4 3 2 1 5 9 8 7 6))
B_Perm:(8 7 4 9 5 1 6 3 2) Sing:((8 4 1 7 5 3 9 6 2)(4 3 2 9 5 1 8 7 6))
B_Perm:(8 7 6 9 5 1 4 3 2) Sing:((8 7 1 6 5 4 9 3 2))
B_Perm:(3 2 6 9 5 1 4 8 7) Sing:((3 2 1 6 5 4 9 8 7))
B_Perm:(3 8 4 1 5 9 6 2 7) Sing:((1 6 3 2 5 8 7 4 9))
B_Perm:(3 8 4 9 5 1 6 2 7) Sing:((1 6 3 8 5 2 7 4 9)(3 4 1 8 5 2 9 6 7))
B_Perm:(3 8 6 1 5 9 4 2 7) Sing:((1 6 3 2 5 8 7 4 9))
B_Perm:(3 8 6 9 5 1 4 2 7) Sing:((3 8 1 6 5 4 9 2 7))
B_Perm:(7 2 4 1 5 9 6 8 3) Sing:((2 1 7 4 5 6 3 9 8))
B_Perm:(7 2 4 9 5 1 6 8 3) Sing:((2 1 7 6 5 4 3 9 8)(4 2 1 7 5 3 9 8 6))
B_Perm:(7 2 6 1 5 9 4 8 3) Sing:((2 1 7 6 5 4 3 9 8))
B_Perm:(7 2 6 9 5 1 4 8 3) Sing:((7 2 1 6 5 4 9 8 3))

B_Perm:(7 8 4 1 5 9 6 2 3)　Sing:((2 7 6 1 5 9 4 3 8)(4 7 2 1 5 9 8 3 6))
B_Perm:(7 8 4 9 5 1 6 2 3)　Sing:((2 7 6 9 5 1 4 3 8)(7 4 1 8 5 2 9 6 3)
　　　　　　　　　　　　　　　　(4 8 1 7 5 3 9 2 6)(4 7 2 9 5 1 8 3 6))
B_Perm:(7 8 6 1 5 9 4 2 3)　Sing:((2 7 6 1 5 9 4 3 8))
B_Perm:(7 8 6 9 5 1 4 2 3)　Sing:((2 7 6 9 5 1 4 3 8)(7 8 1 6 5 4 9 2 3))
B_Perm:(2 4 7 1 5 9 3 6 8)　Sing:((2 1 4 3 5 7 6 9 8))
B_Perm:(2 4 7 9 5 1 3 6 8)　Sing:((2 4 1 7 5 3 9 6 8))
B_Perm:(2 6 3 9 5 1 7 4 8)　Sing:((2 3 1 6 5 4 9 7 8))
B_Perm:(2 6 7 1 5 9 3 4 8)　Sing:((2 3 6 1 5 9 4 7 8))
B_Perm:(2 6 7 9 5 1 3 4 8)　Sing:((2 1 7 6 5 4 3 9 8)(2 3 6 9 5 1 4 7 8)
　　　　　　　　　　　　　　　　(2 6 1 7 5 3 9 4 8))

B_Perm:(8 4 3 9 5 1 7 6 2)　Sing:((4 3 1 8 5 2 9 7 6))
B_Perm:(8 4 7 1 5 9 3 6 2)　Sing:((8 1 4 3 5 7 6 9 2)(4 1 8 7 5 3 2 9 6)
　　　　　　　　　　　　　　　　(4 3 8 1 5 9 2 7 6))
B_Perm:(8 4 7 9 5 1 3 6 2)　Sing:((8 4 1 7 5 3 9 6 2)(4 3 8 9 5 1 2 7 6))
B_Perm:(8 6 3 1 5 9 7 4 2)　Sing:((6 3 2 1 5 9 8 7 4))
B_Perm:(8 6 3 9 5 1 7 4 2)　Sing:((8 3 1 6 5 4 9 7 2)(6 3 2 9 5 1 8 7 4))
B_Perm:(8 6 7 1 5 9 3 4 2)　Sing:((8 3 6 1 5 9 4 7 2)(6 1 8 7 5 3 2 9 4))
B_Perm:(8 6 7 9 5 1 3 4 2)　Sing:((8 1 7 6 5 4 3 9 2)(8 3 6 9 5 1 4 7 2)
　　　　　　　　　　　　　　　　(8 6 1 7 5 3 9 4 2))

B_Perm:(4 2 3 1 5 9 7 8 6)　Sing:((2 1 4 3 5 7 6 9 8))
B_Perm:(4 2 3 9 5 1 7 8 6)　Sing:((2 1 4 7 5 3 6 9 8))
B_Perm:(4 2 7 1 5 9 3 8 6)　Sing:((2 1 4 3 5 7 6 9 8))
B_Perm:(4 2 7 9 5 1 3 8 6)　Sing:((4 2 1 7 5 3 9 8 6))
B_Perm:(4 8 3 1 5 9 7 2 6)　Sing:((1 4 3 2 5 8 7 6 9))
B_Perm:(4 8 3 9 5 1 7 2 6)　Sing:((4 3 1 8 5 2 9 7 6))
B_Perm:(4 8 7 1 5 9 3 2 6)　Sing:((1 8 4 3 5 7 6 2 9))
B_Perm:(4 8 7 9 5 1 3 2 6)　Sing:((4 8 1 7 5 3 9 2 6))
B_Perm:(6 2 3 1 5 9 7 8 4)　Sing:((2 1 6 3 5 7 4 9 8))
B_Perm:(6 2 3 9 5 1 7 8 4)　Sing:((2 1 6 7 5 3 4 9 8)(3 2 1 6 5 4 9 8 7))
B_Perm:(6 2 7 1 5 9 3 8 4)　Sing:((3 2 6 1 5 9 4 8 7))
B_Perm:(6 2 7 9 5 1 3 8 4)　Sing:((2 1 7 6 5 4 3 9 8)(3 2 6 9 5 1 4 8 7)
　　　　　　　　　　　　　　　　(6 2 1 7 5 3 9 8 4))
B_Perm:(6 8 3 1 5 9 7 2 4)　Sing:((3 6 2 1 5 9 8 4 7))
B_Perm:(6 8 3 9 5 1 7 2 4)　Sing:((3 8 1 6 5 4 9 2 7)(3 6 2 9 5 1 8 4 7)
　　　　　　　　　　　　　　　　(6 3 1 8 5 2 9 7 4))
B_Perm:(6 8 7 1 5 9 3 2 4)　Sing:((3 8 6 1 5 9 4 2 7))
B_Perm:(6 8 7 9 5 1 3 2 4)　Sing:((1 8 7 6 5 4 3 2 9)(3 8 6 9 5 1 4 2 7)
　　　　　　　　　　　　　　　　(6 8 1 7 5 3 9 2 4))
B_Perm:(3 4 8 1 5 9 2 6 7)　Sing:((1 3 6 2 5 8 4 7 9)(1 4 3 2 5 8 7 6 9)
　　　　　　　　　　　　　　　　(3 1 4 2 5 8 6 9 7))
B_Perm:(3 4 8 9 5 1 2 6 7)　Sing:((1 3 6 8 5 2 4 7 9)(3 4 1 8 5 2 9 6 7))
B_Perm:(3 6 2 9 5 1 8 4 7)　Sing:((3 2 1 6 5 4 9 8 7))
B_Perm:(3 6 8 1 5 9 2 4 7)　Sing:((1 6 3 2 5 8 7 4 9)(3 2 6 1 5 9 4 8 7))
B_Perm:(3 6 8 9 5 1 2 4 7)　Sing:((3 1 8 6 5 4 2 9 7)(3 2 6 9 5 1 4 8 7)
　　　　　　　　　　　　　　　　(3 6 1 8 5 2 9 4 7))

B_Perm:(7 4 2 9 5 1 8 6 3) Sing:((4 2 1 7 5 3 9 8 6))
B_Perm:(7 4 8 1 5 9 2 6 3) Sing:((2 1 7 6 5 4 3 9 8)(7 1 4 2 5 8 6 9 3)
                                        (4 2 7 1 5 9 3 8 6))
B_Perm:(7 4 8 9 5 1 2 6 3) Sing:((7 4 1 8 5 2 9 6 3)(4 1 8 7 5 3 2 9 6)
                                        (4 2 7 9 5 1 3 8 6))
B_Perm:(7 6 2 9 5 1 8 4 3) Sing:((7 2 1 6 5 4 9 8 3))
B_Perm:(7 6 8 1 5 9 2 4 3) Sing:((7 2 6 1 5 9 4 8 3))
B_Perm:(7 6 8 9 5 1 2 4 3) Sing:((7 1 8 6 5 4 2 9 3)(7 2 6 9 5 1 4 8 3)
                                        (7 6 1 8 5 2 9 4 3))
B_Perm:(4 3 8 1 5 9 2 7 6) Sing:((4 1 3 2 5 8 7 9 6))
B_Perm:(4 3 8 9 5 1 2 7 6) Sing:((4 3 1 8 5 2 9 7 6))
B_Perm:(4 7 2 1 5 9 8 3 6) Sing:((2 4 3 1 5 9 7 6 8))
B_Perm:(4 7 2 9 5 1 8 3 6) Sing:((2 4 3 9 5 1 7 6 8)(4 2 1 7 5 3 9 8 6))
B_Perm:(4 7 8 1 5 9 2 3 6) Sing:((1 7 4 2 5 8 6 3 9)(1 4 8 3 5 7 2 6 9)
                                        (2 1 7 6 5 4 3 9 8)(4 2 7 1 5 9 3 8 6))
B_Perm:(4 7 8 9 5 1 2 3 6) Sing:((4 1 8 7 5 3 2 9 6)(4 2 7 9 5 1 3 8 6)
                                        (4 7 1 8 5 2 9 3 6))
B_Perm:(6 3 2 9 5 1 8 7 4) Sing:((3 2 1 6 5 4 9 8 7))
B_Perm:(6 3 8 1 5 9 2 7 4) Sing:((3 2 6 1 5 9 4 8 7)(6 1 3 2 5 8 7 9 4))
B_Perm:(6 3 8 9 5 1 2 7 4) Sing:((3 1 8 6 5 4 2 9 7)(3 2 6 9 5 1 4 8 7)
                                        (6 3 1 8 5 2 9 7 4))
B_Perm:(6 7 2 1 5 9 8 3 4) Sing:((3 6 2 1 5 9 8 4 7))
B_Perm:(6 7 2 9 5 1 8 3 4) Sing:((2 7 1 6 5 4 9 3 8)(3 6 2 9 5 1 8 4 7)
                                        (6 2 1 7 5 3 9 8 4))
B_Perm:(6 7 8 1 5 9 2 3 4) Sing:((2 7 6 1 5 9 4 3 8)(3 6 8 1 5 9 2 4 7)
                                        (6 2 7 1 5 9 3 8 4))
B_Perm:(6 7 8 9 5 1 2 3 4) Sing:((1 7 8 6 5 4 2 3 9)(2 7 6 9 5 1 4 3 8)
                                        (3 6 8 9 5 1 2 4 7)(6 1 8 7 5 3 2 9 4)
                                        (6 2 7 9 5 1 3 8 4)(6 7 1 8 5 2 9 3 4))

13.4.3. Singular locus for C_4.

C_Perm:(1 2 6 4 5 3 7 8) Sing:((1 2 4 3 6 5 7 8))
C_Perm:(1 7 3 4 5 6 2 8) Sing:((1 3 2 5 4 7 6 8)(1 4 3 2 7 6 5 8))
C_Perm:(1 7 3 5 4 6 2 8) Sing:((1 3 2 5 4 7 6 8))
C_Perm:(1 7 6 4 5 3 2 8) Sing:((1 7 4 3 6 5 2 8))
C_Perm:(8 2 3 4 5 6 7 1) Sing:((2 1 6 4 5 3 8 7)(3 2 1 5 4 8 7 6)
                                        (4 2 3 1 8 6 7 5))
C_Perm:(8 2 3 5 4 6 7 1) Sing:((2 1 6 5 4 3 8 7)(3 2 1 5 4 8 7 6))
C_Perm:(8 2 6 4 5 3 7 1) Sing:((8 2 4 3 6 5 7 1)(2 1 6 5 4 3 8 7))
C_Perm:(8 2 6 5 4 3 7 1) Sing:((2 1 6 5 4 3 8 7))
C_Perm:(8 7 3 4 5 6 2 1) Sing:((8 3 2 5 4 7 6 1)(8 4 3 2 7 6 5 1))
C_Perm:(8 7 3 5 4 6 2 1) Sing:((8 3 2 5 4 7 6 1))
C_Perm:(8 7 6 4 5 3 2 1) Sing:((8 7 4 3 6 5 2 1))
C_Perm:(2 1 6 4 5 3 8 7) Sing:((2 1 4 3 6 5 8 7))
C_Perm:(2 8 3 4 5 6 1 7) Sing:((2 3 1 5 4 8 6 7)(2 4 3 1 8 6 5 7))
C_Perm:(2 8 3 5 4 6 1 7) Sing:((2 3 1 5 4 8 6 7))

C_Perm:(2 8 6 4 5 3 1 7)   Sing:((2 8 4 3 6 5 1 7))
C_Perm:(7 1 3 4 5 6 8 2)   Sing:((3 1 2 5 4 7 8 6)(4 1 3 2 7 6 8 5))
C_Perm:(7 1 3 5 4 6 8 2)   Sing:((3 1 2 5 4 7 8 6))
C_Perm:(7 1 6 4 5 3 8 2)   Sing:((7 1 4 3 6 5 8 2))
C_Perm:(7 8 3 4 5 6 1 2)   Sing:((1 7 6 4 5 3 2 8)(7 3 1 5 4 8 6 2)
                                 (7 4 3 1 8 6 5 2)(3 8 2 5 4 7 1 6)
                                 (3 7 5 1 8 4 2 6)(3 7 4 8 1 5 2 6)
                                 (5 7 1 3 6 8 2 4)(5 3 2 8 1 7 6 4)
                                 (4 8 3 2 7 6 1 5)(4 7 1 6 3 8 2 5))
C_Perm:(7 8 3 5 4 6 1 2)   Sing:((1 7 6 5 4 3 2 8)(7 3 1 5 4 8 6 2)
                                 (3 8 2 5 4 7 1 6)(3 7 5 8 1 4 2 6)
                                 (5 7 1 6 3 8 2 4)(5 3 2 8 1 7 6 4))
C_Perm:(7 8 6 4 5 3 1 2)   Sing:((1 7 6 5 4 3 2 8)(7 8 4 3 6 5 1 2))
C_Perm:(1 3 7 4 5 2 6 8)   Sing:((1 3 4 2 7 5 6 8))
C_Perm:(1 6 2 4 5 7 3 8)   Sing:((1 4 2 3 6 7 5 8))
C_Perm:(1 6 7 4 5 2 3 8)   Sing:((1 2 6 5 4 3 7 8)(1 6 4 2 7 5 3 8)
                                 (1 4 7 3 6 2 5 8))
C_Perm:(8 3 2 4 5 7 6 1)   Sing:((3 2 1 5 4 8 7 6)(4 3 2 1 8 7 6 5))
C_Perm:(8 3 2 5 4 7 6 1)   Sing:((3 2 1 5 4 8 7 6))
C_Perm:(8 3 7 4 5 2 6 1)   Sing:((8 3 4 2 7 5 6 1)(3 2 8 5 4 1 7 6))
C_Perm:(8 3 7 5 4 2 6 1)   Sing:((3 2 8 5 4 1 7 6))
C_Perm:(8 6 2 4 5 7 3 1)   Sing:((8 4 2 3 6 7 5 1)(6 2 1 5 4 8 7 3))
C_Perm:(8 6 2 5 4 7 3 1)   Sing:((6 2 1 5 4 8 7 3))
C_Perm:(8 6 7 4 5 2 3 1)   Sing:((8 2 6 5 4 3 7 1)(8 6 4 2 7 5 3 1)
                                 (8 4 7 3 6 2 5 1)(6 5 8 2 7 1 4 3)
                                 (6 4 8 7 2 1 5 3))
C_Perm:(8 6 7 5 4 2 3 1)   Sing:((6 5 8 7 2 1 4 3))
C_Perm:(3 1 7 4 5 2 8 6)   Sing:((3 1 4 2 7 5 8 6))
C_Perm:(3 8 2 4 5 7 1 6)   Sing:((3 2 1 5 4 8 7 6)(3 4 2 1 8 7 5 6))
C_Perm:(3 8 2 5 4 7 1 6)   Sing:((3 2 1 5 4 8 7 6))
C_Perm:(3 8 7 4 5 2 1 6)   Sing:((3 8 4 2 7 5 1 6))
C_Perm:(6 1 2 4 5 7 8 3)   Sing:((4 1 2 3 6 7 8 5))
C_Perm:(6 1 7 4 5 2 8 3)   Sing:((2 1 6 5 4 3 8 7)(6 1 4 2 7 5 8 3)
                                 (4 1 7 3 6 2 8 5))
C_Perm:(6 8 2 4 5 7 1 3)   Sing:((2 6 5 1 8 4 3 7)(2 6 4 8 1 5 3 7)
                                 (6 2 1 5 4 8 7 3)(6 4 2 1 8 7 5 3)
                                 (4 8 2 3 6 7 1 5))
C_Perm:(6 8 2 5 4 7 1 3)   Sing:((2 6 5 8 1 4 3 7)(6 2 1 5 4 8 7 3))
C_Perm:(6 8 7 4 5 2 1 3)   Sing:((2 8 6 5 4 3 1 7)(6 8 4 2 7 5 1 3)
                                 (4 8 7 3 6 2 1 5))
C_Perm:(2 3 8 4 5 1 6 7)   Sing:((2 3 4 1 8 5 6 7))
C_Perm:(2 6 1 4 5 8 3 7)   Sing:((2 4 1 3 6 8 5 7))
C_Perm:(2 6 8 4 5 1 3 7)   Sing:((2 1 6 5 4 3 8 7)(2 6 4 1 8 5 3 7)
                                 (2 4 8 3 6 1 5 7))
C_Perm:(7 3 1 4 5 8 6 2)   Sing:((3 2 1 5 4 8 7 6)(4 3 1 2 7 8 6 5))
C_Perm:(7 3 1 5 4 8 6 2)   Sing:((3 2 1 5 4 8 7 6))

C_Perm:(7 3 8 4 5 1 6 2) Sing:((7 3 4 1 8 5 6 2)(3 2 8 5 4 1 7 6)
(5 1 7 3 6 2 8 4)(4 1 7 6 3 2 8 5)
(4 3 8 2 7 1 6 5))

C_Perm:(7 3 8 5 4 1 6 2) Sing:((3 2 8 5 4 1 7 6)(5 1 7 6 3 2 8 4))
C_Perm:(7 6 1 4 5 8 3 2) Sing:((7 4 1 3 6 8 5 2))
C_Perm:(7 6 8 4 5 1 3 2) Sing:((7 1 6 5 4 3 8 2)(7 6 4 1 8 5 3 2)
(7 4 8 3 6 1 5 2))

C_Perm:(3 2 8 4 5 1 7 6) Sing:((3 2 4 1 8 5 7 6))
C_Perm:(3 7 1 4 5 8 2 6) Sing:((1 3 2 5 4 7 6 8)(3 4 1 2 7 8 5 6))
C_Perm:(3 7 1 5 4 8 2 6) Sing:((1 3 2 5 4 7 6 8))
C_Perm:(3 7 8 4 5 1 2 6) Sing:((1 5 7 3 6 2 4 8)(1 4 7 6 3 2 5 8)
(3 1 7 5 4 2 8 6)(3 7 4 1 8 5 2 6)
(3 4 8 2 7 1 5 6))

C_Perm:(3 7 8 5 4 1 2 6) Sing:((1 5 7 6 3 2 4 8))
C_Perm:(6 2 1 4 5 8 7 3) Sing:((4 2 1 3 6 8 7 5))
C_Perm:(6 2 8 4 5 1 7 3) Sing:((2 1 6 5 4 3 8 7)(6 2 4 1 8 5 7 3)
(4 2 8 3 6 1 7 5))

C_Perm:(6 2 8 5 4 1 7 3) Sing:((2 1 6 5 4 3 8 7))
C_Perm:(6 7 1 4 5 8 2 3) Sing:((1 6 5 2 7 4 3 8)(1 6 4 7 2 5 3 8)
(2 6 1 5 4 8 3 7)(6 4 1 2 7 8 5 3)
(4 7 1 3 6 8 2 5))

C_Perm:(6 7 1 5 4 8 2 3) Sing:((1 6 5 7 2 4 3 8))
C_Perm:(6 7 8 4 5 1 2 3) Sing:((1 7 6 5 4 3 2 8)(2 6 8 5 4 1 3 7)
(6 1 7 5 4 2 8 3)(6 7 4 1 8 5 2 3)
(6 4 8 2 7 1 5 3)(4 7 8 3 6 1 2 5))

C_Perm:(6 7 8 5 4 1 2 3) Sing:((1 2 6 5 4 3 7 8))
C_Perm:(1 7 4 3 6 5 2 8) Sing:((1 4 3 2 7 6 5 8))
C_Perm:(1 7 4 6 3 5 2 8) Sing:((1 4 3 7 2 6 5 8))
C_Perm:(1 7 5 3 6 4 2 8) Sing:((1 5 3 2 7 6 4 8))
C_Perm:(1 7 5 6 3 4 2 8) Sing:((1 5 3 7 2 6 4 8))
C_Perm:(8 2 4 3 6 5 7 1) Sing:((2 1 6 4 5 3 8 7)(4 2 3 1 8 6 7 5))
C_Perm:(8 2 4 6 3 5 7 1) Sing:((2 1 6 5 4 3 8 7)(4 2 3 8 1 6 7 5))
C_Perm:(8 2 5 3 6 4 7 1) Sing:((2 1 6 5 4 3 8 7)(5 2 3 1 8 6 7 4))
C_Perm:(8 2 5 6 3 4 7 1) Sing:((2 1 6 5 4 3 8 7)(5 2 3 8 1 6 7 4))
C_Perm:(8 7 4 3 6 5 2 1) Sing:((8 4 3 2 7 6 5 1))
C_Perm:(8 7 4 6 3 5 2 1) Sing:((8 4 3 7 2 6 5 1))
C_Perm:(8 7 5 3 6 4 2 1) Sing:((8 5 3 2 7 6 4 1))
C_Perm:(8 7 5 6 3 4 2 1) Sing:((8 5 3 7 2 6 4 1))
C_Perm:(2 8 4 3 6 5 1 7) Sing:((2 4 3 1 8 6 5 7))
C_Perm:(2 8 4 6 3 5 1 7) Sing:((2 4 3 8 1 6 5 7))
C_Perm:(2 8 5 3 6 4 1 7) Sing:((2 5 3 1 8 6 4 7))
C_Perm:(2 8 5 6 3 4 1 7) Sing:((2 5 3 8 1 6 4 7))
C_Perm:(7 1 4 3 6 5 8 2) Sing:((4 1 3 2 7 6 8 5))
C_Perm:(7 1 4 6 3 5 8 2) Sing:((4 1 3 7 2 6 8 5))
C_Perm:(7 1 5 3 6 4 8 2) Sing:((5 1 3 2 7 6 8 4))
C_Perm:(7 1 5 6 3 4 8 2) Sing:((5 1 3 7 2 6 8 4))

C_Perm:(7 8 4 3 6 5 1 2)   Sing:((7 4 3 1 8 6 5 2)(4 8 3 2 7 6 1 5)
                            (4 7 6 1 8 3 2 5))
C_Perm:(7 8 4 6 3 5 1 2)   Sing:((7 4 3 8 1 6 5 2)(4 8 3 7 2 6 1 5)
                            (4 7 6 8 1 3 2 5))
C_Perm:(7 8 5 3 6 4 1 2)   Sing:((7 5 3 1 8 6 4 2)(5 8 3 2 7 6 1 4)
                            (5 7 6 1 8 3 2 4))
C_Perm:(7 8 5 6 3 4 1 2)   Sing:((7 5 3 8 1 6 4 2)(5 8 3 7 2 6 1 4)
                            (5 7 6 8 1 3 2 4))
C_Perm:(1 4 7 3 6 2 5 8)   Sing:((1 4 3 2 7 6 5 8))
C_Perm:(1 5 7 3 6 2 4 8)   Sing:((1 5 3 2 7 6 4 8))
C_Perm:(8 4 2 3 6 7 5 1)   Sing:((6 2 1 4 5 8 7 3)(4 3 2 1 8 7 6 5))
C_Perm:(8 4 2 6 3 7 5 1)   Sing:((6 2 1 5 4 8 7 3)(4 3 2 8 1 7 6 5))
C_Perm:(8 4 7 3 6 2 5 1)   Sing:((8 4 3 2 7 6 5 1)(4 3 8 7 2 1 6 5))
C_Perm:(8 4 7 6 3 2 5 1)   Sing:((4 3 8 7 2 1 6 5))
C_Perm:(8 5 2 3 6 7 4 1)   Sing:((6 2 1 5 4 8 7 3)(5 3 2 1 8 7 6 4))
C_Perm:(8 5 2 6 3 7 4 1)   Sing:((6 2 1 5 4 8 7 3)(5 3 2 8 1 7 6 4))
C_Perm:(8 5 7 3 6 2 4 1)   Sing:((8 5 3 2 7 6 4 1)(5 3 8 7 2 1 6 4))
C_Perm:(8 5 7 6 3 2 4 1)   Sing:((5 3 8 7 2 1 6 4))
C_Perm:(4 1 7 3 6 2 8 5)   Sing:((4 1 3 2 7 6 8 5))
C_Perm:(4 8 2 3 6 7 1 5)   Sing:((2 6 1 4 5 8 3 7)(2 4 3 8 1 6 5 7)
                            (4 2 1 6 3 8 7 5)(4 3 2 1 8 7 6 5))
C_Perm:(4 8 2 6 3 7 1 5)   Sing:((2 6 1 5 4 8 3 7)(2 4 3 8 1 6 5 7)
                            (4 2 1 6 3 8 7 5))
C_Perm:(4 8 7 3 6 2 1 5)   Sing:((4 8 3 2 7 6 1 5))
C_Perm:(5 1 7 3 6 2 8 4)   Sing:((5 1 3 2 7 6 8 4))
C_Perm:(5 8 2 3 6 7 1 4)   Sing:((2 6 1 5 4 8 3 7)(2 5 3 8 1 6 4 7)
                            (5 2 1 6 3 8 7 4)(5 3 2 1 8 7 6 4))
C_Perm:(5 8 2 6 3 7 1 4)   Sing:((2 6 1 5 4 8 3 7)(2 5 3 8 1 6 4 7)
                            (5 2 1 6 3 8 7 4))
C_Perm:(5 8 7 3 6 2 1 4)   Sing:((5 8 3 2 7 6 1 4))
C_Perm:(2 4 8 3 6 1 5 7)   Sing:((2 4 3 1 8 6 5 7))
C_Perm:(2 5 8 3 6 1 4 7)   Sing:((2 5 3 1 8 6 4 7))
C_Perm:(7 4 1 3 6 8 5 2)   Sing:((4 3 1 2 7 8 6 5))
C_Perm:(7 4 1 6 3 8 5 2)   Sing:((4 3 1 7 2 8 6 5))
C_Perm:(7 4 8 3 6 1 5 2)   Sing:((7 4 3 1 8 6 5 2)(4 1 7 6 3 2 8 5)
                            (4 3 8 2 7 1 6 5))
C_Perm:(7 4 8 6 3 1 5 2)   Sing:((4 3 8 7 2 1 6 5))
C_Perm:(7 5 1 3 6 8 4 2)   Sing:((5 3 1 2 7 8 6 4))
C_Perm:(7 5 1 6 3 8 4 2)   Sing:((5 3 1 7 2 8 6 4))
C_Perm:(7 5 8 3 6 1 4 2)   Sing:((7 5 3 1 8 6 4 2)(5 1 7 6 3 2 8 4)
                            (5 3 8 2 7 1 6 4))
C_Perm:(7 5 8 6 3 1 4 2)   Sing:((5 3 8 7 2 1 6 4))
C_Perm:(4 2 8 3 6 1 7 5)   Sing:((2 1 6 4 5 3 8 7)(4 2 3 1 8 6 7 5))
C_Perm:(4 2 8 6 3 1 7 5)   Sing:((2 1 6 5 4 3 8 7))
C_Perm:(4 7 1 3 6 8 2 5)   Sing:((1 4 3 7 2 6 5 8)(4 3 1 2 7 8 6 5))
C_Perm:(4 7 1 6 3 8 2 5)   Sing:((1 4 3 7 2 6 5 8))

C_Perm:(4 7 8 3 6 1 2 5)   Sing:((4 1 7 6 3 2 8 5)(4 7 3 1 8 6 2 5)
                                   (4 3 8 2 7 1 6 5))
C_Perm:(4 7 8 6 3 1 2 5)   Sing:((1 4 7 6 3 2 5 8))
C_Perm:(5 2 8 3 6 1 7 4)   Sing:((2 1 6 5 4 3 8 7)(5 2 3 1 8 6 7 4))
C_Perm:(5 2 8 6 3 1 7 4)   Sing:((2 1 6 5 4 3 8 7))
C_Perm:(5 7 1 3 6 8 2 4)   Sing:((1 5 3 7 2 6 4 8)(5 3 1 2 7 8 6 4))
C_Perm:(5 7 1 6 3 8 2 4)   Sing:((1 5 3 7 2 6 4 8))
C_Perm:(5 7 8 3 6 1 2 4)   Sing:((5 1 7 6 3 2 8 4)(5 7 3 1 8 6 2 4)
                                   (5 3 8 2 7 1 6 4))
C_Perm:(5 7 8 6 3 1 2 4)   Sing:((1 3 7 5 4 2 6 8))
C_Perm:(1 6 4 2 7 5 3 8)   Sing:((1 4 3 2 7 6 5 8))
C_Perm:(1 6 4 7 2 5 3 8)   Sing:((1 4 3 7 2 6 5 8))
C_Perm:(8 3 4 2 7 5 6 1)   Sing:((3 2 8 4 5 1 7 6)(4 3 2 1 8 7 6 5))
C_Perm:(8 3 4 7 2 5 6 1)   Sing:((3 2 8 5 4 1 7 6)(4 3 2 8 1 7 6 5))
C_Perm:(8 3 5 2 7 4 6 1)   Sing:((3 2 8 5 4 1 7 6)(5 3 2 1 8 7 6 4))
C_Perm:(8 3 5 7 2 4 6 1)   Sing:((3 2 8 5 4 1 7 6)(5 3 2 8 1 7 6 4))
C_Perm:(8 6 4 2 7 5 3 1)   Sing:((8 4 3 2 7 6 5 1)(6 5 2 1 8 7 4 3))
C_Perm:(8 6 4 7 2 5 3 1)   Sing:((8 4 3 7 2 6 5 1)(6 5 2 8 1 7 4 3))
C_Perm:(8 6 5 2 7 4 3 1)   Sing:((6 5 2 1 8 7 4 3))
C_Perm:(8 6 5 7 2 4 3 1)   Sing:((6 5 2 8 1 7 4 3))
C_Perm:(3 8 4 2 7 5 1 6)   Sing:((3 4 2 1 8 7 5 6))
C_Perm:(3 8 4 7 2 5 1 6)   Sing:((3 4 2 8 1 7 5 6))
C_Perm:(3 8 5 2 7 4 1 6)   Sing:((3 5 2 1 8 7 4 6))
C_Perm:(3 8 5 7 2 4 1 6)   Sing:((3 5 2 8 1 7 4 6))
C_Perm:(6 1 4 2 7 5 8 3)   Sing:((4 1 3 2 7 6 8 5))
C_Perm:(6 1 4 7 2 5 8 3)   Sing:((4 1 3 7 2 6 8 5))
C_Perm:(6 8 4 2 7 5 1 3)   Sing:((2 6 5 1 8 4 3 7)(6 4 2 1 8 7 5 3)
                                   (4 8 3 2 7 6 1 5))
C_Perm:(6 8 4 7 2 5 1 3)   Sing:((2 6 5 8 1 4 3 7)(6 4 2 8 1 7 5 3)
                                   (4 8 3 7 2 6 1 5))
C_Perm:(6 8 5 2 7 4 1 3)   Sing:((6 5 2 1 8 7 4 3))
C_Perm:(6 8 5 7 2 4 1 3)   Sing:((6 5 2 8 1 7 4 3))
C_Perm:(1 4 6 2 7 3 5 8)   Sing:((1 2 4 3 6 5 7 8))
C_Perm:(1 4 6 7 2 3 5 8)   Sing:((1 2 4 6 3 5 7 8))
C_Perm:(1 5 3 7 2 6 4 8)   Sing:((1 3 2 5 4 7 6 8))
C_Perm:(1 5 6 2 7 3 4 8)   Sing:((1 2 5 3 6 4 7 8))
C_Perm:(1 5 6 7 2 3 4 8)   Sing:((1 2 3 5 4 6 7 8))
C_Perm:(8 4 3 2 7 6 5 1)   Sing:((4 3 2 1 8 7 6 5))
C_Perm:(8 4 3 7 2 6 5 1)   Sing:((4 3 2 8 1 7 6 5))
C_Perm:(8 4 6 2 7 3 5 1)   Sing:((8 2 4 3 6 5 7 1)(6 2 5 1 8 4 7 3)
                                   (6 4 2 1 8 7 5 3)(4 2 8 6 3 1 7 5)
                                   (4 3 8 2 7 1 6 5))
C_Perm:(8 4 6 7 2 3 5 1)   Sing:((8 2 4 6 3 5 7 1)(6 2 5 8 1 4 7 3)
                                   (6 4 2 8 1 7 5 3)(4 3 8 7 2 1 6 5))
C_Perm:(8 5 3 2 7 6 4 1)   Sing:((5 3 2 1 8 7 6 4))
C_Perm:(8 5 3 7 2 6 4 1)   Sing:((8 3 2 5 4 7 6 1))

C_Perm:(8 5 6 2 7 3 4 1) Sing:((8 2 5 3 6 4 7 1)(6 5 2 1 8 7 4 3)
                                (5 2 8 6 3 1 7 4)(5 3 8 2 7 1 6 4))
C_Perm:(8 5 6 7 2 3 4 1) Sing:((8 2 3 5 4 6 7 1)(6 5 2 8 1 7 4 3)
                                (5 3 8 7 2 1 6 4))
C_Perm:(4 1 6 2 7 3 8 5) Sing:((2 1 4 3 6 5 8 7))
C_Perm:(4 1 6 7 2 3 8 5) Sing:((2 1 4 6 3 5 8 7))
C_Perm:(4 8 3 2 7 6 1 5) Sing:((4 3 2 1 8 7 6 5))
C_Perm:(4 8 3 7 2 6 1 5) Sing:((4 3 2 8 1 7 6 5))
C_Perm:(4 8 6 2 7 3 1 5) Sing:((2 8 4 3 6 5 1 7)(2 6 5 1 8 4 3 7)
                                (4 6 2 1 8 7 3 5))
C_Perm:(4 8 6 7 2 3 1 5) Sing:((2 8 4 6 3 5 1 7)(2 6 5 8 1 4 3 7)
                                (4 6 2 8 1 7 3 5))
C_Perm:(5 1 3 7 2 6 8 4) Sing:((3 1 2 5 4 7 8 6))
C_Perm:(5 1 6 2 7 3 8 4) Sing:((2 1 5 3 6 4 8 7))
C_Perm:(5 1 6 7 2 3 8 4) Sing:((2 1 3 5 4 6 8 7))
C_Perm:(5 8 3 2 7 6 1 4) Sing:((5 3 2 1 8 7 6 4))
C_Perm:(5 8 3 7 2 6 1 4) Sing:((3 8 2 5 4 7 1 6)(5 3 2 8 1 7 6 4))
C_Perm:(5 8 6 2 7 3 1 4) Sing:((2 8 5 3 6 4 1 7)(5 6 2 1 8 7 3 4))
C_Perm:(5 8 6 7 2 3 1 4) Sing:((2 8 3 5 4 6 1 7)(5 6 2 8 1 7 3 4))
C_Perm:(3 4 1 2 7 8 5 6) Sing:((1 3 2 4 5 7 6 8))
C_Perm:(3 4 1 7 2 8 5 6) Sing:((1 3 2 5 4 7 6 8))
C_Perm:(3 4 8 2 7 1 5 6) Sing:((3 4 2 1 8 7 5 6))
C_Perm:(3 5 1 2 7 8 4 6) Sing:((1 3 2 5 4 7 6 8))
C_Perm:(3 5 1 7 2 8 4 6) Sing:((1 3 2 5 4 7 6 8))
C_Perm:(3 5 8 2 7 1 4 6) Sing:((3 5 2 1 8 7 4 6))
C_Perm:(6 4 1 2 7 8 5 3) Sing:((4 3 1 2 7 8 6 5))
C_Perm:(6 4 1 7 2 8 5 3) Sing:((4 3 1 7 2 8 6 5))
C_Perm:(6 4 8 2 7 1 5 3) Sing:((2 1 6 5 4 3 8 7)(6 4 2 1 8 7 5 3)
                                (4 3 8 2 7 1 6 5))
C_Perm:(6 4 8 7 2 1 5 3) Sing:((4 3 8 7 2 1 6 5))
C_Perm:(6 5 8 2 7 1 4 3) Sing:((6 5 2 1 8 7 4 3))
C_Perm:(6 5 8 7 2 1 4 3) Sing:((2 1 6 5 4 3 8 7))
C_Perm:(4 3 8 2 7 1 6 5) Sing:((4 3 2 1 8 7 6 5))
C_Perm:(4 6 1 2 7 8 3 5) Sing:((1 4 2 6 3 7 5 8)(1 4 3 2 7 6 5 8)
                                (2 4 1 3 6 8 5 7))
C_Perm:(4 6 1 7 2 8 3 5) Sing:((1 4 3 7 2 6 5 8)(2 4 1 6 3 8 5 7))
C_Perm:(4 6 8 2 7 1 3 5) Sing:((2 1 6 5 4 3 8 7)(2 4 8 3 6 1 5 7)
                                (4 6 2 1 8 7 3 5))
C_Perm:(4 6 8 7 2 1 3 5) Sing:((2 4 8 6 3 1 5 7))
C_Perm:(5 3 1 7 2 8 6 4) Sing:((3 2 1 5 4 8 7 6))
C_Perm:(5 3 8 2 7 1 6 4) Sing:((5 3 2 1 8 7 6 4))
C_Perm:(5 3 8 7 2 1 6 4) Sing:((3 2 8 5 4 1 7 6))
C_Perm:(5 6 1 2 7 8 3 4) Sing:((1 5 2 6 3 7 4 8)(1 5 3 2 7 6 4 8)
                                (2 5 1 3 6 8 4 7))
C_Perm:(5 6 1 7 2 8 3 4) Sing:((1 5 3 7 2 6 4 8)(2 3 1 5 4 8 6 7))
C_Perm:(5 6 8 2 7 1 3 4) Sing:((2 5 8 3 6 1 4 7)(5 6 2 1 8 7 3 4))

C_Perm:(5 6 8 7 2 1 3 4)   Sing:((2 3 8 5 4 1 6 7))
C_Perm:(2 6 4 1 8 5 3 7)   Sing:((2 4 3 1 8 6 5 7))
C_Perm:(2 6 4 8 1 5 3 7)   Sing:((2 4 3 8 1 6 5 7))
C_Perm:(7 3 4 1 8 5 6 2)   Sing:((3 1 7 4 5 2 8 6)(3 2 5 1 8 4 7 6)
                        (5 1 3 2 7 6 8 4)(4 3 2 1 8 7 6 5))
C_Perm:(7 3 4 8 1 5 6 2)   Sing:((3 1 7 5 4 2 8 6)(3 2 5 8 1 4 7 6)
                        (5 1 3 7 2 6 8 4)(4 3 2 8 1 7 6 5))
C_Perm:(7 3 5 1 8 4 6 2)   Sing:((3 1 7 5 4 2 8 6)(3 2 5 1 8 4 7 6)
                        (5 1 3 2 7 6 8 4))
C_Perm:(7 3 5 8 1 4 6 2)   Sing:((3 1 7 5 4 2 8 6)(3 2 5 8 1 4 7 6)
                        (5 1 3 7 2 6 8 4))
C_Perm:(7 6 4 1 8 5 3 2)   Sing:((7 4 3 1 8 6 5 2))
C_Perm:(7 6 4 8 1 5 3 2)   Sing:((7 4 3 8 1 6 5 2))
C_Perm:(3 7 4 1 8 5 2 6)   Sing:((1 5 3 2 7 6 4 8)(3 4 2 1 8 7 5 6))
C_Perm:(3 7 4 8 1 5 2 6)   Sing:((1 5 3 7 2 6 4 8)(3 4 2 8 1 7 5 6))
C_Perm:(3 7 5 1 8 4 2 6)   Sing:((1 5 3 2 7 6 4 8))
C_Perm:(3 7 5 8 1 4 2 6)   Sing:((1 5 3 7 2 6 4 8))
C_Perm:(6 2 4 1 8 5 7 3)   Sing:((2 1 6 4 5 3 8 7)(4 2 3 1 8 6 7 5))
C_Perm:(6 2 4 8 1 5 7 3)   Sing:((2 1 6 5 4 3 8 7)(4 2 3 8 1 6 7 5))
C_Perm:(6 2 5 1 8 4 7 3)   Sing:((2 1 6 5 4 3 8 7))
C_Perm:(6 2 5 8 1 4 7 3)   Sing:((2 1 6 5 4 3 8 7))
C_Perm:(6 7 4 1 8 5 2 3)   Sing:((2 6 5 1 8 4 3 7)(6 4 2 1 8 7 5 3)
                        (4 7 3 1 8 6 2 5))
C_Perm:(6 7 4 8 1 5 2 3)   Sing:((2 6 5 8 1 4 3 7)(6 4 2 8 1 7 5 3)
                        (4 7 3 8 1 6 2 5))
C_Perm:(6 7 5 1 8 4 2 3)   Sing:((1 6 5 2 7 4 3 8))
C_Perm:(6 7 5 8 1 4 2 3)   Sing:((1 6 2 5 4 7 3 8))
C_Perm:(2 4 6 1 8 3 5 7)   Sing:((2 1 4 3 6 5 8 7))
C_Perm:(2 4 6 8 1 3 5 7)   Sing:((2 1 4 6 3 5 8 7))
C_Perm:(2 5 3 8 1 6 4 7)   Sing:((2 3 1 5 4 8 6 7))
C_Perm:(2 5 6 1 8 3 4 7)   Sing:((2 1 5 3 6 4 8 7))
C_Perm:(2 5 6 8 1 3 4 7)   Sing:((2 1 3 5 4 6 8 7))
C_Perm:(7 4 3 1 8 6 5 2)   Sing:((4 3 2 1 8 7 6 5))
C_Perm:(7 4 3 8 1 6 5 2)   Sing:((4 3 2 8 1 7 6 5))
C_Perm:(7 4 6 1 8 3 5 2)   Sing:((7 1 4 3 6 5 8 2)(4 1 7 6 3 2 8 5)
                        (4 3 7 1 8 2 6 5))
C_Perm:(7 4 6 8 1 3 5 2)   Sing:((7 1 4 6 3 5 8 2)(4 3 7 8 1 2 6 5))
C_Perm:(7 5 3 1 8 6 4 2)   Sing:((5 3 2 1 8 7 6 4))
C_Perm:(7 5 3 8 1 6 4 2)   Sing:((7 3 1 5 4 8 6 2)(5 3 2 8 1 7 6 4))
C_Perm:(7 5 6 1 8 3 4 2)   Sing:((7 1 5 3 6 4 8 2)(5 1 7 6 3 2 8 4)
                        (5 3 7 1 8 2 6 4))
C_Perm:(7 5 6 8 1 3 4 2)   Sing:((7 1 3 5 4 6 8 2)(5 3 7 8 1 2 6 4))
C_Perm:(4 2 3 1 8 6 7 5)   Sing:((2 1 4 3 6 5 8 7))
C_Perm:(4 2 3 8 1 6 7 5)   Sing:((2 1 4 6 3 5 8 7))
C_Perm:(4 2 6 1 8 3 7 5)   Sing:((2 1 4 3 6 5 8 7))
C_Perm:(4 2 6 8 1 3 7 5)   Sing:((2 1 4 6 3 5 8 7))

C_Perm:(4 7 3 1 8 6 2 5)   Sing:((4 3 2 1 8 7 6 5))
C_Perm:(4 7 3 8 1 6 2 5)   Sing:((4 3 2 8 1 7 6 5))
C_Perm:(4 7 6 1 8 3 2 5)   Sing:((1 7 4 3 6 5 2 8))
C_Perm:(4 7 6 8 1 3 2 5)   Sing:((1 7 4 6 3 5 2 8))
C_Perm:(5 2 3 1 8 6 7 4)   Sing:((2 1 5 3 6 4 8 7))
C_Perm:(5 2 3 8 1 6 7 4)   Sing:((2 1 5 6 3 4 8 7)(3 2 1 5 4 8 7 6))
C_Perm:(5 2 6 1 8 3 7 4)   Sing:((2 1 5 3 6 4 8 7))
C_Perm:(5 2 6 8 1 3 7 4)   Sing:((2 1 6 5 4 3 8 7))
C_Perm:(5 7 3 1 8 6 2 4)   Sing:((5 3 2 1 8 7 6 4))
C_Perm:(5 7 3 8 1 6 2 4)   Sing:((3 7 1 5 4 8 2 6)(5 3 2 8 1 7 6 4))
C_Perm:(5 7 6 1 8 3 2 4)   Sing:((1 7 5 3 6 4 2 8))
C_Perm:(5 7 6 8 1 3 2 4)   Sing:((1 7 3 5 4 6 2 8))
C_Perm:(3 4 7 1 8 2 5 6)   Sing:((1 3 5 2 7 4 6 8)(1 4 3 2 7 6 5 8)
                                (3 1 4 2 7 5 8 6))
C_Perm:(3 4 7 8 1 2 5 6)   Sing:((1 3 5 7 2 4 6 8)(1 4 3 7 2 6 5 8)
                                (3 1 4 7 2 5 8 6))
C_Perm:(3 5 2 8 1 7 4 6)   Sing:((3 2 1 5 4 8 7 6))
C_Perm:(3 5 7 1 8 2 4 6)   Sing:((1 5 3 2 7 6 4 8)(3 1 5 2 7 4 8 6))
C_Perm:(3 5 7 8 1 2 4 6)   Sing:((1 5 3 7 2 6 4 8)(3 1 2 5 4 7 8 6))
C_Perm:(6 4 2 1 8 7 5 3)   Sing:((4 3 2 1 8 7 6 5))
C_Perm:(6 4 2 8 1 7 5 3)   Sing:((4 3 2 8 1 7 6 5))
C_Perm:(6 4 7 1 8 2 5 3)   Sing:((2 1 6 5 4 3 8 7)(6 1 4 2 7 5 8 3)
                                (4 3 7 1 8 2 6 5))
C_Perm:(6 4 7 8 1 2 5 3)   Sing:((6 1 4 7 2 5 8 3)(4 3 7 8 1 2 6 5))
C_Perm:(6 5 2 8 1 7 4 3)   Sing:((6 2 1 5 4 8 7 3))
C_Perm:(6 5 7 1 8 2 4 3)   Sing:((6 1 5 2 7 4 8 3))
C_Perm:(6 5 7 8 1 2 4 3)   Sing:((6 1 2 5 4 7 8 3))
C_Perm:(4 3 7 1 8 2 6 5)   Sing:((4 1 3 2 7 6 8 5))
C_Perm:(4 3 7 8 1 2 6 5)   Sing:((4 1 3 7 2 6 8 5))
C_Perm:(4 6 2 1 8 7 3 5)   Sing:((2 4 3 1 8 6 5 7))
C_Perm:(4 6 2 8 1 7 3 5)   Sing:((2 4 3 8 1 6 5 7)(4 2 1 6 3 8 7 5))
C_Perm:(4 6 7 1 8 2 3 5)   Sing:((1 6 4 2 7 5 3 8)(1 4 7 3 6 2 5 8)
                                (2 1 6 5 4 3 8 7)(2 4 6 1 8 3 5 7)
                                (4 1 6 2 7 3 8 5))
C_Perm:(4 6 7 8 1 2 3 5)   Sing:((1 6 4 7 2 5 3 8)(1 4 7 6 3 2 5 8)
                                (2 4 6 8 1 3 5 7)(4 1 2 6 3 7 8 5))
C_Perm:(5 3 2 8 1 7 6 4)   Sing:((3 2 1 5 4 8 7 6))
C_Perm:(5 3 7 1 8 2 6 4)   Sing:((3 2 5 1 8 4 7 6)(5 1 3 2 7 6 8 4))
C_Perm:(5 3 7 8 1 2 6 4)   Sing:((3 1 7 5 4 2 8 6)(3 2 5 8 1 4 7 6)
                                (5 1 3 7 2 6 8 4))
C_Perm:(5 6 2 1 8 7 3 4)   Sing:((2 5 3 1 8 6 4 7))
C_Perm:(5 6 2 8 1 7 3 4)   Sing:((2 6 1 5 4 8 3 7)(2 5 3 8 1 6 4 7)
                                (5 2 1 6 3 8 7 4))
C_Perm:(5 6 7 1 8 2 3 4)   Sing:((1 6 5 2 7 4 3 8)(1 5 7 3 6 2 4 8)
                                (2 3 5 1 8 4 6 7)(5 1 6 2 7 3 8 4))

C_Perm:(5 6 7 8 1 2 3 4)　Sing:((1 6 2 5 4 7 3 8)(1 3 7 5 4 2 6 8)
　　　　　　　　　　　　　　　(1 5 3 7 2 6 4 8)(2 1 6 5 4 3 8 7)
　　　　　　　　　　　　　　　(2 3 5 8 1 4 6 7)(5 1 2 6 3 7 8 4))

13.4.4. Singular locus for D_4.

D_Perm:(1 7 3 5 4 6 2 8)　Sing:((1 3 2 4 5 7 6 8))
D_Perm:(8 2 3 5 4 6 7 1)　Sing:((2 1 6 5 4 3 8 7)(3 2 1 4 5 8 7 6))
D_Perm:(8 2 6 4 5 3 7 1)　Sing:((2 1 6 5 4 3 8 7))
D_Perm:(8 7 3 4 5 6 2 1)　Sing:((8 3 2 5 4 7 6 1))
D_Perm:(2 8 3 5 4 6 1 7)　Sing:((2 3 1 4 5 8 6 7))
D_Perm:(7 1 3 5 4 6 8 2)　Sing:((3 1 2 4 5 7 8 6))
D_Perm:(7 8 3 4 5 6 1 2)　Sing:((1 7 6 4 5 3 2 8)(7 3 1 5 4 8 6 2)
　　　　　　　　　　　　　　　(3 8 2 5 4 7 1 6)(3 7 5 1 8 4 2 6)
　　　　　　　　　　　　　　　(3 7 4 8 1 5 2 6)(5 7 1 3 6 8 2 4)
　　　　　　　　　　　　　　　(5 3 2 8 1 7 6 4)(4 7 1 6 3 8 2 5)
　　　　　　　　　　　　　　　(4 3 2 1 8 7 6 5))
D_Perm:(7 8 6 5 4 3 1 2)　Sing:((1 7 6 4 5 3 2 8))
D_Perm:(1 6 7 4 5 2 3 8)　Sing:((1 2 6 5 4 3 7 8))
D_Perm:(8 3 2 5 4 7 6 1)　Sing:((3 2 1 4 5 8 7 6))
D_Perm:(8 3 7 4 5 2 6 1)　Sing:((3 2 8 5 4 1 7 6))
D_Perm:(8 6 2 4 5 7 3 1)　Sing:((6 2 1 5 4 8 7 3))
D_Perm:(8 6 7 5 4 2 3 1)　Sing:((8 2 6 4 5 3 7 1)(6 5 8 7 2 1 4 3)
　　　　　　　　　　　　　　　(6 4 8 2 7 1 5 3))
D_Perm:(3 8 2 5 4 7 1 6)　Sing:((3 2 1 4 5 8 7 6))
D_Perm:(6 1 7 4 5 2 8 3)　Sing:((2 1 6 5 4 3 8 7))
D_Perm:(6 8 2 4 5 7 1 3)　Sing:((2 6 5 1 8 4 3 7)(2 6 4 8 1 5 3 7)
　　　　　　　　　　　　　　　(6 2 1 5 4 8 7 3))
D_Perm:(6 8 7 5 4 2 1 3)　Sing:((2 8 6 4 5 3 1 7))
D_Perm:(2 6 8 4 5 1 3 7)　Sing:((2 1 6 5 4 3 8 7))
D_Perm:(7 3 1 5 4 8 6 2)　Sing:((3 2 1 4 5 8 7 6))
D_Perm:(7 3 8 4 5 1 6 2)　Sing:((3 2 8 5 4 1 7 6)(5 1 7 3 6 2 8 4)
　　　　　　　　　　　　　　　(4 1 7 6 3 2 8 5))
D_Perm:(7 6 8 5 4 1 3 2)　Sing:((7 1 6 4 5 3 8 2))
D_Perm:(3 7 1 5 4 8 2 6)　Sing:((1 3 2 4 5 7 6 8))
D_Perm:(3 7 8 4 5 1 2 6)　Sing:((1 5 7 3 6 2 4 8)(1 4 7 6 3 2 5 8)
　　　　　　　　　　　　　　　(3 1 7 5 4 2 8 6))
D_Perm:(6 2 8 4 5 1 7 3)　Sing:((2 1 6 5 4 3 8 7))
D_Perm:(6 7 1 4 5 8 2 3)　Sing:((1 6 5 2 7 4 3 8)(1 6 4 7 2 5 3 8)
　　　　　　　　　　　　　　　(2 6 1 5 4 8 3 7))
D_Perm:(6 7 8 5 4 1 2 3)　Sing:((1 7 6 4 5 3 2 8)(2 6 8 4 5 1 3 7)
　　　　　　　　　　　　　　　(6 1 7 4 5 2 8 3))
D_Perm:(8 2 4 6 3 5 7 1)　Sing:((2 1 6 5 4 3 8 7))
D_Perm:(8 2 5 3 6 4 7 1)　Sing:((2 1 6 5 4 3 8 7))
D_Perm:(8 7 4 3 6 5 2 1)　Sing:((4 3 2 1 8 7 6 5))
D_Perm:(8 7 5 6 3 4 2 1)　Sing:((5 3 2 8 1 7 6 4))
D_Perm:(7 8 4 3 6 5 1 2)　Sing:((4 7 6 1 8 3 2 5))

D_Perm:(7 8 5 6 3 4 1 2) Sing:((5 7 6 8 1 3 2 4))
D_Perm:(8 4 2 6 3 7 5 1) Sing:((6 2 1 5 4 8 7 3))
D_Perm:(8 4 7 3 6 2 5 1) Sing:((4 3 8 7 2 1 6 5))
D_Perm:(8 5 2 3 6 7 4 1) Sing:((6 2 1 5 4 8 7 3))
D_Perm:(8 5 7 6 3 2 4 1) Sing:((5 3 8 2 7 1 6 4))
D_Perm:(4 8 2 6 3 7 1 5) Sing:((2 6 1 5 4 8 3 7)(2 4 3 1 8 6 5 7)
                                 (4 2 1 3 6 8 7 5))
D_Perm:(5 8 2 3 6 7 1 4) Sing:((2 6 1 5 4 8 3 7)(2 5 3 8 1 6 4 7)
                                 (5 2 1 6 3 8 7 4))
D_Perm:(7 4 8 3 6 1 5 2) Sing:((4 1 7 6 3 2 8 5))
D_Perm:(7 5 8 6 3 1 4 2) Sing:((5 1 7 3 6 2 8 4))
D_Perm:(4 2 8 6 3 1 7 5) Sing:((2 1 6 5 4 3 8 7))
D_Perm:(4 7 1 6 3 8 2 5) Sing:((1 4 3 2 7 6 5 8))
D_Perm:(4 7 8 3 6 1 2 5) Sing:((4 1 7 6 3 2 8 5))
D_Perm:(5 2 8 3 6 1 7 4) Sing:((2 1 6 5 4 3 8 7))
D_Perm:(5 7 1 3 6 8 2 4) Sing:((1 5 3 7 2 6 4 8))
D_Perm:(5 7 8 6 3 1 2 4) Sing:((5 1 7 3 6 2 8 4))
D_Perm:(8 3 4 7 2 5 6 1) Sing:((3 2 8 5 4 1 7 6))
D_Perm:(8 3 5 2 7 4 6 1) Sing:((3 2 8 5 4 1 7 6))
D_Perm:(8 6 4 2 7 5 3 1) Sing:((6 5 2 1 8 7 4 3))
D_Perm:(8 6 5 7 2 4 3 1) Sing:((6 4 2 8 1 7 5 3))
D_Perm:(6 8 4 2 7 5 1 3) Sing:((2 6 5 1 8 4 3 7))
D_Perm:(6 8 5 7 2 4 1 3) Sing:((2 6 4 8 1 5 3 7))
D_Perm:(8 4 6 2 7 3 5 1) Sing:((6 2 5 1 8 4 7 3)(4 2 8 6 3 1 7 5)
                                 (4 3 2 1 8 7 6 5))
D_Perm:(8 5 6 7 2 3 4 1) Sing:((6 2 4 8 1 5 7 3)(5 2 8 3 6 1 7 4)
                                 (5 3 2 8 1 7 6 4))
D_Perm:(4 8 6 2 7 3 1 5) Sing:((2 6 5 1 8 4 3 7))
D_Perm:(5 8 6 7 2 3 1 4) Sing:((2 6 4 8 1 5 3 7))
D_Perm:(3 4 1 2 7 8 5 6) Sing:((1 3 2 4 5 7 6 8))
D_Perm:(3 5 1 7 2 8 4 6) Sing:((1 3 2 4 5 7 6 8))
D_Perm:(6 4 8 2 7 1 5 3) Sing:((2 1 6 5 4 3 8 7))
D_Perm:(6 5 8 7 2 1 4 3) Sing:((2 1 6 5 4 3 8 7))
D_Perm:(4 6 1 7 2 8 3 5) Sing:((1 4 2 3 6 7 5 8))
D_Perm:(4 6 8 2 7 1 3 5) Sing:((2 1 6 5 4 3 8 7))
D_Perm:(5 6 1 2 7 8 3 4) Sing:((1 5 2 6 3 7 4 8))
D_Perm:(5 6 8 7 2 1 3 4) Sing:((2 1 6 5 4 3 8 7))
D_Perm:(7 3 4 8 1 5 6 2) Sing:((3 1 7 5 4 2 8 6)(3 2 5 8 1 4 7 6)
                                 (5 1 3 7 2 6 8 4))
D_Perm:(7 3 5 1 8 4 6 2) Sing:((3 1 7 5 4 2 8 6)(3 2 4 1 8 5 7 6)
                                 (4 1 3 2 7 6 8 5))
D_Perm:(3 7 4 8 1 5 2 6) Sing:((1 5 3 7 2 6 4 8))
D_Perm:(3 7 5 1 8 4 2 6) Sing:((1 4 3 2 7 6 5 8))
D_Perm:(6 2 4 8 1 5 7 3) Sing:((2 1 6 5 4 3 8 7))
D_Perm:(6 2 5 1 8 4 7 3) Sing:((2 1 6 5 4 3 8 7))
D_Perm:(6 7 4 1 8 5 2 3) Sing:((2 6 5 1 8 4 3 7))

D_Perm:(6 7 5 8 1 4 2 3)   Sing:((2 6 4 8 1 5 3 7))
D_Perm:(7 4 6 1 8 3 5 2)   Sing:((4 1 7 6 3 2 8 5))
D_Perm:(7 5 6 8 1 3 4 2)   Sing:((5 1 7 3 6 2 8 4))
D_Perm:(4 2 3 1 8 6 7 5)   Sing:((2 1 4 3 6 5 8 7))
D_Perm:(4 7 6 1 8 3 2 5)   Sing:((1 4 3 2 7 6 5 8))
D_Perm:(5 2 3 8 1 6 7 4)   Sing:((2 1 5 6 3 4 8 7))
D_Perm:(5 7 6 8 1 3 2 4)   Sing:((1 5 3 7 2 6 4 8))
D_Perm:(3 4 7 8 1 2 5 6)   Sing:((1 3 5 7 2 4 6 8))
D_Perm:(3 5 7 1 8 2 4 6)   Sing:((1 3 4 2 7 5 6 8))
D_Perm:(6 4 7 1 8 2 5 3)   Sing:((2 1 6 5 4 3 8 7))
D_Perm:(6 5 7 8 1 2 4 3)   Sing:((2 1 6 5 4 3 8 7))
D_Perm:(4 6 7 1 8 2 3 5)   Sing:((1 4 3 2 7 6 5 8)(2 1 6 5 4 3 8 7))
D_Perm:(5 6 7 8 1 2 3 4)   Sing:((1 5 3 7 2 6 4 8)(2 1 6 5 4 3 8 7))

Bibliography

[1] S. ABEASIS, A. DEL FRA, Degenerations for the representations of a quiver of type A_m, J. Alg., **93** (1985), 376–412.

[2] S. ABEASIS, A. DEL FRA AND H. KRAFT, The geometry of the representations of A_m, Math. Ann., **256** (1981), 401–418.

[3] S.S. ABHYANKAR, Enumerative combinatorics of Young tableaux, *Monographs and Textbooks in Pure and Applied Mathematics*, 115, Marcel Dekker, Inc., New York (1988).

[4] H.H. ANDERSEN, Schubert varieties and Demazure's character formula, *Invent. Math.*, **79** (1985), 611–618.

[5] A. ARABIA, Cohomologie **T**-équivariante de **G**/**B** pour un groupe **G** de Kac–Moody, *C.R.Acad. Sci.*, Paris Sr. I Math., **302** (1986), no. 17, 631–634.

[6] A. ARABIA, Cycles de Schubert et cohomologie équivariante de K/T, *Invent. Math.*, **85** (1986), 39–52.

[7] A. ARABIA, Cohomologie T-équivariante de la variété de drapeaux d'un groupe de Kac–Moody, *Bull. Soc. Math.*, France **117** (1989), 129–165.

[8] A. ARABIA, Classes d'Euler équivariantes et points rationnellement lisses, *Ann. Inst. Fourier*, **48** (1998), 861–912.

[9] A. BEILINSON, J. BERNSTEIN, Localization of g-modules, *C.R. Acad.Sci. Paris*, Ser. I Math, **292** (1981), pp. 15–18.

[10] I. BERNSTEIN, I. GELFAND, S. GELFAND, Structure of representations generated by highest weight vectors, *Funct. Anal. and Appl.*, **5** (1971), 1–8.

[11] I. BERNSTEIN, I. GELFAND, S. GELFAND, Schubert Cells and Cohomology of the Spaces G/P, *Russian Math. Surveys*, **28** (1973), 1–26.

[12] S.C. BILLEY, Kostant polynomials and the cohomology ring for G/B, *Duke Math. J.*, **96** (1999), 205–224.

[13] S.C. BILLEY, Pattern avoidance and rational smoothness of Schubert varieties, *Adv. in Math.*, **139** (1998), 141–156.

[14] S.C. BILLEY, C.K. FAN AND J. LOSONCZY, The parabolic map, *J. of Alg.*, **214** (1999), 1–7.

[15] S.C. BILLEY AND G.S. WARRINGTON, Kazhdan–Lusztig Polynomials for 321-hexagon-avoiding permutations, *J. Alg. Comb*, to appear.

[16] A. BJÖRNER AND F. BRENTI, An improved tableau criterion for Bruhat order, *Electron. J. Combin.*, **3** (1996), no. 1.

[17] B.D. BOE, Kazhdan–Lusztig polynomials for Hermitian symmetric spaces, *Trans. Amer. Math. Soc.*, **309**, (1988), 279–294.

[18] B.D. BOE, T.J. ENRIGHT AND B. SHELTON, Determination of the intertwining operators for holomorphically induced representations of hermitian symmetric pairs, *Pacific J. Math.*, **131** (1998), 39–50.

[19] M. BÓNA, The permutation classes equinumerous to the smooth class, *Elec. J. Combinatorics*, **5** (1998).

[20] A. BOREL, *Linear Algebraic Groups*, second edition, Springer-Verlag, New York, 1991.

[21] A. Borel, *Intersection Cohomology*, Birkhäuser, 1984.

[22] R. BOTT AND H. SAMELSON, Application of the theory of Morse to symmetric spaces, *Amer. J. Math.*, **80** (1958) 964–1029.

[23] N. BOURBAKI, *Groupes et Algèbres de Lie*, Chapitres 4, 5 et 6, Hermann, Paris, 1968.

[24] N. BOURBAKI, *Groupes et Algèbres de Lie*, Chapitres 7 et 8, Hermann, Paris, 1975.

[25] T. BRADEN AND R. MACPHERSON,*From moment graphs to intersection cohomology*, preprint arXiv:math.AG/0008200.

[26] F. BRENTI, Combinatorial expansions of Kazhdan–Lusztig polynomials, *J. London Math. Soc.*, **55** (1997), 448–472.

[27] F. BRENTI, Kazhdan–Lusztig polynomials and *R*-polynomials from a combinatorial point of view, *Discrete Math.*, **193** (1998), no. 1–3, 93–116.

[28] M. BRION, Equivariant cohomology and equivariant intersection theory, in *Representation theories and Algebraic geometries* (A. Broer, ed), 1–37, Kluwer, Dordrecht (1998).

[29] M. BRION, Rational smoothness and fixed points of torus actions, *Transformation Groups*, **4** (1999), 127–156.

[30] M. BRION AND P. POLO, Generic singularities of certain Schubert varieties, preprint (1998).

[31] J.-L. BRYLINSKI AND M. KASHIWARA, Kazhdan–Lusztig conjectures and holonomic systems, *Invent. Math.*, **64** (1981), 387–410.

[32] J. CARRELL, On the smooth points of a Schubert variety, *CMS Conference Proceedings*, 16, 15–24, Proceedings of the conference on "Representations of Groups: Lie, Algebraic, Finite, and Quantum," Banff, Alberta, June 1994.

[33] J.B. CARRELL, The Bruhat graph of a Coxeter group, a conjecture of Deodhar, and rational smoothness of Schubert varieties, *Proceedings of Symposia in Pure Math.*, **56** (1994), 53–61.

[34] J.B. CARRELL AND J. KUTTLER, On the smooth points of T-stable varieties in G/B and the Peterson Map, preprint (1999).

[35] C. CHEVALLEY, Classification de groupes de Lie algébriques, Séminaire, 1956–58, *Secrétariat mathématique*, vol. II, rue Pierre-Curie, Paris (1958).

[36] C. CHEVALLEY, Sur les décompositions cellulaires des espaces G/B, *Proc. Symp. Prue. Math.*, **56** (1994), Part I, 1–25.

[37] M. DEMAZURE, Désingularisation des variétés de Schubert généralisées, *Ann. Sci. E.N.S.*, **7** (1974), 53–88.

[38] V. DEODHAR, Local Poincaré duality and non-singularity of Schubert varieties, *Comm. Algebra*, **13** (1985), 1379–1388.

[39] V. DEODHAR, A combinatorial setting for questions in Kazhdan–Lusztig theory, *Geom. Dedicata*, **36** (1990).

[40] V. DEODHAR, A brief survey of Kazhdan–Lusztig theory and related topics, *Proceedings of Symposia in Pure Math*, **56** (1994), 105–124.

[41] V. DEODHAR, On the Kazhdan–Lusztig conjectures, *Nederl. Akad. Wetensch. Proc. Ser. A.*, **85** (1982), 1–17.

[42] M. DYER, On some generalizations of the Kazhdan–Lusztig polynomials for universal Coxeter systems, *J. Algebra*, **116** (1988).

[43] M. DYER, The nil-Hecke ring and Deodhar's conjecture on Bruhat intervals, *Invent. Math.*, **111** (1993), 571–574.

[44] C. EHRESMANN, Sur la topologie de certains espaces homogènes, *Ann. Math.*, **35** (1934), 396–443.

[45] D. EISENBUD, *Commutative Algebra with a view toward Algebraic Geometry*, Springer-Verlag, GTM, 150.

[46] C.K. FAN, Schubert varieties and short braidedness, *Trans. Groups*, **3** (1998), 51–56.

[47] S. FOMIN AND A. ZELEVINSKY, Recognizing Schubert cells, preprint, (1998).

[48] W. FULTON, *Introduction to Intersection Theory in Algebraic Geometry*, CBMS Regional Conference Series in Mathematics, 54, AMS, Providence, 1984.

[49] W. FULTON, *Young Tableaux: With applications to representation theory and geometry*, volume 35 of London Mathematical Society Student Texts, Cambridge University Press, New York, 1997.

[50] W. FULTON AND P. PRAGACZ, *Schubert varieties and degeneracy loci*, Lect. Notes in Math., Springer-Verlag, 1689.

[51] A. GALLIGO, Computations of certain Hilbert funcrions related with Schubert Calculus, Lecture Notes in Math., 1124, Springer-Verlag (1985).

[52] V. GASHAROV, Factoring the Poincaré polynomials for the Bruhat order on S_n, *Combinatorial Theory*, Series A, **83** (1998), 159–164.

[53] V. GASHAROV, Sufficiency of Lakshmibai–Sandhya's singularity condition for Schubert varieties, to appear in *Compositio Math.*

[54] D. GLASSBRENNER AND K.E. SMITH, Singularities of certain ladder determinantal varieties, *J. of Pure and Applied Alg.*, **100** (1995), 59–75.

[55] N. GONCIULEA, Singular loci of varieties of complexes–II, to appear in *J. Alg.*.

[56] N. GONCIULEA AND V. LAKSHMIBAI, Singular loci of Schubert varieties and Ladder determinantal varieties *J. Alg.*, **229**:2 (2000), 463–497.

[57] N. GONCIULEA AND V. LAKSHMIBAI, Flag varieties, to be published by *Hermann–Actualités Mathematiques*.

[58] M. GORESKY, Tables: Kazhdan–Lusztig polynomials for classical groups.

[59] M. GORESKY AND R. MACPHERSON, Intersection homology–II, *Invent. Math.*, **71** (1983), 77–129.

[60] M. HAIMAN, Smooth Schubert Varieties, unpublished.

[61] H. HANSEN, On cycles in flag manifolds, *Math. Scand.*, **33** (1973), 269–274.

[62] J. HARRIS, *Algebraic Geometry: A First Course*, Graduate Texts in Mathematics 133, Springer-Verlag (1992).

[63] R. HARTSHONE, *Algebraic Geometry*, Graduate Texts in Mathematics 52, Springer-Verlag (1977).

[64] J. HERZOG AND N.V. TRUNG, Gröbner bases and multiplicity of determinantal and Pfaffian ideals, *Adv. Math.*, **96** (1992), 1–37.

[65] H. HILLER, *Geometry of Coxeter Groups*, Pitman Advanced Publishing Program, (1982).

[66] F. HIRZEBRUSCH, *Topological methods in Algebraic Geometry*, Springer-Verlag, 1978.

[67] W.V.D. HODGE AND C. PEDOE, *Methods of Algebraic Geometry*, vol. II, Cambridge University Press (1952).

[68] J.E. HUMPHREYS, *Linear Algebraic Groups*, Graduate Texts in Mathematics 21, Springer-Verlag, (1975).

[69] J.E. HUMPHREYS, *Reflection groups and Coxeter groups*, Cambridge University Press (1990).

[70] J.E. HUMPHREYS, *Introduction to Lie Algebras and Representation Theory*, Graduate Texts in Mathematics 9, Springer-Verlag (1972).

[71] S.P. INAMDAR, A note on Frobenius splitting of Schubert varieties and linear syzygies, *Amer. J. Math.*, **116** (1994), 1587–1590.

[72] S.P. INAMDAR AND V.B. MEHTA, Frobenius splitting of Schubert varieties and linear syzygies, *Amer. J. Math.*, **116** (1994), 1569–1586.

[73] R. IRVING, The socle filtration of a Verma module, *Ann. Scient. Ec. Norm. Sup.*, **21** (1988), 47–65.

[74] J.C. JANTZEN, Moduln mit einem höchsten Gewicht, Lecture Notes in Math., 750, Springer-Verlag.

[75] J.C. JANTZEN, *Einhüllende Algebren halbeinfacher Lie–Algebren*, Springer-Verlag (1983).

[76] J.C. JANTZEN, *Representations of Algebraic Groups*, Academic Press (1987).

[77] A. JOSEPH, On the Demazure character formula, *Ann. Scient. Ec. Norm. Sup.* **18** (1985), 389–419.

[78] D. KAZHDAN AND G. LUSZTIG, Representations of Coxeter groups and Hecke algebras, *Invent. Math.*, **53** (1979), 165–184.

[79] D. KAZHDAN AND G. LUSZTIG, Schubert varieties and Poincaré duality, *Proc. Symp. Pure. Math.*, AMS, **36** (1980), 185–203.

[80] G. KEMPF, Linear systems on homogeneous spaces, *Ann. Math.*, **103** (1976), 557–591.

[81] G. KEMPF AND A. RAMANATHAN, Multicones over Schubert varieties, *Invent. Math.*, **87** (1987), 353–363.

[82] F. KIRWAN, *An Introduction to Intersection Homology Theory*, Longman Scientific and Technical, London (1988).

[83] B. KOSTANT, The principal three–dimensional subgroup and the Betti numbers of a complex simple lie group, *Amer. J. Math.*, **81** (1959), 973–1032.

[84] B. KOSTANT, Groups over $\mathbb{Z}$, Algebraic groups and discontinuous subgroups, *Proc. Symp. Pure. Math*, **9**, AMS (1966).

[85] B. KOSTANT, Structure of the truncated icosahedron (e.g., fullerene or c_{60}, viral coatings) and a 60–element conjugacy class in psl(2, 11), *Selecta Math.*, **1**:1 (1995):163–195.

[86] B. KOSTANT AND S. KUMAR, The Nil-Hecke Ring and Cohomology of G/P for a Kac–Moody Group G^*, *Adv. in Math.*, **62** (1986), 187–237.

[87] B. KOSTANT AND S. KUMAR, T-equivariant K-theory of generalized flag varieties, *J. Diff. Geom.*, **32**(1990), 549–603.

[88] KRATTENHALER AND M. PROHASKA, A remarkable formula for counting non-intersecting lattice paths in a ladder with respect to turns, *Transactions AMS*, **351**(1995),1035–1042.

[89] V. KREIMAN AND V. LAKSHMIBAI, Gröbner bases, Hilbert polynomial, and multiplicity, preprint (2000).

[90] D.M. KULKARNI, Hilbert polynomial of a certain ladder determinantal ideal, *J. Algebraic Combinatorics*, (1993), 57–72.

[91] S. KUMAR, The nil-Hecke ring and singularities of Schubert varieties, *Invent. Math.*, **123** (1996), 471–506.

[92] S. KUMAR, P. LITTELMANN, Frobenius splitting in characteristic 0 and the quantum Frobenius map, preprint (1999).

[93] V. LAKSHMIBAI, Standard monomial Theory for G_2, *J. Alg.*, **98** (1986), 281–318.

[94] V. LAKSHMIBAI, Geometry of G/P–VI, *J. Alg.*, **108** (1987), 355–402.

[95] V. LAKSHMIBAI, Geometry of G/P–VII, *J. Alg.*, **108** (1987), 403–434.

[96] V. LAKSHMIBAI, Geometry of G/P–VIII, *J. Alg.*, **108** (1987), 435–471.

[97] V. LAKSHMIBAI, Bases for quantum Demazure modules, *CMS Conference Proc.*, 16, 199–216, Proceedings of the conference on "Representations of Groups: Lie, Algebraic, Finite, and Quantum," Banff, Alberta, Canada, June 1994.

[98] V. LAKSHMIBAI, Tangent spaces to Schubert Varieties, *Math. Res. Lett.*, **2** (1995), 473–477.

[99] V. LAKSHMIBAI, On Tangent Spaces to Schubert Varieties–I, *J. Alg.*, **230** :1 (2000), 222–244.

[100] V. LAKSHMIBAI, On Tangent Spaces to Schubert Varieties–II, *J. Alg.*, 224 (2000), 167–197.

[101] V. LAKSHMIBAI, Singular Loci of Varieties of Complexes, (to appear in *J. Algebraic combinatorics*).

[102] V. LAKSHMIBAI, Schubert varieties and standard monomial theory, *Topics in Algebra*, Banach Center Publ., 26, Part 2, PWN, Warsaw, (1990), 365–378.

[103] V. LAKSHMIBAI AND P. MAGYAR, Degeneracy schemes, Schubert varieties and Quiver varieties, *International Math. Research Notices*, **12** (1998), 627–640.

[104] V. LAKSHMIBAI, C. MUSILI AND C.S. SESHADRI, Cohomology of line bundles on G/B, *Ann. E.N.S.*, **7** (1974), 89–137.

[105] V. LAKSHMIBAI, C. MUSILI AND C.S. SESHADRI, Geometry of G/P–IV, *Proc. Ind. Acad. Sci.*, 88A (1979), 279–362.

[106] V. LAKSHMIBAI AND K.N. RAJESWARI, Geometry of G/P–IX, *J. Alg.*, **130** (1990), 122–165.

[107] V. LAKSHMIBAI AND B. SANDHYA, Criterion for smoothness of Schubert varieties in $SL(n)/B$, *Proc. Indian Acad. Sci.* (Math. Sci.), **100** (1990), 45–52.

[108] V. LAKSHMIBAI AND C.S. SESHADRI Geometry of G/P–II, *Proc. Ind. Acad. Sci.*, **87A** (1978), 1–54.

[109] V. LAKSHMIBAI AND C.S. SESHADRI, Singular locus of a Schubert variety, *Bull. AMS*, **11** (1984), 363–366.

[110] V. LAKSHMIBAI AND C.S. SESHADRI, Geometry of G/P–V, *J. Alg.*, **100** (1986), 462–557.

[111] V. LAKSHMIBAI AND M. SONG, Criterion for smoothness of Schubert varieties in Sp_{2n}/B, *J. Alg.*, **187** (1997), 332–352.

[112] V. LAKSHMIBAI AND J. WEYMAN, Multiplicities of points on a Schubert variety in a minuscule G/P, *Adv. in Math.*, **84** (1990), 179–208.

[113] A. LASCOUX, Foncteurs de Schur et Grassmanniennes, Thesis, Université Paris–VII (1977).

[114] A. LASCOUX, Polynômes de Kazhdan–Lusztig pour les variétés de Schubert vexillaires. (French) [Kazhdan–Lusztig polynomials for vexillary Schubert varieties],*C.R. Acad. Sci.*, Paris Sér. I Math., **321** (1995), 667–670.

[115] A. LASCOUX, Syzygies des variétés déterminantales, *Adv. Math.*, **30** (1978), 202–237.

[116] A. LASCOUX, Ordonner le groupe symétrique : pourquoi utiliser l'algèbre de Iwahori–Hecke, *Proc. ICM Berlin*, Doc. Math. 1998 Extra Vol. III, 355–364 (electronic).

[117] A. LASCOUX AND M.P. SCHÜTZENBERGER, Polynômes de Kazhdan and Lusztig pour les grassmanniennes. (French) [Kazhdan–Lusztig polynomials for Grassmannians], *Astérisque*, 87–88 (1981), 249–266, Young tableaux and Schur functions in algebra and geometry (Toruń, 1980).

[118] A. LASCOUX AND M.P. SCHÜTZENBERGER, Polynômes de Schubert, *C.R. Acad. Sci Paris.*, **294** (1982), 447–450.

[119] A. LASCOUX AND M.P. SCHÜTZENBERGER, Schubert polynomials and Littlewood–Richardson rule, *Lett. in Math. Phys.*, **10** (1985), 111–124.

[120] A. LASCOUX AND M.P. SCHÜTZENBERGER, Treillis et bases des groupes de Coxeter. *Electron. J. Combin.*, **3**:2 (1996).

[121] P. LITTELMANN, Contracting modules and standard monomial theory for symmetrizable Kac–Moody algebras (to appear in *JAMS*).

[122] I.G. MACDONALD, The Poincaré series of a Coxeter group, *Math. Ann.*, **199** (1972), 161–174.

[123] I.G. MACDONALD, Schubert polynomials, *C.U.P.*, (1991).

[124] I.G. MACDONALD, Symmetric functions and Hall polynomials, Second edition, *Clarendon Press*, Oxford (1995).

[125] L. MANIVEL, Fonctions symétriques, polynômes de Schubert et lieux de dégénerescence, *Cours Spécialisés, Vol.3, Soc. Math., France*, (1998) (MR 99k:05159).

[126] C. MCCRORY, A characterization of homology manifolds, *J. London Math. Soc.*, 16(2) (1977), 146–159.

[127] V.B. MEHTA AND A. RAMANATHAN, Frobenius splitting and cohomology vanishing for Schubert varieties, *Annals of Math.*, **122** (1985), 27–40.

[128] V.B. MEHTA AND V. SRINIVAS, Normality of Schubert varieties, *Amer. J. Math.*, 109 (1987), 987–989.

[129] S.B. MULAY, Determinantal loci and the flag variety, *Adv. Math.*, **74** (1989), 1–30.

[130] D. MUMFORD, *Complex projective varieties*, Grundlehren der math. Wissenschaften 221, Springer-Verlag (1976).

[131] C. MUSILI, Postulation formula for Schubert varieties, *J. Indian Math. Soc.*, **36** (1972), 143–171.

[132] C. MUSILI, Some properties of Schubert varieties, *J. Indian Math. Soc.*, **38** (1974), 131–145.

[133] C. MUSILI AND C.S. SESHADRI, Schubert varieties and the variety of complexes, *Arithmetic and Geometry, II*, Prog. Math. 36, Birkhäuser (1983), 329–359.

[134] M.S. NARASIMHAN AND S. RAMANAN, Moduli of vector bundles on a compact Riemann surface, *Ann. Math.*, **89** (1969), 14–51.

[135] P. POLO, On Zariski tangent spaces of Schubert varieties, and a proof of a conjecture of Deodhar, *Indag. Math.*, **5** (1994), 483–493.

[136] P. POLO, Construction of arbitrary Kazhdan–Lusztig polynomials, *Representation Theory* (an electronic journal of the AMS), Vol.3 (1999), 90–104.

[137] R. PROCTOR, Classical Bruhat orders are lexicographic shellable, *J. Alg.*, **77** (1982), 104–126.

[138] S. RAMANAN AND A. RAMANATHAN, Projective normality of Flag varieties and Schubert varieties, *Invent. Math.*, **79** (1985), 217–224.

[139] A. RAMANATHAN, Schubert varieties are arithmetically Cohen–Macaulay, *Invent. Math.*, **80** (1985), 283–294.

[140] A. RAMANATHAN, Equations defining Schubert varieties and Frobenius splitting of diagonals, *Publ. Math. I.H.E.S.*, **65** (1987), 61–90.

[141] J. Rosenthal, Schubert varietäten und deren Singularitäten, Diplom thesis, University of Basel, Switzerland (1986).

[142] J. ROSENTHAL AND ZELEVINSKY, An explicit formula for the multiplicity of points on a classical Schubert variety, preprint (1998).

[143] P. SANKARAN AND P. VANCHINATHAN, Small resolutions of Schubert varieties in symplectic and orthogonal Grassmannians, *Publ. RIMS*, **30** (1994), 443–458.

[144] P. SANKARAN AND P. VANCHINATHAN, Small resolutions of Schubert varieties and Kazhdan–Lusztig polynomials, *Publ. RIMS*, **31** (1995), 465–480.

[145] H. SCHUBERT, *Kalkül der abzählenden geometrie*, Teubner, Leipzig, (1879), reprinted, Springer-Verlag, 1979.

[146] C.S. SESHADRI, Geometry of G/P–I, *C.P. Ramanujam: A Tribute, 207*, published by Tata Institute, Bombay, 207–239, Springer-Verlag (1978).

[147] C.S. SESHADRI, Line bundles on Schubert varieties, *Proceedings of the Bombay Colloquium on Vector bundles on Algebraic Varieties* (1984).

[148] M. SONG, Schubert varieties in $Sp(2n)/B$, Ph.D. Thesis, Northeastern University (1996).

[149] T.A. SPRINGER, Quelques applications de la cohomologie d'intersection, *Astérisque*, 92–93 (1982), 249–273, Bourbaki Seminar, Col 1981/1982.

[150] Z. STANKOVA, Forbidden subsequences, *Discrete Math.*, **132** (1994), 291–316.

[151] R.P. STANLEY, Some combinatorial aspects of the Schubert calculus, Combinatoire et Représentation du Groupe Symétrique, Lecture Notes in Math 579, Springer-Verlag (1977), 217–251.

[152] P. SVANES, Coherent cohomology on Schubert subschemes of flag schemes, *Adv. Math.*, **14** (1974), 369–453. 291–316.

[153] S.J. TELLER, Computing the antipenumbra of an area light source. *Computer Graphics*, **26**:2) (1992), 139–148.

[154] S.J. TELLER AND M.E. HOHMEYER, Computing the lines piercing four lines, Technical report (1991), U.C. Berkeley, Computer Science Department.

[155] A. VAN DEN HOMBERGH, Note on a paper by Bernstein, Gelfand, Gelfand on Verma modules, *Proc. Konin. Neder. Aka. Amsterddam Ser. A*, **77** (1974), 352–356.

[156] D.-N. VERMA, Structure of certain induced representations of complex semisimple Lie algebras, *Bull. Amer. Math. Soc.*, **74** (1968), 160–166.

[157] A. ZELEVINSKY, Small resolutions of singularities of Schubert varieties, *Funct. Anal. Appl.*, **17** (1983), 142–144.

Index